Chemical Signals in Vertebrates 14

AF896325

Christina D. Buesching
Editor

Chemical Signals in Vertebrates 14

Springer

Editor
Christina D. Buesching
Wildlife Conservation Research Unit,
Department of Zoology
University of Oxford
Oxford, Oxfordshire, UK

ISBN 978-3-030-17615-0 ISBN 978-3-030-17616-7 (eBook)
https://doi.org/10.1007/978-3-030-17616-7

© Springer Nature Switzerland AG 2019
This work is subject to copyright. All rights are reserved by the Publisher, whether the whole or part of the material is concerned, specifically the rights of translation, reprinting, reuse of illustrations, recitation, broadcasting, reproduction on microfilms or in any other physical way, and transmission or information storage and retrieval, electronic adaptation, computer software, or by similar or dissimilar methodology now known or hereafter developed.
The use of general descriptive names, registered names, trademarks, service marks, etc. in this publication does not imply, even in the absence of a specific statement, that such names are exempt from the relevant protective laws and regulations and therefore free for general use.
The publisher, the authors and the editors are safe to assume that the advice and information in this book are believed to be true and accurate at the date of publication. Neither the publisher nor the authors or the editors give a warranty, expressed or implied, with respect to the material contained herein or for any errors or omissions that may have been made. The publisher remains neutral with regard to jurisdictional claims in published maps and institutional affiliations.

This Springer imprint is published by the registered company Springer Nature Switzerland AG
The registered company address is: Gewerbestrasse 11, 6330 Cham, Switzerland

It's the soul of life
It's chemistry
It's the poetry of understanding

Mammals, reptiles, birds, and fish
All lives leave pieces of soul
Creeping, swimming, running, fearing
Loving, fighting, hating, greeting, nursing
Looking for new and old horizons

It's chemistry
It's the soul of life

Carlos Grau Paricio

Preface

This 14th volume of Chemical Signals in Vertebrates (CSiV) is based on a mixture of conference proceedings from the CSiV XIV meeting held in Cardiff, Wales in September 2017 and additional papers, often from authors who could not attend the meeting itself. The meeting was attended by over 80 delegates, many of them still at the start of their scientific career, and the manuscripts in this book have been written by over 50 co-authors. I would like to take this opportunity to express my gratitude to our local host, Dr. Carsten Mueller, for the excellent organisation and smooth running of the conference, and all his work, time and effort in the lead up. In addition, I would like to thank the staff of the Sir Martin Evans Building, School of Biosciences, Cardiff University, as well as Dr. Nadine Sugianto and Ms. Tanesha Allen from the Wildlife Conservation Research Unit, University of Oxford, for their help 'behind the scenes' with the logistics on the ground. We are particularly grateful for the support from Markes for the conference pens and notepads and for the sponsorship from the Monell Chemical Senses Center to cover travel costs for the speakers in the Scent of Disease session.

As we have come to expect from previous CSiV meetings, the talks and posters presented were again of high quality, and the ensuing discussions are certain to spark new ideas and future collaborations. I particularly want to thank all contributors to this volume, where-in the era of the 'Research Excellence Framework[1]' criteria and the need to demonstrate one's publication record in high impact factor journals to Research Council funding agencies-book chapters are in danger of becoming an anachronism. My hope is that the type of synthesis offered by this series of books continues to stimulate new ideas and trans-disciplinary research, which has so benefitted the growing depth and breadth of our field.

As in previous CSiV conferences, the range of presentations at this 14th meeting embraced an incredibly broad range, spanning from behavioural observations on intra-specific communication of raccoon dogs in Japan and inter-specific information exchange between predator species in Botswana, to the chemical composition of urinary scent marks of badgers in England, and the evolution of scent proteomics

[1] https://www.ref.ac.uk/about/what-is-the-ref/.

in the world's salamanders. This variety of topics and species is reflected in the contributions to this book, where we have, alongside the established sections on (animal) intra-specific communication and inter-specific cues and signals, also a dedicated section on olfaction in humans; a field that is likely to feature substantially at the next CSiV 15 organised by Dr. Benoist Schaal and his team at the Centre des Sciences du Goût at the Université de Bourgogne in Dijon, France in summer 2020.

For the previous CSiV 13 volume, Dr. Dietland Müller-Schwarze, one of the veterans in the field who has been on the forefront of semio-chemistry and olfactory communication for the past 40 years, was invited to write a perspective of the future of research into chemical signals in vertebrates. He predicted that technological advances would continue to shape semio-chemical research, and this prescience has certainly proven to be the case. Studies combining advanced molecular techniques, such as next-generation sequencing, with traditional gas chromatographic analysis of scent characteristics are becoming ever more common; evidenced particularly by the rapidly growing field of proteomics and olfactory speciation.

New insights from these inter-disciplinary approaches permit evolutionary and ecological insights, many with implications for applied species management and conservation, and I have thus dedicated a section on the role of scent in evolution in this book and develop some testable hypotheses for the future in the final chapter.

Oxford, UK
January 2019

Christina D. Buesching

Contents

Part I Intra-specific Communication in Non-Human Vertebrates

Perspectives on Chemical Signals Conveying Information in Rodents 3
Michael H. Ferkin

Latrines as Potential Communication Centres in Short-Beaked Echidnas 13
Rachel L. Harris, Jenny Sprent, and Stewart C. Nicol

Do Urinary Volatiles Carry Communicative Messages in Himalayan Snow Leopards [*Panthera uncia*, (Schreber, 1775)]? 27
Subhadeep Das, Sourav Manna, Sandipan Ray, Payel Das, Upashna Rai, Biswatosh Ghosh, and Mousumi Poddar Sarkar

Encoded Information Within Urine Influences Behavioural Responses Among European Badgers (*Meles meles*) 38
Tanesha M. Allen, Nadine A. Sugianto, Charlotte Ryder, Chris Newman, David W. Macdonald, and Christina D. Buesching

LPS-Induced Immune System Stimulation Alters Urinary Volatiles and Behaviour in Growing Pigs 60
Sankarganesh Devaraj, Anoosh Rakhshandeh, Edgar Aviles-Rosa, and John J. McGlone

A Field Study of Wild Echidna Responses to Conspecific Odour 71
Rachel L. Harris, Elissa Z. Cameron, and Stewart C. Nicol

How Diet Affects Vertebrate Semiochemistry 81
Jan Havlíček, Jitka Fialová, and S. Craig Roberts

The Social Function of Latrines: A Hypothesis-Driven Research Approach 94
Christina D. Buesching and Neil Jordan

Part II Olfaction in Humans

The Effects of Artificial Fragrances on Human Olfactory Communication . . . 107
Caroline Allen, Jan Havlíček, and S. Craig Roberts

The Human Mammary Odour Factor: Variability and Regularities in Sources and Functions . . . 118
Benoist Schaal, Sébastien Doucet, Robert Soussignan, Magali Klaey-Tassone, Bruno Patris, and Karine Durand

Cross-Cultural Approaches to Better Understand Chemical Communication in Humans . . . 139
Camille Ferdenzi

Adaptation of the University of Pennsylvania Smell Identification Test for the Population of Central Russia . . . 153
Vera V. Voznessenskaya, Maria A. Klyuchnikova, Elena I. Rodionova, and Anna Voznesenskaya

Part III Inter-specific Cues and Signals

House Mouse (*Mus musculus*) Avoidance of Olfactory Cues from Ferrets and Other Mammalian and Reptilian Predators: Preliminary Results . . . 165
Carlos Grau, Eva Teruel, Julius Leclercq, and Patrick Pageat

Do Carnivores Have a World Wide Web of Interspecific Scent Signals? . . . 182
Peter Apps, Kasim Rafiq, and J. Weldon McNutt

Part IV Semio-chemistry and Evolution

Chemistry Between Salamanders: Evolution of the SPF Courtship Pheromone System in Salamandridae . . . 205
Franky Bossuyt, Margo Maex, Dag Treer, Lisa M. Schulte, Ines Van Bocxlaer, and Sunita Janssenswillen

Comparative Structural Modeling of Bovine Vomeronasal Type-1 Receptor 1 (VN1R1) and Elucidation of Molecular Interactions with Pheromones Using *in silico* Approaches . . . 221
Rajesh Durairaj, Cécile Bienboire-Frosini, and Patrick Pageat

Detecting the Smell of Disease and Injury: Scoping Evolutionary and Ecological Implications . . . 238
Chris Newman and Christina D. Buesching

Index . . . 251

Contributors

Caroline Allen School of Psychology, Newcastle University, Newcastle-upon-Tyne, UK

Tanesha M. Allen Wildlife Conservation Research Unit, Department of Zoology, Recanati-Kaplan Centre, University of Oxford, Abingdon, Oxfordshire, UK

Peter Apps Botswana Predator Conservation Trust, Maun, Botswana

Edgar Aviles-Rosa Laboratory of Animal Behavior, Physiology and Welfare, Department of Animal and Food Sciences, Texas Tech University, Lubbock, TX, USA

Cécile Bienboire-Frosini Department of Behavioural and Physiological Mechanisms of Adaptation (D-MPCA), Research Institute in Semiochemistry and Applied Ethology (IRSEA), Apt, France

Franky Bossuyt Amphibian Evolution Lab, Biology Department, Vrije Universiteit Brussel, Brussels, Belgium

Christina D. Buesching Wildlife Conservation Research Unit, Department of Zoology, Recanati-Kaplan Centre, University of Oxford, Abingdon, Oxfordshire, UK

Elissa Z. Cameron School of Biological Sciences, University of Tasmania, Hobart, TAS, Australia;
School of Biological Sciences, University of Canterbury, Christchurch, New Zealand

Payel Das Chemical Signal & Lipidomics Laboratory, Department of Botany, Centre of Advanced Study, University of Calcutta, Kolkata, India

Subhadeep Das Chemical Signal & Lipidomics Laboratory, Department of Botany, Centre of Advanced Study, University of Calcutta, Kolkata, India

Sankarganesh Devaraj Laboratory of Animal Behavior, Physiology and Welfare, Department of Animal and Food Sciences, Texas Tech University, Lubbock, TX, USA

Sébastien Doucet Developmental Ethology and Cognitive Psychology Group, Centre for Smell, Taste and Food Science, CNRS (UMR 6265), Université de Bourgogne-Franche Comté, Dijon, France

Rajesh Durairaj Department of Behavioural and Physiological Mechanisms of Adaptation (D-MPCA), Research Institute in Semiochemistry and Applied Ethology (IRSEA), Apt, France

Karine Durand Developmental Ethology and Cognitive Psychology Group, Centre for Smell, Taste and Food Science, CNRS (UMR 6265), Université de Bourgogne-Franche Comté, Dijon, France

Camille Ferdenzi Centre de Recherche En Neurosciences de Lyon, CNRS UMR 5292, INSERM U1028, Université Claude Bernard Lyon 1, Bron Cedex, France

Michael H. Ferkin Department of Biological Sciences, University of Memphis, Memphis, TN, USA

Jitka Fialová Department of Zoology, Charles University, Prague, Czech Republic

Biswatosh Ghosh Department of Zoology, Bidhannagar College, Salt Lake, Kolkata, India

Carlos Grau Department of Behavioural and Physiological Mechanisms of Adaptation (DMPCA), Research Institute in Semiochemistry and Applied Ethology (IRSEA), Quartier Salignan, Apt, France

Rachel L. Harris School of Biological Sciences, University of Tasmania, Hobart, TAS, Australia;
Department of Evolutionary Anthropology, Duke University, Durham, NC, USA

Jan Havlíček Department of Zoology, Charles University, Prague, Czech Republic

Sunita Janssenswillen Amphibian Evolution Lab, Biology Department, Vrije Universiteit Brussel, Brussels, Belgium

Neil Jordan Centre for Ecosystem Science, University of New South Wales, Sydney, Australia;
Taronga Conservation Society Australia, Taronga Western Plains Zoo, Wildlife Reproduction Centre, Dubbo, Australia;
Botswana Predator Conservation Trust, Maun, Botswana

Magali Klaey-Tassone Developmental Ethology and Cognitive Psychology Group, Centre for Smell, Taste and Food Science, CNRS (UMR 6265), Université de Bourgogne-Franche Comté, Dijon, France

Maria A. Klyuchnikova A.N. Severtsov Institute of Ecology and Evolution, Russian Academy of Sciences, Moscow, Russia

Julius Leclercq Animal Experimentation Unit, Research Institute in Semiochemistry and Applied Ethology (IRSEA), Quartier Salignan, Apt, France

David W. Macdonald Wildlife Conservation Research Unit, Department of Zoology, Recanati-Kaplan Centre, University of Oxford, Abingdon, Oxfordshire, UK

Margo Maex Amphibian Evolution Lab, Biology Department, Vrije Universiteit Brussel, Brussels, Belgium

Sourav Manna Chemical Signal & Lipidomics Laboratory, Department of Botany, Centre of Advanced Study, University of Calcutta, Kolkata, India

John J. McGlone Laboratory of Animal Behavior, Physiology and Welfare, Department of Animal and Food Sciences, Texas Tech University, Lubbock, TX, USA

J. Weldon McNutt Botswana Predator Conservation Trust, Maun, Botswana

Chris Newman Wildlife Conservation Research Unit, Department of Zoology, Recanati-Kaplan Centre, University of Oxford, Abingdon, Oxfordshire, UK

Stewart C. Nicol School of Biological Sciences, University of Tasmania, Hobart, TAS, Australia

Patrick Pageat Department of Chemicals Identification and Development of Analogues, Research Institute in Semiochemistry and Applied Ethology (IRSEA), Quartier Salignan, Apt, France

Bruno Patris Developmental Ethology and Cognitive Psychology Group, Centre for Smell, Taste and Food Science, CNRS (UMR 6265), Université de Bourgogne-Franche Comté, Dijon, France

Kasim Rafiq Botswana Predator Conservation Trust, Maun, Botswana;
School of Natural Science and Psychology, Liverpool John Moores University, Liverpool, UK

Upashna Rai Padmaja Naidu Himalayan Zoological Park, Darjeeling, West Bengal, India

Anoosh Rakhshandeh Department of Animal and Food Sciences, Texas Tech University, Lubbock, TX, USA

Sandipan Ray Chemical Signal & Lipidomics Laboratory, Department of Botany, Centre of Advanced Study, University of Calcutta, Kolkata, India

S. Craig Roberts Division of Psychology, University of Stirling, Stirling, UK

Elena I. Rodionova A.N. Severtsov Institute of Ecology and Evolution, Russian Academy of Sciences, Moscow, Russia;
Kharkevich Institute for Information Transmission, Russian Academy of Sciences, Moscow, Russia

Charlotte Ryder Wildlife Conservation Research Unit, Department of Zoology, Recanati-Kaplan Centre, University of Oxford, Abingdon, Oxfordshire, UK

Mousumi Poddar Sarkar Chemical Signal & Lipidomics Laboratory, Department of Botany, Centre of Advanced Study, University of Calcutta, Kolkata, India

Benoist Schaal Developmental Ethology and Cognitive Psychology Group, Centre for Smell, Taste and Food Science, CNRS (UMR 6265), Université de Bourgogne-Franche Comté, Dijon, France

Lisa M. Schulte Amphibian Evolution Lab, Biology Department, Vrije Universiteit Brussel, Brussels, Belgium

Robert Soussignan Developmental Ethology and Cognitive Psychology Group, Centre for Smell, Taste and Food Science, CNRS (UMR 6265), Université de Bourgogne-Franche Comté, Dijon, France

Jenny Sprent School of Biological Sciences, University of Tasmania, Hobart, TAS, Australia

Nadine A. Sugianto Wildlife Conservation Research Unit, Department of Zoology, Recanati-Kaplan Centre, University of Oxford, Abingdon, Oxfordshire, UK

Eva Teruel Statistical Analysis Department, Research Institute in Semiochemistry and Applied Ethology (IRSEA), Quartier Salignan, Apt, France

Dag Treer Amphibian Evolution Lab, Biology Department, Vrije Universiteit Brussel, Brussels, Belgium

Ines Van Bocxlaer Amphibian Evolution Lab, Biology Department, Vrije Universiteit Brussel, Brussels, Belgium

Anna Voznesenskaya A.N. Severtsov Institute of Ecology and Evolution, Russian Academy of Sciences, Moscow, Russia;
Kharkevich Institute for Information Transmission, Russian Academy of Sciences, Moscow, Russia

Vera V. Voznessenskaya A.N. Severtsov Institute of Ecology and Evolution, Russian Academy of Sciences, Moscow, Russia

Part I

Intra-specific Communication in Non-Human Vertebrates

Perspectives on Chemical Signals Conveying Information in Rodents

Michael H. Ferkin

Department of Biological Sciences, University of Memphis, Ellington Hall, Memphis, TN 38152, USA
mhferkin@memphis.edu

Abstract. Social cues must be detectable by receivers or bystanders before they can respond to the information contained in the cue. Information about conspecifics and heterospecifics can be extracted from social cues. At that point, the information in the cue becomes a fact and the cue becomes a signal. Chemical signals, such as scent marks, provide accurate information that can be used by conspecifics in sexual and social contexts. Scent marks of many conspecifics distributed in an area may be akin to a chemical bulletin board that provides information about the identity of senders and how recently they were in that area. These scent marks deposited by senders can provide both private and public information to receivers. The manner in which receivers use this information to make decisions will depend on the context and manner in which it encounters these scent marks. Thus, a learned rule of thumb that will allow the animal to respond in a predictable way to a scent mark and when to scent mark would not suffice. The sender and the receiver must identify the phenotype and genotype of their counterparts, recollect interactions with them, be aware of their own condition and the context of the communication. Higher level cognitive processing would likely be required for receivers and senders to make potentially informed decisions that reflect tradeoffs in their survival and fitness.

1 Introduction

Chemical signals from scent marks and odour-producing tissues provide a great deal of information about the sender to receivers (Ferkin 2011; Roberts 2007; Wyatt 2014). The manner in which an animal uses this information to make decisions will depend on the context and manner in which it encounters these scent marks (Franklin and Ferkin 2006, 2008; Roberts et al. 2014). It is unlikely that a learned rule of thumb that will allow the animal to respond in a predictable way to a scent mark would suffice. To select the appropriate response, an individual will need to identify the phenotype and genotype of the sender, recall any interactions with that sender, be aware of its own condition and respond in a manner that will presumably increase the individual's survival and fitness (Ferkin 2015; Green et al. 2015; Tinnesand et al. 2015). Doing so will involve the receivers learning of spatial and temporal associations among senders and their marks (Ferkin and Pierce 2007; Roberts et al. 2014), storing them, and later recalling these associations (Franklin and Ferkin 2006, 2008). Receivers will also need to learn other features associated with the sender such as when the scent mark was

© Springer Nature Switzerland AG 2019
C. D. Buesching (Ed.): *Chemical Signals in Vertebrates 14*, pp. 3–12, 2019.
https://doi.org/10.1007/978-3-030-17616-7_1

deposited, how long it has been since the receiver encountered the sender, and where that encounter occurred (Franklin and Ferkin 2006, 2008). Higher level cognitive processing involving procedural memory, episodic memory, autobiographical memory and making judgements of numerical discrimination would be required for receivers to make potentially informed decisions, when they encounter the scent marks of particular conspecifics (Ferkin 2011).

Information about conspecifics and heterospecifics can be extracted from social cues. The information can reflect features of the sender, such as condition-dependent traits related to its phenotype, genotype and its interactions with the environment or with other individuals (Valone 1989; Dall et al. 2005 ; Danchin et al. 2008). The social cues may contain personal information of the sender that is usually relevant to particular conspecifics, and usually private. Alternatively, the information can be detected by non-targeted individuals or bystanders and become public information, and possibly provide receivers with inadvertent social information (Galef and Giraldeau 2001; Danchin et al. 2004; Valone 2007). The key issue is that these social cues must be detectable by receivers or bystanders before they can respond to the information contained in the cue. At that point, the information in the cue becomes a fact and the cue becomes a signal (Danchin et al. 2008). A signal may impact on the fitness of the sender and the receiver. Most work has focused on visual and auditory cues and signals as providers of information for receivers (Danchin et al. 2008). Less work has focused on how scent marks can also serve as cues and signals that provide information to receivers in a manner that benefits them (Johnston 2009; Ferkin and Hobbs 2014; Wyatt 2014; Buesching and Stankowich 2017; Ferkin and Ferkin 2017; Ferkin et al. 2017).

For many terrestrial mammals, scent marks convey information about the sender to nearby conspecifics and heterospecifics (Johnston and delBarco-Trillo 2009; Kaur et al. 2014; Roberts et al. 2014; Wyatt 2014). These scent marks are viewed as honest signals of the sender's quality or condition (Roberts 2007; Ferkin 2011; Thonhauser et al. 2013). That is, scent marks from sources such as urine (Zala et al. 2004; Charlton 2014, 2015), saliva (Block et al. 1981), faeces (Palphramand and White 2007), and those from specialized glands, such as the preorbital gland (Ceacero et al. 2015; Pluháček et al. 2015), anal gland (Bills et al. 2013; Tinnesand et al. 2015), foot pads (Owen et al. 2015), submandibular glands (Mykytowycz 1965; Camacho-Arroyo et al. 1999) and the integument (Martin et al. 2014; McLean 2014) can accurately reflect the condition and phenotype (Sabau and Ferkin 2013; Sharpe 2015) or genotype of the sender (Green et al. 2015). Scent marks from these multiple sources likely provide unique and overlapping information about the sender (Johnston 2003, 2009; Hurst and Beynon 2004; Kaur et al. 2014). To this point, Johnston and del-Barco-Trillo (2009) hypothesized that scent marks may be viewed as a mosaic of chemicals. These mosaics are also individually distinct (Halpin 1986; Johnston 2003, 2009; Thom and Hurst 2004; Kaur et al. 2014).

Studies have also shown that some subsets of major urinary proteins found in scent marks of male house mice (*Mus domesticus*) and rats *(Rattus norvegicus)* provide information about the identity of sender, essentially a 'bar code' (Hurst et al. 2001; Gómez-Baena et al. 2014). This information is stable over time, and individual as well as species-specific, and appears to be based on the individual's genotype and not its phenotype (Hurst et al. 2001; Cheetham et al. 2007; Green et al. 2015). A recent discovery revealed that wild male house mice (*M. musculus*) expressed their own unique combination and ratio of MUPs in their urine (Kaur et al. 2014). The subsets of MUPs were individually distinct and distinguishable from those found in the urine of other male house mice. Male house mice used this information to avoid counter-marking their own scent marks and those of male conspecifics whose subset of major urinary proteins is similar to their own (Kaur et al. 2014). However, another study on mice challenges the fingerprint properties of MUPs (Thoß et al. 2015). The MUP profile of a given individual changed over the course of several weeks, supporting more a 'dynamic expression' hypothesis than a 'barcode' hypothesis. Thus, Thoß et al. (2015) suggested that animals may recognize individuals by investigating relatively dynamic signals instead of fixed ones.

Scent marks of many conspecifics distributed in an area may be akin to a chemical bulletin board that provides information about the identity of senders and how recently they were in that area (Woodward et al. 2000; Johnston 2003). Consequently, odours and scent marks produced and deposited by senders can provide both private and public information (Ferkin and Ferkin 2017). The spatial and temporal distribution of these marks, as well as the fact that they can remain in the environment for hours and days may provide individuals that encounter these bulletin boards additional time-sensitive information about particular senders (Buesching et al. 2002; Johnston 2009; Ferkin 2011, 2015; Kean et al. 2015). Such information can augment or update that which may be acquired through direct interactions with the senders. The scent marks in such a chemical bulletin board may include signals from rivals and potential mates. Thus, the response of individuals when they encounter scent marks from different senders or under different contexts should be flexible, allowing receivers to adjust their responses accordingly, depending on the identity of the sender, past associations and context (Ferkin 2011). The response chosen will be modulated by the receiver's ability and capacity for discrimination, categorization, judgments of relative numerousness, and recollection of the emotional valence attached to the previous interaction with the sender/s (Franklin and Ferkin 2006, 2008; Johnston and delBarco-Trillo 2009; Ferkin 2011; Roberts et al. 2014). The particular response should represent a balance of the costs and benefits associated with that choice (Ferkin 2015) as well as the likelihood of eavesdropping or copying by conspecifics (Danchin et al. 2004; Valone 2007; Ferkin and Ferkin 2017). The response in question would have been modified over time through evolutionary pressures placed on individuals in that species (Roberts 2007; Ferkin et al. 2017). In this way, receivers in species that face similar ecological and social challenges may display responses to scent marks that are more similar compared to those from species facing different challenges (Wyatt 2014), indicating some degree of convergence and stability over time (Ferkin 2015).

There appear to be a few general rules of thumb regarding how decisions are made by receivers. Typically, individuals spend more time investigating the scent marks of opposite-sex conspecifics than those of same-sex conspecifics (Brown and Macdonald 1985). Individuals may also spend more time investigating the scent marks of more attractive than those of less attractive opposite-sex conspecifics. Differences among the attractiveness of the scent marks of opposite-sex conspecifics can be detected and may affect decisions about mate selection (Johnston 2009; Wyatt 2014). However, mate choice may be affected by temporal shifts in the attractiveness of odours and the responses of receivers to those odours (Ferkin 2011; Sabau and Ferkin 2013). The scent marks of female house mice and Syrian hamsters (*Mesocricetus auratus*) are attractive to male conspecifics only when the female is in oestrus (Kavaliers et al. 1994; Lai et al. 1996; delBarco-Trillo et al. 2009). Postpartum oestrus female mammals produce odours and scent marks that are more attractive to males than are those produced by females not in postpartum oestrus (Ferkin and delBarco-Trillo 2014). However, senders and receivers may not always behave in a congruent fashion and may not follow a simple rule of thumb when it comes to senders depositing scent marks and to receivers responding preferentially to particular senders. Many studies have shown that the response to scent marks is context-dependent, thereby allowing receivers to adjust their responses accordingly, depending on the identity of the sender, past associations, and previous information (Ferkin 2011). Importantly, the production of scent marks and the responses to them likely evolved under the pressures of sexual selection (Roberts 2007; Danchin et al. 2008; Ferkin 2015; Petty and Drea 2015).

2 Sources of Scents

Some sources of scent produce marks that convey sexual information about the sender.

For example, meadow voles, *Microtus pennsylvanicus*, have a highly localized pattern of sexual information on their bodies during the breeding season (Ferkin and Johnston 1995). The faeces, anogenital area and urine scent marks of meadow voles are attractive to opposite-sex conspecifics but not to same-sex conspecifics. Saliva/mouth secretions of female voles are attractive to male but not to female conspecifics. In contrast, the posterolateral region scent marks of male voles are attractive to both male and female conspecifics. Thus, in meadow voles, the secretions from the mouth and posterolateral region likely encode different discernable information than that from the faeces, anogenital area and urine (Ferkin and Johnston 1995).

Similarly, the chemical signals for discrimination and recognition of individuals can come from many different specific sources, such as a variety of specialized scent glands (e.g. sebaceous glands, apocrine and eccrine sweat glands, etc.) as well as urine and faeces. In Syrian hamsters, for example there are five different sources of individually distinctive scents, namely flank gland, vaginal secretions, ear glands (inside the pinna), urine and faeces (Johnston et al. 1991). Six other potential sources of odours (fur from the midline ventral surface, fur from the dorsal surface between the shoulders, saliva, feet, fur behind the ears, and the flank-gland area from flank-gonadectomized males) were tested in Syrian hamsters but were not found to be individually distinctive, as measured by habituation–dishabituation tests (Johnston et al. 1991). Similar results were found with Djungarian hamsters (*Phodopus campbelli*). Male hamsters were able to discriminate individual differences between other males using scents from the midventral gland, urine, faeces, mouth and the sacculi from the corner of the mouth (Lai and Johnston 1994). These different odour sources that provide information about individual identity may contain some redundant information, but the combined information from of all of these sources may facilitate the creation of stronger memories about the identity of a given conspecific (Johnston 2003, 2009; Johnston and delBarco-Trillo 2009).

Receivers gain specific information about the scent marks of individuals that they encounter because scent marks and over-marks provide individually distinct information about senders (Johnston and delBarco-Trillo 2009; Kaur et al. 2014; Ferkin 2015). Indeed, this information may be viewed as public information. Public information as defined by Valone (2007) is the 'indirect information used by individuals to estimate the quality of environmental parameters'. It may be gleaned by individuals noting the behaviour of others or by the overt decisions made by them (Danchin et al. 2004). Nevertheless, this social information provided may not be intentional. For example, scent marks may contain inadvertent social information. To examine this possibility, Ferkin and Ferkin (2017) first exposed male voles to the scented bedding of a sexually receptive female and the bedding or a varying number of male conspecifics that were placed next to the bedding of that female. After the exposure, male voles were tested to assess if they spend different amounts of time investigating the fresh scent mark of the female from their exposure phase and the scent mark of a novel female. The tests revealed that male voles preferred the scent mark of a female whose bedding was associated with the bedding of three or five males compared to that of a novel female. Male voles showed no preferences for the scent mark of a novel female or that of the female who was associated with the bedding of one male or two males or zero males (clean bedding). These findings suggest that male voles may use social information provided by the scent marks of both rival males and that of a female conspecific to influence their odour preferences for a particular female. Our results are similar to those reported for male guppies that preferred to associate with females that they observed to have had more suitors compared to females they observed to have had no suitors or fewer suitors (Witte and Ryan 2002; Auld and Godin 2015). Thus, preferences for the 'popular' female vole may facilitate mate copying among male meadow voles (Ferkin and Ferkin 2017).

This speculation is intriguing in that male meadow voles are promiscuous, mate multiply, (Berteaux et al. 1999), and produce litters of mixed paternity (Boonstra et al. 1993). Thus, mating with a female that may have mated with three other males would lower that male's paternity assurance as well as increase the likelihood of sperm competition (Boonstra et al. 1993; Berteaux et al. 1999). Studies have shown that male meadow voles paired with a female while exposed to the cotton bedding of male conspecifics increase the number of spermatozoa in their ejaculate relative to that of males meadow voles paired with a female and not exposed to the bedding of male conspecifics (delBarco-Trillo and Ferkin 2004, 2006; Vaughn et al. 2008). Thus, it is likely that individuals may have to balance the costs of sending and receiving potentially public information, which may reduce their survival or attract competitors, with the benefits of attempting to mate with a 'popular' (Morand-Ferron et al. 2010).

3 Concluding Thoughts

For a scent mark to be informative it must provide a reliable, honest signal that allows individuals detecting it to predict fitness trade-offs if they choose or not choose to respond to it. The reliability of the scent mark will depend on how likely it conveys context-appropriate information. Moreover, because scent marks can be deposited in the absence of receivers and persist in the environment for days or longer (Wyatt 2014), it can be detected and provide information to receivers over a prolonged period of time. Thus, scent marks would be a source of public information, which may be inadvertent and/or have negative consequences for the sender and the receiver it was attempting to target (Ferkin and Ferkin 2017). For scent marks to function as sources of information the sender must gain a benefit that outweighs the survival or fitness cost of alerting others to its identity and its presence in an area. Likewise, the receiver that encounters the scent mark must not incur costs that are greater than what the gain in terms of survival and fitness by attending to the scent mark and responding to the information conveyed (Roberts 2007; Roberts et al. 2014; Ferkin 2015). The type of information, the amount and value of the information that is sent and received may likely vary with the life history of the participants. For example, the behaviour and physiology of senders and receivers may differ as they age and gain experience in when, where and under what conditions they may scent mark and the fitness tradeoff involved. For receivers, detecting the scent marks, choosing how they may respond, what information they may choose to convey, and what sources they may use would have to be reconciled with fitness costs and benefits to do so, and this would depend on the context and constraints.

Gosling and Roberts (2001) and Roberts (2007) provided us with excellent flowcharts for the processing of scent marks in house mice. Franklin and Ferkin (2006, 2008) and Ferkin (2011, 2015) extended such flowcharts using results from empirical studies to incorporate key cognitive elements that would be used by voles and potentially other mammals when they encounter scent marks (Ferkin et al. 2010; Vlautin and Ferkin 2012; Petrulis 2013a, b; Kaur et al. 2014; Roberts et al. 2014). Briefly, the schema is as follows: The first step in the schema involves the detection of the scent mark, which requires sensory processes. After the scent mark is detected, it must be attended to. This involves neuroendocrine and cognitive processes that underlie or mediate the interpretation of the scent mark by the individual. Interpreting the information transmitted by the scent mark will involve discrimination, perceptual, procedural, spatial memory, and episodic memory, categorization and the weighting or assessment of valence to it by the receiver. The number of senders, their identity, age and sex and eavesdroppers may affect the value of the information to the receiver. At this point, the receiver may choose whether to respond to the information. If the information induces a response, the simplest response may depend on the receiver's determination of whether the scent mark provides information about the sender posing some type of threat or a mating opportunity. If the scent mark is from a potential mate, the response would likely be to scent mark or over-mark and seek out the sender. If the scent mark is from a familiar same-sex conspecific and the information is not novel, it may be ignored by the receiver. If, however, the scent mark is from an unfamiliar same-sex conspecific and the information is new or salient, it may be investigated by the receiver. If the scent mark is from a heterospecific, the response would depend on whether or not the sender was perceived as a threat. If the information conveys a threat the receiver would avoid scent marking or seeking out interactions with the particular sender; the receiver may also shift its nest or area of activity away from the location of that sender's marks. If the information is from a heterospecific that does not pose a threat to the receiver, the receiver may ignore it, deposit its own scent marks or continue its activity in that area.

The schema highlights that the response of the receiver to the information provided by the sender should represent a tradeoff in the costs and benefits to the receiver. The actual trade-off should be context dependent. However, the context for the sender may not always correspond to that for the sender. This would depend in part on the amount of time that has passed between depositing the scent mark and it being detected. Scent marks can persist in the environment and may be detected by the target as well as eavesdroppers and bystanders. Thus, the decision to be a sender and providing reliable and accurate personal information in a scent mark must represent a fitness tradeoff.

Acknowledgements. I thank Adam C. Ferkin, Drs. Javier delBarco-Trillo, Christina D. Buesching, and Lyndsey Pierson, and two anonymous reviewers for reading earlier versions of this manuscript. The writing of this manuscript was supported by funds from the Jack H. Morris Distinguished Professorship.

References

Auld HL, Godin JGJ (2015) Sexual voyeurs and copiers: social copying and the audience effect on male mate choice in the guppy. Behav Ecol Sociobiol 69:1795–1807

Berteaux D, Bety J, Rengifo E, Bergeron J (1999) Multiple paternity in meadow voles (*Microtus pennsylvanicus*): investigating the role of the female. Behav Ecol Sociobiol 45:283–291

Bills KL, Samuelson DA, Larkin ILV (2013) Anal glands of the Florida manatee, *Trichechus manatus latirostris*: a potential source of chemosensory signal expression. Marine Mamm Sci 29:280–292

Block ML, Volpe LC, Hayes MJ (1981) Saliva as a chemical cue in the development of social behavior. Science 211:1062–1064

Boonstra R, Xia X, Pavone L (1993) Mating system of the meadow vole, *Microtus pennsylvanicus*. Behav Ecol 4:83–89

Brown RE, Macdonald DW (eds) (1985) Social odours in mammals. Clarendon Press, Oxford, UK

Buesching CD, Stankowich T (2017) Communication amongst the musteloids: signs, signals, and cues. In: Macdonald DW, Newman C, Harrington LA (eds) Biology and conservation of wild musteloids. Oxford University Press, Oxford, UK, pp 158–177

Buesching CD, Waterhouse JS, Macdonald DW (2002) Gas-chromatographic analyses of the subcaudal gland secretion of the European Badger (*Meles meles*) Part II: time-related variation in the individual-specific composition. J Chem Ecol 28:57–69

Camacho-Arroyo I, Cerbón M, Gamboa-Domínguez A, González-Agüero G, González-Mariscal G (1999) Immunocytochemical detection of estrogen and progesterone receptors in the rabbit submandibular gland. Comp Biochem Physiol 123:179–186

Ceacero F, Pluháček J, Zábransky M (2015) Pre-orbital gland opening during aggressive interactions in rusa deer (*Rusa timorensis*). Behav Proc 111:51–54

Charlton BD (2014) Discrimination of sex and reproductive state in koalas, *Phascolarctos cinereus*, using chemical cues in urine. Anim Behav 91:118–124

Charlton BD (2015) Chemosensory discrimination of identity and familiarity in koalas. Behav Proc 119:38–43

Cheetham SA, Thom MD, Jury F, Ollier WER, Beynon RJ, Hurst JL (2007) The genetic basis of individual-recognition signals in the mouse. Curr Biol 17:1771–1777

Danchin E, Giraldeau L-A, Valone TJ, Wagner RH (2004) Public information: from nosy neighbors to cultural evolution. Science 305:487–491

Danchin E, Giraldeau L-A, Cézilly F (eds) (2008) Behavioural ecology. Oxford University Press, Oxford, UK

Dall SRX, Giraldeau L-A, Olsson O, McNamara JM, Stephens DW (2005) Information and its use by animals in evolutionary ecology. Trends Ecol Evol 20:187–193

delBarco-Trillo J, Ferkin MH (2004) Male mammals respond to a risk of sperm competition conveyed by odours of conspecific males. Nature 431:446–449

delBarco-Trillo J, Ferkin MH (2006) Male meadow voles respond differently to risk and intensity of sperm competition. Behav Ecol 17:581–585

delBarco-Trillo J, LaVenture AB, Johnston RE (2009) Male hamsters discriminate estrous state from vaginal secretions and individuals from flank marks. Behav Proc 82:18–24

Ferkin MH, Ferkin DA, Ferkin BD, Vlautin CT (2010) Olfactory experience affects the response of meadowvoles to the opposite-sex scent donor of mixed-sex over-marks. Ethol 116:821–831

Ferkin MH (2011) Odor-related behavior and cognition in meadow voles, *Microtus pennsylvanicus* (Arvicolidae, Rodentia). Folia Zool 60:262–276

Ferkin MH (2015) The response of rodents to scent marks: Four broad hypotheses. Horm Behav 68:43–52

Ferkin MH, delBarco-Trillo J (2014) The behavior of female meadow voles, *Microtus pennsylvanicus*, during postpartum estrus and the responses of males to them. Mamm Biol 79:81–89

Ferkin MH, Ferkin AC (2017) The number of male conspecifics affects the odor preferences and the copulatory behavior of male meadow voles, *Microtus pennsylvanicus*. Behaviour 154:413–433

Ferkin MH, Hobbs NJ (2014) Female meadow voles, *Microtus pennsylvanicus,* respond differently to the scent marks of multiple male conspecifics. Anim Cogn 17:715–722

Ferkin MH, Johnston RE (1995) Meadow voles, *Microtus pennsylvanicus*, use multiple sources of scent for sex recognition. Anim Behav 49:37–44

Ferkin MH, Pierce AA (2007) Perspectives on over-marking: is it good to be on top? J Ethol 25:107–116

Ferkin MH, delBarco-Trillo J, Petrulis A (2017) Communication by chemical signals: physiological mechanisms, ontogeny and learning, function, evolution and cognition. In: Pfaff DW, Joëls M (eds) Hormones, brain, and behavior, 3rd edn. Elsevier Press, Oxfordshire, UK, pp 285–327

Franklin S, Ferkin MH (2006) An ontology for comparative cognition: a functional approach. Comp Cogn Behav Rev 1:36–52

Franklin S, Ferkin MH (2008) Using broad cognitive models to apply computational intelligence to animal cognition. In: Smolinski TG, Milanova MM, Hassanien A-E (eds) Applications of computational intelligence in biology: current trends and open problems. Springer, Berlin, Germany, pp 363–394

Galef BG, Giraldeau LA (2001) Social influences on foraging in vertebrates: causal mechanisms and adaptive functions. Anim Behav 51:3–15

Gosling LM, Roberts SC (2001) Scent-marking by male mammals: cheat-proof signals to competitors and mates. Adv Study Behav 30:169–217

Green JP, Holmes AM, Davidson AJ, Paterson S, Stockley P, Beynon RJ, Hurst JL (2015) The genetic basis of kinrecognition in a cooperatively breeding mammal. Curr Biol 25:2631–2641

Halpin ZT (1986) Individual odors among mammals: origins and functions. Adv Study Behav 16:39–70

Hurst JL, Beynon RJ (2004) Scent wars: the chemobiology of competitive signalling in mice. BioEssays 26:1288–1298

Hurst JL, Payne CE, Nevison CM, Marie AD, Humphries RE, Robertson DHL, Cavaggioni A, Beynon RJ (2001) Individual recognition in mice mediated by major urinary proteins. Nature 414:631–634

Johnston RE (2003) Chemical communication in rodents: from pheromones to individual recognition. J Mamm 84:1141–1162

Johnston RE (2009) Individual odors and social communication, individual recognition, kin recognition, and scent over-marking. Adv Study Behav 30:439–505

Johnston RE, delBarco-Trillo J (2009) Communication by chemical signals: behavior, social recognition, hormones and the role of the vomeronasal and olfactory systems. In: Pffaf D, Arnold AP, Fahrbach SE, Etgen AM, Rubin RT (eds) Hormones, brain, and behavior, vol 2. Elsevier Science, New York, USA, pp 797–870

Johnston RE, Derzie A, Chiang G, Jernigan P, Lee H-C (1991) Sources of individual odors in golden hamsters. Chem Senses 16:540–541

Kaur AW, Ackels T, Kuo T-H, Cichy A, Dey S, Hays C, Kateri M, Logan DW, Marton TF, Spehr M, Stowers L (2014) Murine pheromone proteins constitute a context-dependent combinatorial code governing multiple social behaviors. Cell 157:676–688

Kavaliers M, Wiebe JP, Galea LA (1994) Male preference for the odors of estrous female mice is enhanced by the neurosteroid 3 alpha-hydroxy-4-pregnen-20-one (3-alpha HP). Brain Res 646:140–144

Kean EF, Chadwick EA, Müller CT (2015) Scent signals individual identity and country of origin in otters. Mamm Biol 80:99–105

Lai S-C, Johnston RE (1994) Individual odors in Djungarian hamsters (*Phodopus campbelli*). Ethology 96:117–126

Lai S-C, Vasilieva NY, Johnston RE (1996) Odors providing sexual information in Djungarian hamsters: evidence for an across-odor code. Horm Behav 30:26–36

Martin J, Carranza J, López P, Alarcoa S, Pedro-González J (2014) A new sexual signal in rutting male red deer: age related chemical scent constituents in the belly black spot. Mamm Biol 79:362–368

McLean S (2014) Scent glands of the common bushtail possum (*Trichosurus vulpecula*). New Zealand J Zool 41:193–202

Morand-Ferron J, Doligez B, Dall SRX, Reader SM (2010) Social information use. Encyclopedia of Anim Behav 3:242–250

Mykytowycz R (1965) Further observations on the territorial function and histology of the submandibular cutaneous (chin) glands in the rabbit, *Oryctolagus cuniculus* (L.). Anim Behav 13:400–412

Owen MA, Swasigood RR, Slocomb A, Amstrup SC, Dumer GM, Simac K, Pessier AP (2015) An experimental investigation of chemical communication in the polar bear. J Zool 295: 36–43

Palphramand KL, White PCL (2007) Badgers, *Meles meles*, discriminate between neighbour, alien and self scent. Anim Behav 74:429–436

Petrulis A (2013a) Chemosignals, hormones and mammalian reproduction. Horm Behav 63: 723–741

Petrulis A (2013b) Chemosignals and hormones in the neural control of mammalian sexual behavior. Front Neuroendocrinol 34:255–267

Petty JMA, Drea CM (2015) Female rule in lemurs is ancestral and hormonally mediated. Scient Reports 5:e9631

Pluháček J, Ceacero F, Lupták P (2015) First records of preorbital gland opening in rare wild barasingha (*Rucervus duvaucelii*). Behav Proc 119:28–31

Roberts SA, Davidson SJ, Beynon RJ, Hurst JL (2014) Female attraction to mate scent and associative learning: the house mouse as a mammalian model. Anim Behav 97:313–321

Roberts SC (2007) Scent marking. In: Wolff JO, Sherman PW (eds) Rodent societies: an ecological and evolutionary perspective. University of Chicago Press, Chicago, USA, pp 255–266

Sabau RM, Ferkin MH (2013) Food deprivation and restriction during late gestation affect the sexual behavior of postpartum female meadow voles, *Microtus pennsylvanicus*. Ethology 119:29–38

Sharpe LL (2015) Handstand scent marking: height matters to dwarf mongooses. Anim Behav 105:173–179

Thom MD, Hurst JL (2004) Individual recognition by scent. Ann Zool Fennici 41:765–787

Thonhauser KE, Raveh S, Hettyey A, Beissmann H, Penn DJ (2013) Why do male mice mate with multiple mates? Behav Ecol Sociobiol 67:1961–1970

Thoß M, Luzynski K, Ante M, Miller I, Penn DJ (2015) Major urinary protein (MUP) profiles show dynamic changes rather than individual "barcode" signatures. Front Ecol Evol 3:71. https://doi.org/10.3389/fevo.2015.00071

Tinnesand HV, Buesching CD, Noonan MJ, Newman C, Zedrosser A, Rosell F, Macdonald DW (2015) Will trespassers be prosecuted or assessed according to their merits? A consilient interpretation of territoriality in a group-living carnivore, the European badger (*Meles meles*). PLoS ONE e0132432

Valone TJ (1989) Group foraging, public information and patch estimation. Oikos 56:357–363

Valone TJ (2007) From eavesdropping on performance to copying the behavior of others: a review of public information use. Behav Ecol Sociobiol 62:1–14

Vaughn AA, delBarco-Trillo J, Ferkin MH (2008) Sperm investment in male meadow voles is affected by the condition of the nearby male conspecifics. Behav Ecol 19:1159–1164

Vlautin CT, Ferkin MH (2012) The influence of predator and conspecific odor on sex differences in path choice in meadow voles. Behaviour 149:133–152

Witte K, Ryan MJ (2002) Mate choice copying in the sailfin molly, *Poecilia latipinna*, in the wild. Anim Behav 63:943–949

Woodward RL, Bartos K, Ferkin MH (2000) Meadow voles (*Microtus pennsylvanicus*) and prairie voles (*M. ochrogaster*) differ in their responses to over-marks from opposite- and same-sex conspecifics. Ethology 106:979–992

Wyatt TD (2014) Pheromones and animal behavior: chemical signals and signatures. Cambridge University Press, Cambridge, UK

Zala SM, Potts WK, Penn DJ (2004) Scent-marking displays provide honest signals of health and infection. Behav Ecol 15:338–344

Latrines as Potential Communication Centres in Short-Beaked Echidnas

Rachel L. Harris[1,2](✉), Jenny Sprent[1], and Stewart C. Nicol[1]

[1] School of Biological Sciences, University of Tasmania, Hobart TAS 7001, Australia
rachelharris138@outlook.com
[2] Department of Evolutionary Anthropology, Duke University, Durham NC 27708, USA

Abstract. Localised defecation sites ('latrines') are an important method of intraspecific olfactory communication in mammals. As efficient, low-cost methods of exchanging information, latrines might be especially important for solitary or widely distributed species, or those with low metabolic rates or low-energy diets. Latrines can be difficult to find and closely monitor, but information on their spatial distribution and patterns of use can provide useful insights into their possible functions. We used geographic mapping and remote cameras to describe spatial, temporal and behavioural patterns in latrine use in a normally solitary monotreme, the short-beaked echidna (*Tachyglossus aculeatus*; 'echidna'). We found 64 echidna latrines, usually under logs in scrub habitat and in areas of intermediate (rather than low or high) areas of overlap between resident echidna home ranges. Latrines were most often visited through July–November, and fresh scats were found between August and February. When visiting latrines, echidnas showed olfactory behaviours such as scanning and nose-poking at the soil, suggesting latrines are not used purely for defecation and likely perform a role in communication. We suggest that latrines play a role in signalling habitat use, particularly among females, whereas males might use latrines to monitor the location and activities of potential mates and competitors.

1 Introduction

Latrines play a central role in olfactory communication in a variety of mammalian species (Brown and Macdonald 1985). Mammals often rely on olfactory cues present in faeces, urine and other secretions to exchange information with conspecifics (Bradbury and Vehrencamp 2011; Wyatt 2014). Once deposited at latrine sites, such odours can persist in the environment long after the signalling animal has left, potentially reaching a wide audience (Müller-Schwarze 2006; Wyatt 2014) without the need for direct interaction between communicating individuals (Darden et al. 2008; Wronski et al. 2013). Consequently, latrines may function as 'information centres' or 'bulletin boards' both within and between social groups, or between widely distributed individuals (Darden et al. 2008; Eppley et al. 2016). Owing to the often cryptic nature of latrines and the animals using them, the specific communicatory functions of latrines are often not well understood. Researchers have variously used a combination of methods, such

© Springer Nature Switzerland AG 2019
C. D. Buesching (Ed.): *Chemical Signals in Vertebrates 14*, pp. 13–26, 2019.
https://doi.org/10.1007/978-3-030-17616-7_2

as geographic mapping and remote camera monitoring, to describe possible functions of latrine use in a wide variety of both social and asocial taxa, including many carnivores (Jordan et al. 2007; Hulsman et al. 2010; Ruibal et al. 2010; Crowley et al. 2012; Rodgers et al. 2015; Buesching et al. 2016), social primates (Irwin et al. 2004; Dröscher and Kappeler 2014; Eppley et al. 2016), rodents (Sneddon 1991; Ziege et al. 2016) and ungulates (Wronski et al. 2006; Wronski and Plath 2010).

Latrines may perform a variety of roles in animal communication, depending on their spatial distribution, how they are used, and by which individuals within a population. For example, concentrations of latrines along territorial borders suggest a role in defence (Hogan et al. 2013; Buesching et al. 2016), whereas increased density of latrines in core areas of home ranges may demarcate areas of exclusive use (Jordan et al. 2007; Wronski and Plath 2010) or signal food availability (Begg et al. 2003). Furthermore, latrines may help to coordinate social and reproductive behaviour by allowing conspecifics to monitor the activity and condition of competitors and potential mates (Sneddon 1991; Rostain et al. 2004; Wronski et al. 2006; Darden et al. 2008).

Here, we integrated two approaches to describe latrine use in the Tasmanian subspecies of short-beaked echidna (*T. a. setosus*). Echidnas are found throughout Australia (Augee 2008) and, like other myrmecophagous species, have low metabolic rates (McNab 1984) and low-energy diets (Sprent and Nicol 2016). In Tasmania, at the southernmost extent of the echidna's range, echidna spatial organisation appears largely driven by females and their response to resource availability: home ranges of females are smaller than those of males (Nicol et al. 2011; Sprent and Nicol 2012), decrease in size with increasing availability of high-quality 'scrub' habitat (Sprent and Nicol 2012) and tend to be more exclusive than those of males (Nicol et al. 2011; Sprent and Nicol 2012). In contrast, the home ranges of male echidnas are much larger than those of females (Nicol et al. 2011), and overlap with those of many other individuals, suggesting male echidnas distribute themselves to maximise access to mates (Nicol et al. 2011; Sprent and Nicol 2012). Echidna distribution is likely mediated, at least in part, by olfactory communication. Their olfactory apparatus is strikingly large and complex, including an enormous olfactory surface and a folded olfactory bulb, in addition to a large prefrontal cortex, which is likely to function as an integration centre for behaviourally important chemosensory information (Ashwell 2013). Furthermore, putative cloacal scent glands produce secretions containing compounds of varying molecular weight and volatility (Harris et al. 2012) which, when combined with cloacal scent-marking behaviour (Dobroruka 1960; Beard et al. 1992), could allow for communication of both short- and long-term signals to conspecifics.

It seems likely that latrines perform a communicatory function among solitary echidnas: for example, males might use latrines to optimise the placement of their core home ranges in relation to those of females (Sprent and Nicol 2012) or allow females to maintain exclusivity in their resource use. In their earlier pilot study, Sprent and colleagues (2006) found that Tasmanian echidnas use latrines continuously between August and February (i.e. from the end of the mating period until the end of the Austral summer, when many individuals begin to enter hibernation; Nicol and Morrow 2012). Both males and females were found near latrines (Sprent et al. 2006), but we lack basic information on how latrines are distributed within a population, and how individual animals use them. For example, of the nearly 6,000 observations from a single

Tasmanian study population made over 15 years, only 13 observations are of echidnas in or near latrines. Using two complementary approaches, geographic mapping and remote camera monitoring, we aimed to describe (1) where latrines are located with respect to habitat and echidna distribution; (2) echidna behaviour at latrines; and (3) temporal variations in latrine use. If latrines are used to advertise resource use, we hypothesise that echidna latrines will be more common in areas of 'high-quality' habitat and in areas with more overlapping home ranges. We expected that if latrines are used for communication, we would observe scent-marking and other olfactory behaviours, and that latrines would be used by multiple individuals. We also anticipated that latrine use would be minimal during the hibernation period (March–May), increase during the breeding season when more individuals are active (June–September), and peak during spring–summer, when maximum foraging and dispersal occurs (October–January).

2 Materials and Methods

2.1 Study Site and Animals

Fieldwork was conducted in a 12 km^2 section of grazing property in the Tasmanian southern midlands (42°28′S, 147°14′E), where the resident echidna population has been the subject of ongoing physiological and ecological studies since 1996. The study site consists of remnant *Eucalyptus amygdalina* woodland interspersed with improved and native pasture, along with patches of *Acacia dealbata*, *Casuarina verticilata* and *Lomandra longifolia*. The site is intersected with numerous gullies, caves and sandstone outcrops; altitude is 200–400 m asl. On the first capture, all echidnas at this site are fitted with subcutaneous passive implantable transponder (PIT) tags (LifeChip, Destron Fearing, MN, USA) for identification. At the time of this study, a total of 249 individuals had been found at the site, including 115 females, 95 males and 39 juveniles (sex unknown). Of these animals, 15 females, 5 males and one juvenile are known to have died. Population size and density are difficult to estimate in this cryptic species, particularly as density seems to vary by habitat type (Nicol et al. 2011; Sprent and Nicol 2012), but all animals found in this site occupy a total area of approximately 900 ha. Up to approximately 15 adults in any one year are fitted with radiofrequency transmitters (Holohil Systems, Ontario, Canada) for serial monitoring, and have colour-coded straws attached to their spines for identification on remote cameras. Further details on site climate and radio-tracking protocols are outlined by Nicol et al. (2011).

2.2 Latrine Location in Relation to Habitat Type and Core Home Ranges

In October 2010, we mapped latrine locations by searching an approximately 62-hectare section in the centre of the field site, known to contain the home ranges of resident adult male and female echidnas. Echidna latrines are typically located under fallen logs, trees or large slabs of rock, with a loose soil base and sufficient space for an echidna to enter (Sprent et al. 2006). Scats are found on or near the latrine surface layer and may be buried within the top 5–10 cm of soil (Grove et al. 2006; Sprent et al. 2006). We located and identified echidna latrines by systematically searching the survey area for windrows, logs and large rocks, then checking for the presence of echidna scats, which are distinctively cylindrical and comprised of soil and invertebrate exoskeleton (Triggs 2004). If no scats were visible on the soil surface of an otherwise probable latrine, we checked for buried scats by probing the top 5–10 cm of soil by hand. After discovering a latrine, we recorded its surrounding habitat features and global positioning coordinates. The vegetation structure was sufficiently open that few logs and rocks would have been inadvertently missed during the latrine survey.

To evaluate how latrine locations might be related to habitat type, we categorised vegetation according to the density of tree cover, based on a high-resolution QuickBird image of the survey area in ArcGIS (version 9.2). To estimate tree density across the survey area, we drew 100 m transects across blocks of vegetation on the image, then counted the number of trees intersected by each transect: 'open' areas contained 0–4 trees on a 100 m transect, 'scrub' areas contained 5–12 trees, and 'thick' areas contained ≥ 13 trees. We digitised the blocks of tree densities, then calculated latrine density for each vegetation category. Next, we compared latrine locations with the number of overlapping core home ranges of individual echidnas in the surrounding area. Echidna location data were collected between February 1996 and October 2010 (Nicol et al. 2011). Echidnas have a high degree of home range fidelity (Nicol et al. 2011) and probably spend most of their adult life within the same home range (Abensperg-Traun 1991; Augee et al. 1992), although they may make occasional excursions out of their home range (see Fig. 1 in Nicol et al. 2011). To ensure that the home ranges overlapping with the latrine survey area and therefore included in our final analyses ($n = 20$) reflected the current population, we extracted home range data only for those individuals that had been observed alive within the previous five years, and calculated core home ranges for only those individuals with 20 or more locations recorded throughout the year (Nicol et al. 2011). We used a fixed kernel method and the reference (h_{ref}) smoothing parameter (Nicol et al. 2011). We used the 50% kernel area, which is generally considered to be the core of an animal's home range and is likely to be used exclusively by an individual (Cimino and Lovari 2003; Cavalcanti and Gese 2009).

For each latrine, we calculated the tree density and the number of core home ranges which overlapped the latrine location. We first used a Chi-squared test to determine if latrines were randomly distributed across the three different vegetation patch classes (open, scrub, thick). Then, we used an analysis of variance (ANOVA) to test for the effects of both vegetation type and sex of home range owners against the number of core home ranges associated with each latrine. Analyses were performed using Statistica (version 6.1; Statsoft, Tulsa, OK, USA). Results were deemed significant at $P < 0.05$.

2.3 Temporal Patterns of Latrine Visits and Echidna Behaviour

To describe seasonal and daily patterns of latrine use and echidna behaviour, we used remote motion-sensing cameras to simultaneously monitor seven latrines for 52 weeks between November 2011 and November 2012. The latrines were distributed across two sections of scrub (approx. 750 m apart) within the latrine survey site. Individual latrines within each section were approximately 80 m apart (mean: 80.9 m, min. 30 m, max. 153 m). At each latrine, we attached two cameras to branches or wooden stakes: one camera recorded motion video (Scoutguard SC550, HuntingCamOnline Gadsden, SC, USA), while a second camera (Reconyx PC800, Holmen, WI, USA) recorded still photographs from an additional angle. We visited cameras each week to download camera footage, check equipment, replace batteries, and check for fresh echidna scats. Because it is not possible to assign individual scats to a donor animal without genotyping techniques, we did not attempt to differentiate between individual scat samples found in latrines (i.e. we noted presence/absence of fresh scats but did not speculate on whether they came from multiple animals or derived from a single, fragmented scat).

We reviewed all videos and identified those containing footage of echidnas. Where possible, we identified known individuals by the location of radio transmitters and colour-coded plastic tubing attached to their spines. We recorded animal behaviour using an ethogram (Table 1) and noted whether a recorded visit coincided with fresh scats being found in the latrine. We used the camera timestamps to calculate the time and duration of latrine visits. Sometimes it was not possible to accurately record echidna behaviour (e.g. poor camera flash, animal partly obscured, cameras only triggered as the animal exited the latrine), and we collectively defined these recordings as 'obstructed view' (OV). We excluded OV records from our calculations of visit duration and behaviour frequency but included them for analyses on visit frequencies and timing. We examined both daily and seasonal patterns in latrine use. First, we classified echidna activity according to the time of day of each latrine visit: 'diurnal' visits occurred between sunrise and sunset, 'twilight' visits occurred either in the mornings (after astronomical twilight and before sunrise) or in the evenings (after sunset and before astronomical twilight), and 'nocturnal' visits occurred overnight, between astronomical twilights (GeoScience Australia). To test for differences in latrine use (number of visits or number of scats) by month, we ran linear mixed effects models using the *glmmADMB* package (version 0.8.3.3; Fournier et al. 2012) in R Studio (version 1.0.136; RStudio Team 2016). For each model, we used a Poisson distribution,

Table 1. Ethogram of echidna behaviours recorded at latrines and frequency with which each behaviour was observed (*N*)

Behaviour	*N*	Description
Dig	10	Using forelegs to move soil or objects, creating depression
Nose-poke	11	Nostrils inserted below soil surface
Roll	6	Lying on back, rolling from side to side
Scent-mark	3	Dragging cloaca along objects or ground
Scrape/ backfill	5	Using forelegs to scrape and move soil or objects underneath the animal's body, usually to fill a small depression
Scratch/groom	11	Using hindlegs to scratch or groom spines
Sniff (ground)	30	Beak towards or touching ground, with/without head movement
Sniff/scan (air)	20	Beak raised above horizontal, with/without head movement
Tongue-flick	3	Tongue visible, moving rapidly in and out of mouth
Walk	31	Travelling either through or within latrine

and included 'month' as a fixed variable and 'latrine ID' as a random variable. For interpretative purposes, we also describe seasonal patterns of latrine use by dividing the year into three seasons according to echidna behaviour and physiology at the time (Nicol and Morrow 2012): 'hibernation' (February to May); 'mating' (June to mid-September); and 'active' (mid-September to January).

3 Results

3.1 *Latrine Location in Relation to Habitat Type and Core Home Ranges*

We found 64 latrines (overlapping with at least 20 core home ranges) in the survey area (Fig. 1a,b); all latrines were located under logs and fallen trees. Latrine density varied significantly between the three patch types $\left(\chi_2^2 = 34.83, P < 0.005\right)$: latrines were approximately four times more densely distributed in scrub patches than in patches of open habitat or thick bush (Table 2). Notably, the presence of latrines did not appear to be a simple function of the availability of log shelters, since not all log sites contained latrines, and log sites were distributed across all three vegetation types.

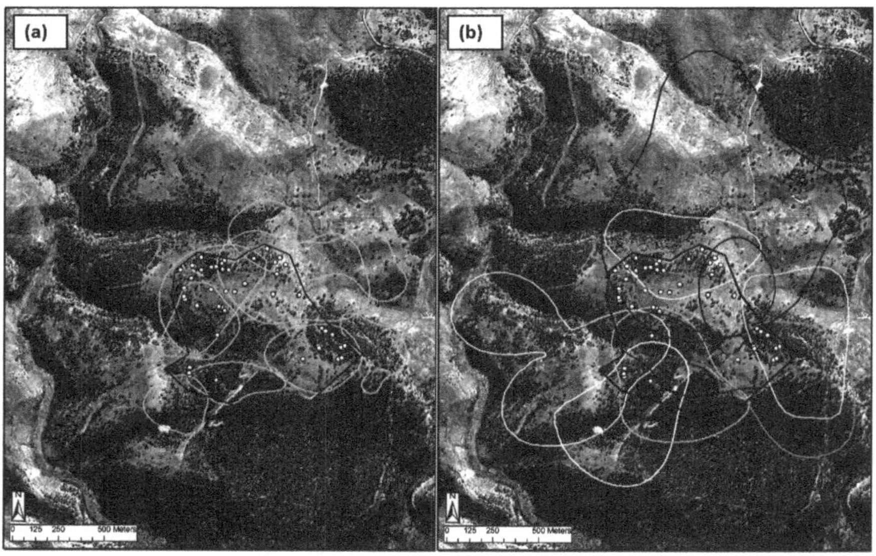

Fig. 1. Location of latrines (white circles) and core home ranges (lines) of known **a** female and **b** male Tasmanian echidnas (*Tachyglossus aculeatus setosus*). The black line indicates the latrine survey area (centre of both images). A colour version of this figure is available from the authors

Over the five years leading up to the 2010 survey, 32 echidnas were observed within the survey area. Twenty adults (9 females, Fig. 1a; 11 males, Fig. 1b) had their core home range overlap with the location of latrines. An additional 12 individuals had core home ranges that either fell outside of the survey area (2 males, 6 females) or could not

Table 2. Total area of each vegetation patch type (hectares, 'ha'), numbers of echidna latrines (*N*) found within each patch type, and density of echidna latrines. Patch types are categorised according to the number of trees on a 100 m transect: 'open' areas contained 0–4 trees, 'scrub' areas contained 5–12 trees, 'thick' areas contained ≥ 13 trees

Patch type	Area (ha)	*N*	Latrine density/ha
Open	31.94	18	0.56
Scrub	16.56	34	2.05
Thick	25.37	12	0.47
Total	**73.87**	**64**	**0.87**

be calculated due to insufficient location data (4 juveniles). Most (62/64) latrines fell within at least one known core home range, but the number of latrines did not increase linearly with the number of overlapping core home ranges: for example, most latrines overlapped with the core home ranges of two females (30 latrines) or two males (35 latrines), whereas very few latrines overlapped with the core home ranges of more than three individuals (e.g. only one latrine overlapped with four male home ranges, and three latrines overlapped with four female home ranges). The number of core home ranges that overlapped with any specific latrine did not vary significantly with vegetation patch type ($F_{2,122} = 0.59$, $P = 0.55$), the sex of the home range owner(s) ($F_{1,122} = 0.0006$, $P = 0.98$), nor the interaction between sex and patch type ($F_{2,122} = 2.32$, $P = 0.10$; Table 3).

Table 3. Echidna core home ranges containing latrines (*N*) in patches of different vegetation type: 'open' areas contained 0–4 trees on a 100 m transect, 'scrub' areas contained 5–12 trees, and 'thick' areas contained ≥ 13 trees

Patch type	*N*	Mean number of known core home ranges (females)	Mean number of core known home ranges (males)
Open	19	2.53 ± 0.18	2.11 ± 0.13
Scrub	33	2.27 ± 0.20	2.18 ± 0.12
Thick	12	1.83 ± 0.11	2.33 ± 0.26

3.2 Echidna Behaviour at Latrines

We recorded 73 visits and 126 min of echidna activity at seven latrines over the 52-week survey period (Fig. 2). All latrines were visited at least once (mean: 10.4 visits; range: 1–20) and all visits were by lone individuals. We positively identified six adults (3 males, 15 visits; 3 females, 4 visits) and one juvenile male from the camera footage. Except for one latrine which was only visited once during the study period, all latrines were visited by more than one individual animal. After excluding OV footage where echidna behaviour could not be recorded accurately ($n = 31$ visits), and footage of a juvenile male that took shelter in a latrine while being radio-tracked, we recorded 130 observations of different echidna behaviours (Table 1) in the remaining 109 min of footage from 41 visits. The most commonly observed behaviours were walking (recorded in 23.8% of visits) and sniffing at the ground (23.1%), along with sniffing or

Fig. 2. Echidna behaviour recorded at a latrine in the Tasmanian southern midlands. Behaviours observed included: **a** sniff (ground); **b** nose-poke; **c** and **d** scrape/backfill. Fresh scats recovered following this visit. Visit duration was 6 min 14 s. The sex and identity of the individual shown are not known. Stills taken from video camera footage recorded at 5.25 pm on 23 November 2011. Video footage is available from the authors

scanning the air (15.4%), scratching or grooming (8.5%), nose-poking (8.5%) and digging (7.7%). We also recorded less frequent behaviours such as rolling (4.6%), scraping or backfilling (3.8%), tongue-flicking (2.3%) and scent-marking (2.3%). It was not possible to directly observe an animal defecating due to the camera angle and the animal's proximity to the ground.

Latrine use varied significantly by month, both in terms of the number of echidna visits ($z = 4.20$, $P < 0.0001$, peaking in July–November; Fig. 3) and the number of times that fresh scats were found ($z = 3.56$, $P = 0.00037$, peaking in December; Fig. 3). Nearly half of all latrine visits occurred during the mating season (June to mid-September, $n = 36$, 49.3%; non-breeding season, mid-September to January, $n = 32$, 43.8%; hibernation season, February to May, $n = 5$, 6.8%). Of the visits occurring during the mating season, most were brief (≤ 30 s, $n = 29$), approximately two-thirds occurred at night, and all positively identified individuals which visited latrines during the mating season were male (3 individuals, $n = 8$ visits, 7 visits between late September and January, i.e. after the breeding season). We only positively identified females in latrines during the non-breeding season between October and December, (3 individuals, $n = 4$ visits). Overall, we recorded more visits at night ($n = 36$) than during the day ($n = 27$) or at twilight ($n = 10$). Across the Tasmanian echidna's active

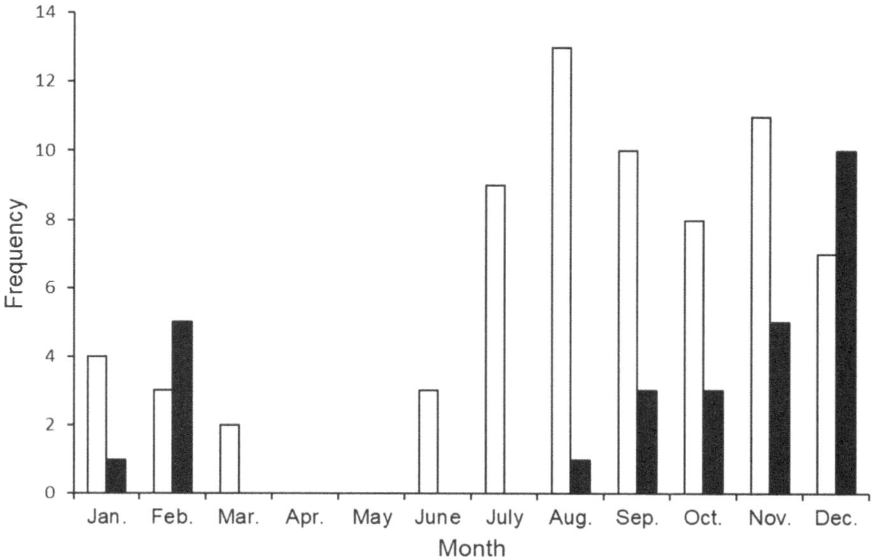

Fig. 3. Total number of echidna visits recorded on camera (white bars) and number of times scat deposits were found (black bars) per month across seven latrine sites between November 2011 and November 2012

period (mating season and non-breeding seasons combined, June–January), the average time between two echidna visits at a given latrine was 12 days (min 200 min, max 63 days). We missed at least 10 echidna visits, as indicated by the presence of fresh echidna scat in the latrine but no corresponding footage, and we only found fresh scats after approximately 38% (28/73) of echidna visits. When fresh scats were found, we also noticed (1) a strong odour in the latrine, and (2) observed behaviours such as rolling, nose-poking, scraping and digging in the camera footage, suggesting these behaviours are associated with defecation events. Latrine visits also tended to be longer when fresh scats were subsequently found (mean: 243 s, range: 30–1144 s) compared with when no scats were found (mean: 112 s, range: 2–1210 s).

4 Discussion

Our results emphasise a role for latrines as communication centres among short-beaked echidnas. Echidna latrines were not randomly distributed throughout the study area: instead, they were concentrated under large logs, which are important echidna shelter sites (Wilkinson et al. 1998) and would probably extend the life of scent marks by protecting them from weather (Alberts 1992; Stewart et al. 2002). Many nearby sites that also appeared to offer good protection from weather and disturbance had no evidence of scats, suggesting that latrines did not simply occur in all echidna shelter sites. Because some latrines can accumulate large amounts of scat with continuous use (e.g. 1–2 kg each; Sprent et al. 2006), yet echidnas did not leave fresh scats at every visit, it appears that latrines are the result of animals inspecting and sometimes deliberately over-marking existing scats, and selectively leaving new scats. Most latrines were used by multiple individuals throughout the camera monitoring period, and we observed several types of olfactory behaviours, suggesting that echidnas could use odour cues in latrines to communicate information on individual presence and resource use to conspecifics.

Rather than being evenly distributed across different types of vegetation, echidna latrines were concentrated in areas of high-quality 'scrub' habitat. If, as in other species, latrines function to defend resources such as refuge or feeding sites, then the significantly heterogeneous landscape may make it more profitable for echidnas to mark most frequently in more valuable areas and not defend areas of marginal value (McLoughlin et al. 2000; Delahay et al. 2007). Consistent with an earlier finding that female home range sizes decline with increasing proportion of scrub habitat (Sprent and Nicol 2012), the greater density of latrines in scrub patches suggests that these habitats may be more valuable, perhaps due to spatial or temporal heterogeneity in food availability or distribution of refuge sites. Moreover, we observed that echidnas of both sexes visited latrines during the nonbreeding season, and latrine use (as measured by scat deposition) peaked during the period when echidna feeding activity at this site was at its annual maximum (Nicol and Morrow 2012). As in other species that experience seasonal changes in food availability, the period of maximal feeding may be the most important time of year for signalling resource use and maintaining optimal spacing between individuals (Pigozzi 1990; Zhou et al. 2015). In female echidnas, such signalling is probably important for maintaining relative home range exclusivity, and perhaps to deter potential immigrants from establishing home ranges in areas that already support an optimal number of residents.

Surprisingly, the number of latrines did not increase in areas of relatively high echidna density: instead, most were found in areas of intermediate density (as indicated by the numbers of individuals with overlapping core home ranges). In other species, territorial (or home range) defence may be abandoned or modified depending upon the relative benefits of maintaining an exclusive area versus the energetic costs of defending it (e.g. brown bears: McLoughlin et al. 2000; European badgers: Buesching et al. 2016). For example, in areas of low animal density, there may be minimal benefit of maintaining latrine sites if no other animals live close by, whereas in high-density areas, latrine use and territorial behaviour may be abandoned due to the greater energetic costs of frequently needing to overmark scats of other animals. We did not observe any similar patterns of scat deposition across latrines in different areas of echidna density, but it is possible that other olfactory cues might be deposited in latrines as urine or scent marks, which could not be measured in this study. Genetic techniques could help to uncover more detailed information on how specific individuals use latrine sites for communication.

Latrine visits peaked during the breeding season, when only a portion of the total population has emerged from hibernation, then continued until November–December. Visits during the breeding season were typically brief and occurred at night, and all positively identified animals were male. These observations are consistent with our understanding of sex differences in activity patterns (Nicol and Morrow 2012; Morrow et al. 2016) and male searching behaviour during the breeding season (Nicol et al. 2011; Nicol and Morrow 2012). GPS tracking shows that roving male echidnas cover large distances, predominately at night, as they search for potential mates (Nicol unpublished data). In contrast, female Tasmanian echidnas are not typically active during the breeding season, as they often remain in hibernation until they are found by searching males (Morrow and Nicol 2009; Nicol and Morrow 2012). Consequently, it seems unlikely that latrines have a role in signalling reproductive status to potential mates. However, male echidnas could use information on sex or individual identity (present in scent secretions and possibly also in faeces; Harris et al. 2014) to optimise the placement of their home ranges with respect to females, thus maximising male reproductive opportunities. Frequent visits to latrines during the breeding season might also allow males to deter and monitor the activity of potential competitors (Eppley et al. 2016). Although we did not detect scats in latrines until August, over halfway through the breeding season, male echidnas might also monitor cues present in urine or scent marks (not measured here).

Although latrines appear to have several potential functions as communication centres in short-beaked echidnas, it is unclear whether the patterns we observed are consistent with other myrmecophagous species: for example, latrines may function in territory defence in captive numbats (*Myrmecobius fasciatus*; Hogan et al. 2013), but not in giant anteaters (*Myrmecophaga tridactyla*; Möcklinghoff 2008). Latrine use in tamanduas, armadillos and pangolins remains to be described. Additionally, investigating latrine use in other echidna subspecies will be beneficial, as they will be subject to different degrees of seasonal food availability and population densities.

Acknowledgements. We thank Gemma Morrow for her assistance in the field, the McShane family for allowing us access to their property and Menna Jones for loan of cameras. Research was carried out with approvals from the University of Tasmania Animal Ethics Committee and the Tasmanian Department of Primary Industries, Parks, Water and the Environment. Research was supported by the Holsworth Wildlife Research Fund, the MA Ingram Trust and the National Geographic Committee for Research and Exploration.

References

Abensperg-Traun M (1991) A study of home-range, movements and shelter use in adult and juvenile echidnas, *Tachyglossus aculeatus* (Monotremata: Tachyglossidae), in Western Australian wheat belt reserves. Aust Mammal 14:13–22

Alberts AC (1992) Constraints on the design of chemical communication systems in terrestrial vertebrates. Am Nat 139(Suppl.):S62–S89

Ashwell KWS (2013) Chemical senses: olfactory and gustatory systems. In: Ashwell KWS (ed) Neurobiology of monotremes: brain evolution in our distant mammalian cousins. CSIRO Publishing, Collingwood, Australia, pp 235–250

Augee ML (2008) Family Tachyglossidae: echidnas. In: Van Dyck S, Strahan R (eds) The mammals of Australia. New Holland Publishers, Sydney, Australia, pp 36–39

Augee ML, Beard LA, Grigg GC et al (1992) Home range of echidnas in the Snowy Mountains. In: Augee ML (ed) Platypus and echidnas. The Royal Zoological Society of New South Wales, Mosman, Australia, pp 225–231

Beard LA, Grigg GC, Augee ML (1992) Reproduction by echidnas in a cold climate. In: Augee ML (ed) Platypus and echidnas. The Royal Zoological Society of New South Wales, Mosman, Australia, pp 93–100

Begg CM, Begg KS, Du Toit JT et al (2003) Scent-marking behaviour of the honey badger, *Mellivora capensis* (Mustelidae), in the southern Kalahari. Anim Behav 66(5):917–929

Bradbury JW, Vehrencamp SL (2011) Principles of animal communication. Sinauer Associates Inc, Sunderland, USA

Brown RE, Macdonald DW (1985) Social odours in mammals. Clarendon Press, Oxford, UK

Buesching CD, Newman C, Service K et al (2016) Latrine marking patterns of badgers (*Meles meles*) with respect to population density and range size. Ecosphere 7(5):e01328

Cavalcanti SM, Gese EM (2009) Spatial ecology and social interactions of jaguars (*Panthera onca*) in the southern Pantanal. Brazil. J Mammal 90(4):935–945

Cimino L, Lovari S (2003) The effects of food or cover removal on spacing patterns and habitat use in roe deer (*Capreolus capreolus*). J Zool 261(3):299–305

Crowley S, Johnson CJ, Hodder D (2012) Spatial and behavioral scales of habitat selection and activity by river otters at latrine sites. J Mammal 93(1):170–182

Darden SK, Steffensen LK, Dabelsteen T (2008) Information transfer among widely spaced individuals: latrines as a basis for communication networks in the swift fox? Anim Behav 75 (2):425–432

Delahay R, Ward A, Walker N et al (2007) Distribution of badger latrines in a high-density population: habitat selection and implications for the transmission of bovine tuberculosis to cattle. J Zool 272(3):311–320

Dobroruka VLJ (1960) Einige Beobachtungen an Ameisenigeln, *Echidna aculeata* Shaw (1792). Z Tierpsychol (trans: Polymeropoulos E) 17:178–181

Dröscher I, Kappeler PM (2014) Maintenance of familiarity and social bonding via communal latrine use in a solitary primate (*Lepilemur leucopus*). Behav Ecol Sociobiol 68(12):2043–2058

Eppley TM, Ganzhorn JU, Donati G (2016) Latrine behaviour as a multimodal communicatory signal station in wild lemurs: the case of *Hapalemur meridionalis*. Anim Behav 111:57–67

Fournier D, Skaug H, Ancheta J et al (2012) AD Model Builder: using automatic differentiation for statistical inference of highly parameterized complex nonlinear models. Optim Method Softw 27:233–249

Grove S, Richards K, Spencer C et al (2006) What lives under large logs in Tasmanian eucalypt forest? Tasmanian Nat 128:86–93

Harris RL, Davies NW, Nicol SC (2012) Chemical composition of odorous secretions in the Tasmanian short-beaked echidna (*Tachyglossus aculeatus setosus*). Chem Senses 37(9):819–836

Harris RL, Holland BR, Cameron EZ et al (2014) Chemical signals in the echidna: differences between seasons, sexes, individuals and gland types. J Zool 293(3):171–180

Hogan LA, Lisle AT, Johnston SD (2013) Eliminative behavior of captive numbats, *Myrmecobius fasciatus* (Mammalia; Marsupialia): pattern and identification of fecal deposits. Zoo Biol 32(3):324–331

Hulsman A, Dalerum F, Swanepoel L et al (2010) Patterns of scat deposition by brown hyaenas *Hyaena brunnea* in a mountain savannah region of South Africa. Wildlife Biol 16(4):445–451

Irwin MT, Samonds KE, Raharison JL et al (2004) Lemur latrines: observations of latrine behavior in wild primates and possible ecological significance. J Mammal 85(3):420–427

Jordan NR, Cherry MI, Manser MB (2007) Latrine distribution and patterns of use by wild meerkats: implications for territory and mate defence. Anim Behav 73:613–622

McLoughlin PD, Ferguson SH, Messier F (2000) Intraspecific variation in home range overlap with habitat quality: a comparison among brown bear populations. Evol Ecol 14(1):39–60

McNab BK (1984) Physiological convergence amongst ant-eating and termite-eating mammals. J Zool 203:485–510
Möcklinghoff L (2008) Social organization and habitat use of the giant anteater (*Myrmecophaga tridactyla* L., 1758) in timber plantations in northern Brazil. Diploma thesis, Julius-Maximilians Universität Würzburg
Morrow G, Nicol SC (2009) Cool sex? Hibernation and reproduction overlap in the echidna. PLoS ONE 4(6):e6070
Morrow GE, Jones SM, Nicol SC (2016) Interaction of hibernation and male reproductive function in wild Tasmanian short-beaked echidnas *Tachyglossus aculeatus setosus*. J Mammal 97(3):852–860
Müller-Schwarze D (2006) Chemical ecology of vertebrates. Cambridge University Press, Cambridge
Nicol SC, Morrow G (2012) Sex and seasonality: reproduction in the echidna (*Tachylossus aculeatus*). In: Ruf T, Bieber C, Arnold W et al (eds) Living in a seasonal world: thermoregulatory and metabolic adaptations. Springer, Heidelberg, Germany, pp 143–153
Nicol SC, Vanpé C, Sprent JA et al (2011) Spatial ecology of a ubiquitous Australian anteater, the short-beaked echidna (*Tachyglossus aculeatus*). J Mammal 92(1):101–110
Pigozzi G (1990) Latrine use and the function of territoriality in the European badger, *Meles meles*, in a Mediterranean coastal habitat. Anim Behav 39(5):1000–1002
Rodgers TW, Giacalone J, Heske EJ et al (2015) Communal latrines act as potentially important communication centers in ocelots *Leopardus pardalis*. Mammal Biol 80:380–384
Rostain RR, Ben-David M, Groves P et al (2004) Why do river otters scent-mark? An experimental test of several hypotheses. Anim Behav 68(4):703–711
RStudio Team (2016) RStudio: Integrated Development Environment for R. Boston, USA. URL http://www.rstudio.com/
Ruibal M, Peakall R, Claridge A (2010) Socio-seasonal changes in scent-marking habits in the carnivorous marsupial *Dasyurus maculatus* at communal latrines. Aust J Zool 58(5):317–322
Sneddon IA (1991) Latrine use by the European rabbit (*Oryctolagus cuniculus*). J Mammal 72 (4):769–775
Sprent J, Nicol SC (2012) Influence of habitat on home-range size in the short-beaked echidna. Aust J Zool 60(1):46–53
Sprent JA, Andersen NA, Nicol SC (2006) Latrine use by the short-beaked echidna, *Tachyglossus aculeatus*. Aust Mammal 28:131–133
Sprent JA, Nicol SC (2016) Diet of the short-beaked echidna (*Tachyglossus aculeatus*) in the Tasmanian Southern Midlands. Aust Mammal 38(2):188–194
Stewart PD, MacDonald DW, Newman C et al (2002) Behavioural mechanisms of information transmission and reception by badgers, *Meles meles*, at latrines. Anim Behav 63(5):999–1007
Triggs B (2004) Tracks, scats and other traces: a field guide to Australian mammals. Oxford University Press Australia, Melbourne, Australia
Wilkinson DA, Grigg GC, Beard LA (1998) Shelter selection and home range of echidnas, *Tachyglossus aculeatus*, in the highlands of south-east Queensland. Wildlife Res 25(3):219–232
Wronski T, Apio A, Plath M (2006) The communicatory significance of localised defecation sites in bushbuck (*Tragelaphus scriptus*). Behav Ecol Sociobiol 60(3):368–378
Wronski T, Apio A, Plath M et al (2013) Sex difference in the communicatory significance of localized defecation sites in Arabian gazelles (*Gazella arabica*). J Ethol 31(2):129–140
Wronski T, Plath M (2010) Characterization of the spatial distribution of latrines in reintroduced mountain gazelles: do latrines demarcate female group home ranges? J Zool 280(1):92–101
Wyatt TD (2014) Pheromones and animal behaviour: chemical signals and signatures. Cambridge University Press, Cambridge, UK

Zhou Y, Chen W, Buesching CD et al (2015) Hog badger (*Arctonyx collaris*) latrine use in relation to food abundance: evidence of the scarce factor paradox. Ecosphere 6(1):19
Ziege M, Bierbach D, Bischoff S et al (2016) Importance of latrine communication in European rabbits shifts along a rural-to-urban gradient. BMC Ecol 16:29

Do Urinary Volatiles Carry Communicative Messages in Himalayan Snow Leopards [*Panthera uncia*, (Schreber, 1775)]?

Subhadeep Das[1], Sourav Manna[1], Sandipan Ray[1], Payel Das[1], Upashna Rai[2], Biswatosh Ghosh[3], and Mousumi Poddar Sarkar[1](✉)

[1] Chemical Signal & Lipidomics Laboratory, Department of Botany, Centre of Advanced Study, University of Calcutta, Kolkata 700019, India
mousumipsarkar1@gmail.com
[2] Padmaja Naidu Himalayan Zoological Park, Darjeeling 734101, West Bengal, India
[3] Department of Zoology, Bidhannagar College, Salt Lake, Kolkata 700064, India

Abstract. Felids urinate and spray 'Marking Fluid' for territorial maintenance and to transmit messages of their reproductive status. The very rare Himalayan snow leopard also utilises these two primary modes for chemical communication. The present paper is the first report on the volatiles in urine of snow leopards which were analysed with the help of headspace solid phase micro extraction gas chromatography mass spectrometry. Chemical profiles revealed the presence of numerous low molecular weight compounds with different functional groups like alcohols, aldehydes, ketones, sulphur containing compounds. Many monoterpene alcohols, which are common secondary metabolites of plants, are abundant in the urine collected during the months of October to December, the typical reproductive season of the snow leopard in the Darjeeling hills of the Eastern Himalaya. 6-Methyl-5-hepten-2-one was identified from this felid which has a characteristic odour perceptible by the human nose. Among many sulphur containing compounds, Dimethyl disulfide and Dimethyl trisulfide were common in all urine samples of both sexes. Saturated, monounsaturated and polyunsaturated fatty acids were also identified from the lipid fraction of the urine which, in nature, may play an important role by increasing the durability of the volatiles.

1 Introduction

'Chemical signals' which regulate a variety of physiological phenomena in many felids are the primary mode of information transfer related to the reproductive behaviour of these carnivores. (Albone 1984; Brahmachary and Dutta 1981, 1984; Wyatt 2014). All cat species, in general, have two modes of pheromonal communication, ordinary Urination and the spraying of Marking Fluid (MF) (Brahmachary and Dutta 1979, 1984, 1987; Brahmachary 1996; Brahmachary and Poddar-Sarkar 2015; Poddar-Sarkar and Brahmachary 2014). Alongside visual, auditory and tactile cues, members of the cat family predominantly use these two behavioural modes to mark their territory and to

© Springer Nature Switzerland AG 2019
C. D. Buesching (Ed.): *Chemical Signals in Vertebrates 14*, pp. 27–37, 2019.
https://doi.org/10.1007/978-3-030-17616-7_3

inform other individuals about their reproductive status. Thus, 'scent marking'—differently termed as MF by us, plays a significant role in social interactions among snow leopards. Controversy regarding the origin of urine and MF in big cats existed for many years in international literature, however, it is now concluded that both these secretions are ejected through the urinary tract of these feline species (Poddar-Sarkar and Brahmachary 2014). In the present article, we try to identify those volatile and less/non-volatile chemical compounds in the urine of snow leopard, *Panthera uncia* syn. *Uncia uncia* (Schreber 1775) which may act as putative pheromones, although substantial evidence is yet to be required at this stage. Due to the extreme climatic conditions of the Darjeeling hills, unusual logistic constraints for work and strict zoo regulations hinder the authors' intention for exhaustive work on this highly threatened animals. However, the authors intend to project their findings based on urinary volatiles which might be the first report on chemical communication of snow leopard in the Himalayas.

P. uncia is a crepuscular felid and the native to the North Eastern Himalayan mountain range of the Indian subcontinent (Fig. 1). Snow leopards generally lead a solitary lifestyle, rarely use audible sounds and exist in a very low population density in the Eastern Himalaya. In order to communicate with each other, snow leopards scrape the ground with their hind legs and spray urine against rocks to leave markings on the landscape (Sharma et al. 2006). Although the chemistry of urinary volatiles from other big cats such as lion, *Panthera leo* (Andersen and Vulpius 1999), tiger, *Pathera tigris tigris* (Poddar-Sarkar and Brahmachary 2014), bobcat, *Lynx rufus* (Mattina et al. 1991) and cheetah, *Acinonyx jubatus* (Poddar-Sarkar and Brahmachary 1997; Burger et al. 2006) were reported by many authors, little is known about the composition of snow leopard urine and the investigation of volatiles by headspace might be informative in the context of this species' olfactory communication.

Fig. 1. Snow leopard *Panthera uncia* [Zoo name-Tista; f2—Studbook No: 2399]

2 Materials and Methods

2.1 Collection of Samples

Urine was collected from snow leopards kept in open-air enclosures in the Padmaja Naidu Himalayan Zoological park (PNHZP), Darjeeling, West Bengal, India (27° 03W′30.1″N 88°15′14.4″E). PNHZP is situated within the Darjeeling hill range (altitude 6,700 ft.) in the designated Eastern Himalaya Biodiversity Hotspot that has an undulated topography, an average temperature range during winter to summer of ~4–15.6 °C, a humid climate during the monsoon season with an average rainfall of ~309 cm and occasional snowfall during winter. After close observation of snow leopard behaviour, the schedule of urine collection was decided to be at dawn and dusk when the animals were fully active. Urine was collected from three female snow leopards (f1 = Studbook No: 2540, DOB 25.05.2004, f2 = Studbook No: 2399, DOB 29.03.2002, f3 = Studbook No: 2538, DOB 11.03.2004) on 17 occasions and from one male (m = Studbook No: 2404. DOB 08.07.2002) on 10 occasions following the procedure previously adopted by our team (Brahmachary and Dutta 1987; Poddar-Sarkar and Brahmachary 2014) over the years. Just before collection, animals were moved to a closed enclosure. Sampling schedule was rationalised for maximum collection opportunity and to maintain uniformity in experimental design covering both reproductive and non-reproductive seasons throughout the year during 2017–2018. Samples were pipetted out from the precleaned floor (only with distilled water) of the enclosure into 10 ml airtight Teflon-coated glass vials (Agilent, India), crimped immediately and transported to the laboratory under the ice. Samples for headspace volatiles (HSVs) were processed at the earliest convenience (i.e. between 72 and 96 h after collection) and samples for lipid work were kept at −20 °C for future analysis.

2.2 Chemical Analysis

Absorption of headspace volatiles (HSV) was optimised by attaching a 1 cm 50/30 μm divinylbenzene/carboxen/polydimethylsiloxane [(DVB/CAR/PDMS); (Supelco, USA) stableflexTM, 24 Ga] SPME fibre with manual assembly holder (Supelco, USA) over the sample vials. Equilibration for absorbing the vapour phase was maintained at room temperature for 2 h in each case. Chemical analysis of HSV was performed using gas chromatography mass spectrometry [GCMS; Agilent 7890A, USA- triple axis MS-5975C] with a DB-WAX (30 m × 0.25 mm × 0.25 μm) column. Samples were desorbed for 10 min at the injector port at a temperature of 260 °C. The column temperature was maintained at 35 °C for 2 min initial hold, then ramping at a rate of 4 °C/min up to 210 °C with 3 min hold. The identity of the compounds was assigned by matching their retention time with authentic standards, whenever available [Sigma (USA) and by Dr. Ehrenstorfer GmbH, (Germany)] (Table 1) as well as by co-chromatography and by comparison of their respective mass spectral data obtained from the NIST (2011) library, by calculating Linear Retention Index (LRI) in relation to n-alkanes of C11–C19, considering the EI-MS fragmentation pattern and from previous records of this laboratory. The flow rate of the carrier gas (Helium) was maintained at 1 ml/min. The front inlet temperature was 250 °C. The temperature of

Table 1. Head Space Volatiles (HSV) identified by SPME-GCMS from urine of male (m) and three females (f1, f2 and f3) of Snow leopards

Volatile compounds	Rt	LRI (DB-WAX)	Identification mode[a]
Alcohols			
1-pentanol$^{f2, f3, m}$	13.206		i, ii, iv, v, vi
1-hexanol$^{f1, f2, m}$	16.457	1359	iii, iv, v, vi
3-Octanol$^{f3, m}$	17.711		i, ii, iv, v
2-Hexen-1-ol, (Z)-$^{f2, m}$	18.041		iv, vi
1-Heptanol$^{f1, f2, f3, m}$	19.677		i, ii, iv, v, vi
2-Ethyl-1-hexanol$^{f1, f2, f3, m}$	20.707	1496	iii, iv, v
Linalool^{f1}	22.348		iv, vii
cis-Dihydro-alpha-terpineol$^{f3, m}$	22.787		iv, vi
1-octanol$^{f1, m}$	22.788	1565	i, ii, iii, iv, v, vi
Cyclohexanol,3,5-dimethyl-$^{f2, m}$	23.856		iv, vii
Terpineol-4-ol$^{f1, m}$	23.914		iv, vii
1-nonanol^{f1}	25.854		i, ii, iv, v, vi
alpha-terpineol$^{f1, f3, m}$	26.738	1652	iii, iv, vi
gamma- terpineol$^{f1, f3, m}$	26.827	1655	iii, iv, vii
Trans-2-pinanol$^{f1, m}$	26.834		iv, vii
Phenethyl alcohol$^{f1, f2, f3, m}$	32.413	1913	i, ii, iii, iv, v
Phenol$^{f1, f2, f3, m}$	34.735		iv, v
p-cresol$^{f3, m}$	36.574		i, iii, iv, v
Ketones			
4-Heptanone$^{f1, f2, f3, m}$	8.561	1118	i, ii, iii, iv, v
3-octanone$^{f1, f2, f3, m}$	12.748		i, ii, iv, v
2-octanone^{f1}	13.987	1284	iii, iv, v
8-Hydroxy-2-octanone$^{f1, f2}$	14.091		iv, vii
2-hexanone,3,4-dimethyl-$^{f1, f2, f3, m}$	14.313		iv, v
Cyclohexanone, 2,2,6- trimethyl-$^{f2, f3, m}$	14.682		iv, viii
3-Ethylcyclopentanone^{f2}	15.28		iv, vii
2,3-octanedione^{f1}	15.385	1323	iii, iv, v
6-Methyl-5-hepten-2- one$^{f1, f2, f3, m}$	15.568	1337	i, ii, iii, iv, v
2-nonanone$^{f1, f2, f3, m}$	17.227	1387	i, ii, iii, iv, v, vi
Trans-3-Nonen-2-one^{f1}	19.207		iv, vii
2-decanone^{f1}	20.707		i, ii, iii, iv, v, vi
2-Undecanone^{f1}	23.882		iv, v
Acetophenone$^{f1, f2, f3}$	25.186	1645	i, ii, iii, iv, v
2-piperidinone, 1-methyl-$^{f2, f3}$	29.957	1815	iii, iv, vii
Aldehydes			
Pentanal$^{f2, f3}$	4.974		iv, v
Hexanal$^{f2, m}$	9.911	1186	iii, iv, v, vi
Heptanal^{f1}	10.738	1182	iii, iv, v, vi

(*continued*)

Table 1. (*continued*)

Volatile compounds	Rt	LRI (DB-WAX)	Identification mode[a]
Octanal[f1, f2, m]	13.855	1287	i, ii, iii, iv, v, vi
Nonanal[f1, m]	17.195	1391	i, ii, iii, iv, v, vi
Decanal[m]	20.548		i, ii, iv, v, vi
Benzaldehyde[f1, f2, f3, m]	21.362	1516	iii, iv, v, vi
Phenylacetaldehyde[f2]	23.373		iv, v
Sulphur containing compounds			
Dimethyl disulphide[f1, f2, f3, m]	7.156		i, ii, iv, v
2,4-Dithiapentane[m, f3]	13.639	1280	iii, iv, vii
Dimethyl trisulfide[f1, f2, f3, m]	16.667	1371	i, ii, iii, iv, v
Thiophene, 2-pentyl-[f1, f2, f3]	19.231		iv, vii
Ethanol,2-(metylthio)-[m, f3]	21.852	1534	iii, iv, vii
2,4,5-Trithiahexane[f2]	25.377		iv, vii
Nitrogen containing compounds			
Hexanenitrile[f2, m]	14.218		iv, v
Pyridine,2,4,6- trimethyl-[f2]	17.036		iv, vii
Pyrazine,tri methyl[f2, f3]	17.94	1405	iii, iv, vii
Oxime-, methoxy- phenyl-_[f1, f2, f3, m]	28.92		iv, vii
Indole[f2, f3, m]	44.386		iv, vii
Hydrocarbons			
Ethylbenzene[f1, m]	8.473	1120	iii, iv, v
beta-ocimene[f1, f2]	8.836		iv, vii
p-xylene[m]	9.189	1134	i, ii, iii, iv, v
Benzene, tert-butyl-[f1, f2, m]	12.869		iv, vii
Terpinolene[f1]	12.92		iv, vii
p-cymene[f3, m]	12.984		iv, vii
benzene,1,2,3- trimethyl[f1, m]	13.156		iv, vii
p-cymenene[f1, f3, m]	18.64		iv, vii
Benzene, 1,3-dichloro-[f1, f2, f3, m]	18.779	1436	iii, iv, vii
1-Phenyl-1-butene[f1, f3]	18.865		iv, vii
Azulene[f1, f2, f3, m]	27.546	1729	iii, iv, v
Acids			
Acetic acid[f1, f3, m]	19.562	1427	i, ii, iii, iv, v
Butanoic acid, 4- hydroxy-[f1, m]	24.55		iv, v

[a]Compounds were identified by (i) comparing retention time (Rt) with authentic compounds; (ii) Co-chromatography with authentic compounds; (iii) Linear retention index (LRI) relative to C11–C19 n-alkanes and compared with published LRI data (PubChem); (iv) comparing mass fragmentation pattern with NIST library (2011); (v) analysis of characteristic features of MS fragments; (vi) Comparison with published mass spectrometric data from our group; (vii) Absolute configuration not determined

the MS source, quadrupole and auxiliary heater were set at 230 °C, 150 °C and 280 ° C, respectively. The electron energy was 70 eV (vacuum pressure-2.21 e-0.5 torr). The mass fragment scan range was 50–450 amu at 0.5 s/scan.

Lipids were extracted from the urine by using the Bligh and Dyer's (1959) method. An aliquot of chloroform extract was taken in a pre-weighed glass vial for gravimetric estimation of lipids. The solvent was evaporated to dryness with a stream of N_2 and the residue weighed again by precision balance. For the analysis of fatty acids (FA), the chloroform phase was used. The fatty acids present in urine were derivatized to fatty acid methyl ester (FAME) by acid catalysed esterification (Poddar-Sarkar 1996). FAME was recovered with n-hexane and finally dried over anhydrous sodium sulphate. The volume of the n-hexane was reduced under a stream of nitrogen and subjected to GCMS. 1 μl of hexane extract of FAME was injected to HP5-MS column (30 m 0.25 mm × 0.25 μm) of GCMS (Agilent Technologies, USA; 7890A GC system with 5975C triple axis detector MS) apparatus. The programme was set at 70 °C initial hold for 1 min for column temperature, ramping at 4 °C/min. up to 260 °C with a final hold for 3 min. FAME were identified by calculating their relative retention time (RRt) and comparison with authentic mixture of 37 FAME and PUFA (Supelco, Lot No: LB80556 and LB77207, USA). Identification was confirmed by comparing mass fragmentation pattern of the compounds from the NIST (2011) data base.

2.3 Statistical Data Analysis

A total number of 27 samples from three females and one male were analysed by GCMS. For quantitative analysis, each peak was normalised by calculating the relative percentage considering total ion count from the chromatogram. Chemical compounds identified by mass fragmentation pattern were grouped into different classes on the basis of their functional group or nature of backbone. Summation of nine classes of compounds from all females during the reproductive season (RS) and non-reproductive season (NRS) were plotted in Fig. 2. In addition, a comparative assessment on the basis of such 35 identified volatile compounds which were present in all urine samples of four leopards were done by heat map (Fig. 3). For heat map generation, successive steps were followed: Step (i) the sampling events were segmented into two seasons: RS and NRS for each animal; (ii) A table was developed by considering average amount for each identified compounds taking all females in a pool and for data in male in separate pool; (iii) Values were converted to the percentage of the row sum of each compound considering NRS & RS separately for female pool as well as for male pool; (iv) Derived values were used to form the final matrix (Fig. 3). Statistical data were processed using past software (3.21 version) for generating the heat map.

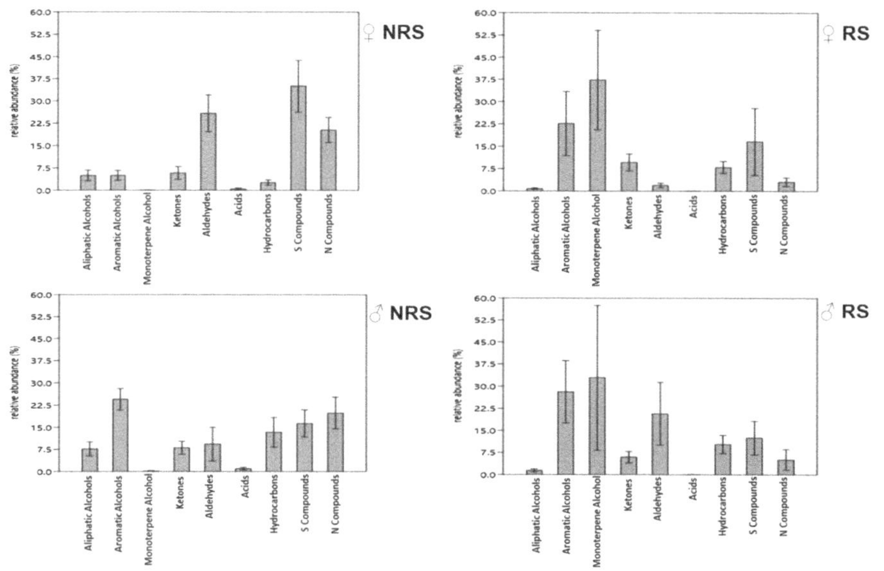

Fig. 2. HSV profile of nine classes of compounds distinctively separated from urine of male and female during non-reproductive (NRS) and reproductive seasons (RS) of snow leopard; S—sulphur containing compounds, N—nitrogen containing compounds

3 Results

A number of volatile organic compounds (VOCs) with different functional groups like alcohols, aldehydes and ketones were identified from the HSVs of urine of snow leopard of both the sexes (Table 1) The most interesting compound *6-Methyl-5-hepten-2-one* which has a characteristic aroma, perceptible by the human nose was emitted from the fresh urine of snow leopard. However, no such distinctive aroma was perceptible from the distilled water washings of the floor processed in the same manner as a control, and 2-acetyl-1-pyrroline, the aroma molecule responsible for the characteristic smell of 'Basmati rice' present in MF of tigers and Indian leopards (Brahmachary et al. 1990; Brahmachary 1996; Poddar-Sarkar and Brahmachary 2014) was not detected in the urine of snow leopards. *Dimethyl disulphide* and *Dimethyl trisulfide*, two sulphur compounds were also identified from urine of snow leopard (Table 1). Low boiling straight chain alcohol of carbon number 5, 6, 7, 8 and aldehyde of 6, 8, 9 were common HSVs present in the urine of both male and females. Two carboxylic acids, such as acetic acid and 4-hydroxy butanoic acid were identified from both sexes. Some urinary constituents like phenol, benzaldehyde, p-cresol, acetophenone were also identified from urine. Some compounds which are very common secondary metabolites of plants like azulene, 1-methyl-2-piperidone, beta-ocimene, p-cymene, p cymenene were also identified in urine of snow leopard. HSV profile of the urine collected from reproductive season showed significant presence of some terpenoids such as alpha-terpineol, gamma-terpineol, terpineol-4-ol, cis-dihydro-alpha-terpineol and terpinolene

(Table 1). We found distinctive variations in the relative abundance of some compounds in both sexes during RS and NRS. During RS, high amounts of monoterpene alcohols and aromatic alcohols were identified in contrast to the lower amounts of sulphur and nitrogen containing compounds as well as aliphatic alcohols (Fig. 2). RS and NRS differed between male and female urine (Fig. 3).

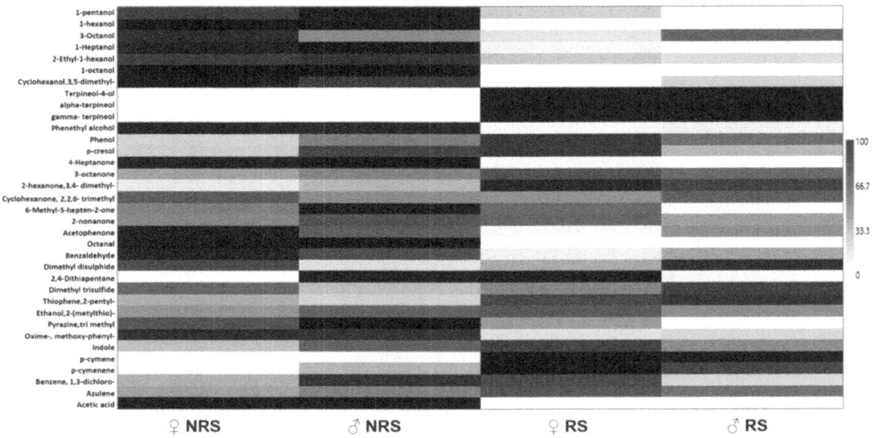

Fig. 3. Heat map generated on the basis of the statement mentioned in Sect. 2.3 of the text. Thirty five compounds of the Y-axis with different functional groups and structural backbones segregated the chemical profile of urine during NRS and RS from both sexes

Lipids present in the urine of snow leopard of both sexes ranged from 0.95 to 1.42 mg/ml. The Lipid fraction of snow leopard urine mostly contained Saturated Fatty Acids (SFA) of even and odd carbon number. Palmitic acid (16:0) is the most abundant in all cases (Fig. 4). In addition to even carbon number SFA such as Decanoic acid (10:0), Dodecanoic acid (12:0), Tetradecanoic acid (14:0), Octadecanoic acid (18:0), Eicosanoic acid (20:0), Docosanoic acid (22:0) and Tetracosanoic acid (24:0) some SFA with odd carbon number, such as Tridecanoic acid (13:0), Pentadecanoic acid (15:0), Heptadecanoic acid (17:0), Nonadecanoic acid (19:0) and Heneicosanoic acid (21:0) were also identified. In addition, Benzeneacetic acid is also detected in the urine of snow leopard. Five monounsaturated FAs (7-Hexadecenoic acid, 9-Hexadecenoic acid, 9-Octadecenoic acid,11-Eicosenoic acid and 13-Docosenoic acid) and two polyunsaturated FAs (9,12-Octadecadienoic acid and 5,8,11-Eicosatrienoic acid) were identified (Fig. 4).

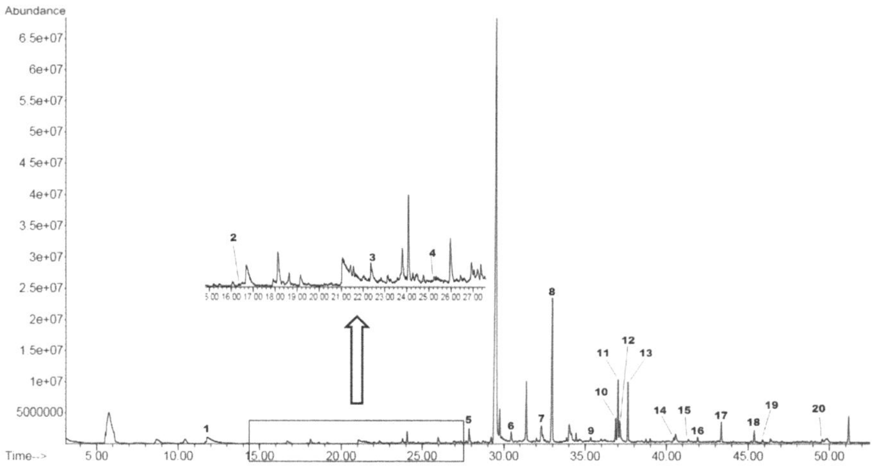

Fig. 4. Fatty Acids identified from lipid part of Urine of Snow leopard: Peak No 1. Benzeneacetic acid methyl ester 2. Decanoic acid methyl ester 3. Dodecanoic acid methyl ester 4. Tridecanoic acid methyl ester 5. Methyl tetradecanoate 6. Pentadecanoic acid methyl ester 7. 9-Hexadecenoic acid methyl ester 8. Hexadecanoic acid methyl ester 9. Heptadecanoic acid methyl ester 10. 9,12-Octadecadienoic acid methyl ester 11. 13-Octadecenoic acid methyl ester 12. 9-Octadecenoic acid methyl ester 13. Methyl stearate 14. 5,8,11-Eicosatrienoic acid methyl ester 15. 11-Eicosenoic acid methyl ester 16. Eicosanoic acid methyl ester 17. Heneicosanoic acid methyl ester 18. 13-Docosenoic acid methyl ester 19. Docosanoic acid methyl ester 20. Tetracosanoic acid methyl ester

4 Discussion

One of the volatiles of fresh snow leopard urine is sulcatone (*6-Methyl-5-hepten-2-one*), which imparts a characteristic odour perceptible to the human nose. Urine of snow leopards contains many characteristic low molecular weight compounds with diverse functional groups such as pentanol, hexanol, heptanol, 3-octanone, nonanal, indole, etc. which might play a role in chemical communication. Similar types of compounds have been shown to moderate and govern a variety of specialised behaviours related to kin recognition, choosing potential partners and maintaining social standings in many mammals including other felids and canids (Andersen and Vulpius 1999; Burger et al. 2006; Raymer et al. 1984; Soso and Koziel 2017; Wilson 1980). Dimethyl disulphide (DMDS), a male attractant compound of Hamster Vaginal Secretions (Singer et al. 1976) and one of the most important constituents of MF in lions (Soso and Koziel 2017) as well as Cheetah urine (Burger et al. 2006) was also confirmed in the urine of Snow leopard. Many secondary metabolites of plant systems were identified from the urine of snow leopard, and we observed fragmented leaves in their scats. Interestingly, we found significant variation in urinary terpenoidal compounds during their reproductive season. Hexanal, Octanal, Nonanal, 4-heptanone and benzaldehyde, identified from urine of both female and male snow leopard were considered as common urinary volatiles of many mammals such as the house mouse

(Novotny et al. 1999), white-tailed deer (Miller et al. 1998) and elephant (Rasmussen and Greenwood 2003), coyote (Schultz et al. 1988), ferret (Zhang et al. 2005) and MF of lions (Soso and Koziel 2017). Other low carbon alcohols and aldehydes are also common urinary volatiles of many mammals (Albone 1984). Phenethyl alcohol, detected in urine of snow leopard, is one of the major volatiles of lion MF (Soso and Koziel 2017). Fatty acids, identified from the urine of snow leopard are similar in nature to other big cats such as tiger, leopard, lion and cheetah (Poddar-Sarkar 1996; Poddar-Sarkar and Brahmachary 2014). It can be assumed that lipids may be delaying the dissipation of urinary volatile constituents which may facilitate animals to mark vast areas for territorial maintenance (Brahmachary and Dutta 1987; Poddar-Sarkar and Brahmachary 1996; Poddar-Sarkar 1996; Poddar-Sarkar and Brahmachary 2004). Brahmachary and Dutta observed previously that steam distillation separates the smell of volatiles which rapidly vanishes after being liberated from the heavier lipids (Brahmachary unpublished; Brahmachary and Choudhuri unpublished; Brahmachary and Dutta 1979; Poddar-Sarkar and Brahmachary 2014). As urine is one of the major sources of pheromone in other felids, it can be presumed that it might play a similar role in snow leopards. Nevertheless, extension of this work may add some new findings in the future. Therefore, by analysing the VOCs of urine throughout the year, the physiological status of the animal can be assessed and could form an important basis for the planning and management of future breeding programmes of this rare species as well as being utilised for zoo management and conservation purposes.

Acknowledgements. Author SD [CSIR sanction no-09/0289(0996)/2017-EMR-1], SM [09/0289(1004)/2017-EMR-1] and PD [108(Sanc.)/ST/P/S&T/1G-24/2014] are grateful to Council of Scientific & Industrial Research (CSIR), Government of India and Govt. of West Bengal respectively for providing their fellowships during this work. We would also like to acknowledge the Department of Science and Technology (Fund for Infrastructure development in Science and Technology programme) Govt. of India for extending GCMS facility in the Department of Botany, University of Calcutta. We also acknowledge the kind help and assistance from Principal Chief Conservator of Forest (wildlife), Govt. of West Bengal, and Director of Padmaja Naidu Himalayan Zoological Park, Darjeeling, West Bengal.

Note:
We dedicate this paper to the memory of our mentor Late Prof. R.L. Brahmachary with our deep grief and sorrow. He corrected our initial draft of this manuscript but passed away on 13 February 2018 when we were submitting the final version of this paper.

References

Albone E (1984) Mammalian semiochemicals. Wiley, Chichester, UK

Andersen KF, Vulpius T (1999) Urinary volatile constituents of the lion. Chem Senses 24: 179–189

Bligh EG, Dyer WJ (1959) A rapid method of total lipid extraction and purification. Can J Biochem Physiol 37:911–917

Brahmachary RL (1996) The expanding world of 2-acetyl-1-pyrroline. Curr Sci 71:257–258

Brahmachary RL, Dutta J (1979) Phenylethylamine as a biochemical marker of tiger. Zeitschrift für Naturforschung C 34:632–633
Brahmachary RL, Dutta J (1981) On the pheromones of tigers: experiments and theory. Am Nat 118:561–567
Brahmachary RL, Dutta J (1984) Pheromones of leopards: facts and theory. Tiger Paper 11:18–23
Brahmachary RL, Dutta J (1987) Chemical communication in the tiger and leopard. In: Tilson RL, Seal US (eds) Tigers of the world: the biology, biopolitics, management, and conservation of an endangered species. Noyes Publications, Park Ridge, NJ
Brahmachary RL, Poddar-Sarkar M (2015) Fifty years of tiger pheromone research. Curr Sci 108:2178–2185
Brahmachary RL, Poddar-Sarkar M, Dutta J (1990) The aroma of rice … and tiger. Nature 344:26
Burger BV, Visser R, Moses A, Le Roux M (2006) Elemental sulfur identified in urine of cheetah, *Acinonyx jubatus*. J Chem Ecol 32:1347–1352
Mattina MJI, Pignatello JJ, Swihart RK (1991) Identification of volatile components of bobcat (*Lynx rufus*) urine. J Chem Ecol 17(2):451–462
Miller LA, Johns BE, Elias DJ (1998) Immunocontraception as a wildlife management tool: some perspectives. Wildl Soc Bull 26:237–243
Novotny MV, Ma W, Wiesler D, Zidek L (1999) Positive identification of the puberty-accelerating pheromone of the house mouse: the volatile ligands associating with the major urinary protein. Proc R Soc Lond B 266:2017–2022
Poddar-Sarkar M (1996) The fixative lipid of tiger pheromone. J Lipid Mediat Cell Signal 15:89–101
Poddar-Sarkar M, Brahmachary RL (1997) Putative semiochemicals in the African cheetah. J Lipid Mediat Cell Signal 15:285–287
Poddar-Sarkar M, Brahmachary RL (2004) Putative chemical signals of leopard. Anim Biol 54(3):255–259
Poddar-Sarkar M, Brahmachary RL (2014) pheromones of tiger and other big cats. In: Mucignat-Caretta C (ed) Neurobiology of chemical communication. CRC Press/Taylor & Francis, Italy
Rasmussen LEL, Greenwood DR (2003) Frontalin: a chemical message of musth in Asian elephant (*Elephas maximus*). Chem Senses 28(5):433–446
Raymer J, Wiesler D, Novotny M, Asa C, Seal US, Mech LD (1984) Volatile constituents of wolf (*Canis lupus*) urine as related to gender and season. Experientia 40(7):707–709
Schultz TH, Flath RA, Stern DJ, Richard MT, Teranishi R, Kruse SK, Butler B, Howard WE (1988) Coyote estrous urine volatiles. J Chem Ecol 14:701–712
Sharma S, Dutta T, Veer bhatnagar Y (2006) Marking site selection by free ranging snow leopard (*Uncia uncia*). In: McNeely JA, McCarthy TM, Smith A, Olsvig-Whittaker L, Wikramanayake ED (eds) Conservation biology in Asia. Nepal: Society for Conservation Biology Asia Section and Resources Himalaya
Singer AG, Agosta WC, O'Connell RJ, Pfaffmann C, Bowen DV, Field FH (1976) Dimethyl disulfide: an attractant pheromone in hamster vaginal secretion. Science 191(4230):948–950
Soso SB, Koziel JA (2017) Characterizing the scent and chemical composition of *Panthera leo* marking fluid using solid-phase microextraction and multidimensional gas chromatography–mass spectrometry-olfactometry. Sci Rep 7(1):513
Wilson EO (1980) Caste and division of labor in leaf-cutter ants (hymenoptera: formicidae: Atta): I. The overall pattern in *A. sexdens*. Behav Ecol Sociobiol 7:143–156
Wyatt TD (2014) Pheromones and animal behavior: chemical signals and signatures, 2nd edn. Cambridge University Press, Cambridge, UK
Zhang JX, Soini HA, Bruce KE, Wiesler D, Woodley SK, Baum MJ, Novotny MV (2005) Putative chemosignals of the ferret (*Mustela furo*) associated with individual and gender recognition. Chem Senses 30(9):727–737

Encoded Information Within Urine Influences Behavioural Responses Among European Badgers (*Meles meles*)

Tanesha M. Allen(✉), Nadine A. Sugianto, Charlotte Ryder, Chris Newman, David W. Macdonald, and Christina D. Buesching

Wildlife Conservation Research Unit, Department of Zoology, Recanati-Kaplan Centre, University of Oxford, Tubney, Abingdon, Oxfordshire, UK
tanesha.allen@zoo.ox.ac.uk

Abstract. Metabolic by-products like sex-steroid metabolites convey information regarding an individual's physiological fitness and reproductive state. Urine and faeces containing these excreted metabolites can function as cues for conspecifics within social contexts, particularly for mate selection. Our pilot study investigated in scent provisioning experiments how donor- and responder-related characteristics influenced behavioural responses towards excreted sex-steroid metabolites. We used 51 urine samples with known urinary hormone concentrations (determined by EIA corrected for specific gravity) collected from known individuals in a high-density population of European badgers (*Meles meles*) in Wytham Woods, Oxfordshire, UK during the spring mating season. Samples were provided in Eppendorf tubes pushed into the ground near badger setts to assess how response likelihood, sniff duration, and countermarking responses varied according to donor-related traits (age, sex, group familiarity and hormone concentrations). From 143 videos containing 26 identifiable adult responders, we observed adult badgers modifying their behaviour towards urine samples based on the perceived relevance to their own traits with badgers investigating and countermarking urine from opposite-sex donors more often than urine from same-sex donors. This suggests that urine-marking can serve as an adaptive signal for individual advertisement. Additionally, all badgers were more likely to investigate male urine (containing testosterone) than female urine (containing oestradiol); however, no other clear effects of sex-steroid metabolites were noticed. We consider that this could be a feature of the social evolution of badgers or arise because badgers signal reproductive status through other olfactory secretions, notably their unique subcaudal gland marking behaviour.

1 Introduction

Mammals generate metabolic by-products that individuals ultimately excrete via urination and defecation (Behringer and Deschner 2017). Since these metabolic by-products result directly from biological pathways (e.g. digestion, muscle metabolism, endocrinological processes), their presence in urinary/faecal compositional profiles can convey information regarding an individual's physiological fitness to conspecifics [e.g. age (Linklater et al. 2013), health (Penn and Potts 1998; Zala et al. 2004), relative body

© Springer Nature Switzerland AG 2019
C. D. Buesching (Ed.): *Chemical Signals in Vertebrates 14*, pp. 38–59, 2019.
https://doi.org/10.1007/978-3-030-17616-7_4

size (Sharpe 2015) and reproductive state (Converse et al. 1995)]. Social information (e.g. social group membership) may also be conveyed (Henkel et al. 2015), particularly if individuals or populations reside in different habitats and/or exploit different dietary sources.

Sex-steroid levels indicate sexual maturity (Lu et al. 1979), with lower levels present during adolescence (Courant et al. 2010) and senescence (Lu et al. 1979; badgers: Sugianto et al. in press a). In females, sex-steroids also reveal oestrus cycle stage, where higher oestradiol levels promote ovulation, alter cervical mucus consistency for improved sperm transport and prepare the uterine lining for implantation (Findlay et al. 2010; Wira et al. 2015; badgers: Sugianto et al. in press b). In males, higher testosterone levels promote spermatogenesis and secondary sex characteristics (Preston et al. 2012; badgers: Buesching et al. 2009, Sugianto et al. in press a, b). Once sex-steroids fulfil their biological function, species-specific filtration pathways through the kidneys and/or liver (Campbell et al. 2001; Snoj et al. 2014) transform these into metabolites (Amendolagine et al. 2018), in conjugated and/or unconjugated form (James et al. 1972; Thomas et al. 1990) excreted via urine or faeces. Consequently, excreted levels generally correlate with circulating plasma titres, and can provide reliable information concerning an individual's reproductive status (Brown et al. 1994; Snoj et al. 2014. Nevertheless, excretion filtration pathways are highly species-specific (Wasser et al. 1988)—for instance, linked to metabolic rate (Goymann 2012)—and thus need to be validated carefully for each study species (Touma and Palme 2005).

Importantly, these metabolites expose olfactory cues about the status of that individual to conspecifics (reviewed in Wyatt 2014), and thus purposeful urinating/defecating behaviours have evolved to function as intentional signals for individual advertisement; for example, to elicit mating opportunities and enhance mate competition (Albone 1984; Macdonald 1979; Thonhauser et al. 2013; Vogt et al. 2014). Responses stimulated by deposited marks may affect mate selection decisions, therefore ultimately influencing social and reproductive partners. Visual signals are less important for nocturnal mammals, making them particularly reliant on olfactory signals for conspecific assessment (Fobes and King 1977; Wong et al. 2010).

Here we undertook a pilot study to begin to examine the behavioural responses of nocturnal European badgers (*Meles meles*; henceforth badgers) to the scent of conspecific urine in a high-density population (40 individuals/km^2), where badgers live in groups, sharing dens (termed setts) in which more than 20 individuals can reside (see Macdonald et al. 2015).

Badgers are macrosmatic and rely heavily on olfactory communication using urine-marking, faecal deposits coated with anal gland secretions, interdigital gland secretions and unique subcaudal gland secretions, originating from a specially evolved subcaudal pouch situated between the tail and anus (reviewed by Buesching and Stankowich 2017; Stubbe 1971). This subcaudal gland secretion plays an important role in badger olfactory communication (Buesching et al. 2002a, b, c), with badgers engaging in social allo-marking (Buesching et al. 2003) and object marking (Buesching and Macdonald 2001), including over-marking of scents left by conspecifics (Buesching and Macdonald 2001). Faeces and these specialised gland secretions encode individual- and group-specific information (reviewed in Buesching and Macdonald 2001;

Buesching and Stankowich 2017; Tinnesand et al. 2015), yet thus far relatively little is known regarding the information encoded in urine (but see Service and Harris 2001). Badger urine has been established to contain hormone metabolites indicative of reproductive status, albeit with a time lag (Sugianto et al. in press b).

Urination generally occurs on the ground during roaming and foraging (Buesching and Macdonald 2001) with no alterations in position (e.g. handstands) to elevate the position of urine marks to signify dominance as observed in other species (Sharpe 2015; White et al. 2002). Hutchings et al. (2002) observed that badgers urinated predominantly at latrines, which are also shared sites used for defecation (Buesching and Jordan, this volume; Stewart et al. 2002), although 38% of urinations occurred at non-latrine locations (e.g. associated with linear features such as fencing and thick undergrowth; see also White et al. 1993). Stewart et al. (2002) found that latrine visits last an average of 23 s, with nearly 54% of that duration involving sniffing. High sniffing activity along with preferences for latrines under tree cover (Stewart et al. 2002) indicate that urine and other latrine scents are important enough to warrant protection from rainfall and other forms of environmental degradation.

While it has been established that badgers can distinguish sex and age from urine marks (Service and Harris 2001), as it does in some other species (Converse et al. 1995; Hradecký 1986; SankarGanesh et al. 2014), this was done without consideration of the responder's identity, sex, age, etc. The possibility that badger urine encodes reproductive status and conveys this to the responder has thus far not been fully explored. Here, we use scent-playback experiments (following Tinnesand et al. 2015) to investigate how known, identified responders react to the provisioning of badger urine collected from known donors, a proportion of which was also analysed for sex-steroid metabolite levels. We analyse how donor- and responder-specific traits moderate responder behaviour.

2 Methods

2.1 Study Population and Sample Collection

Our pilot study used 51 urine samples collected from 29 females and 22 males (42 adults and 9 cubs; Table 1) between 22 May and 3 June 2017 inhabiting a high-density population at Wytham Woods, Oxfordshire, UK (GPS reference: 51° 46′26″N; 1° 19′ 19″W; see Macdonald et al. 2009 for details of the badger population), an area of 424-ha of semi-natural mixed woodland (Savill et al. 2011). Cage traps were baited with

Table 1. Age, sex, and familiarity distribution of urine sample donors (n = 51). Urine samples collected during spring 2017 and presented during scent-playback experiments were distributed across sex and social group. All cub urine was from non-residents

	Adults (n = 42)		Cubs (n = 9)	
	Females (n = 24)	Males (n = 18)	Females (n = 5)	Males (n = 4)
Residents (n = 17)	8	9	NA	NA
Non-residents (n = 34)	16	9	5	4

peanuts and placed near sett entrances as part of a long term (1987–2018) capture–mark–recapture study, following the methodology described by Macdonald et al. (2015).

Badgers were sedated with an intramuscular injection of 0.2 ml ketamine hydrochloride/kg body weight (McLaren et al. 2005) prior to being identified by their unique tattoo number received at first capture (usually as cubs). Individuals were sexed and weighed (in kg to the nearest 100 g). Exact ages were known for individuals first captured as cubs ($n = 41$) while age of those individuals first captured as adults ($n = 10$) was estimated based on tooth wear (categorised on a 1–5 scale) using the following formula (Macdonald et al. 2009; see da Silva and Macdonald 1989 for limitations):

$$Age = -2.51 + 2.10x\,(toothwear)$$

To identify individuals during scent-playback experiments, all adults from the three focal setts were given unique clip marks by cutting a pattern into the black guard hair to expose the white under-fur (Stewart and Macdonald 1997). Cubs were not clipped to minimise thermal compromise.

Badgers were assigned to social groups based on the location (social den, termed a 'sett') at which they were captured, and from their former capture history (following the rules in Annavi et al. 2014). This enabled us to experiment with scents provisioned from unfamiliar (i.e. donor badgers residing at other social groups from the responders) and familiar badgers (i.e. donor and responder residing in the same social group).

Urine was collected in glass vials with Teflon-coated screw lids (Sigma-Aldrich, Dorset, UK) by gentle manual palpation of the bladder (Vicente et al. 2007). Urine samples were then divided between 5–10 Eppendorf tubes (2 ml) filled to 0.5 ml each using plastic pipettes (one pipette per urine sample) for hormonal (enzyme immunoassay: EIA) analysis and scent-playback experiments. Subsamples were frozen at −20 °C until further analysis.

Ethical review and approval were obtained for all trapping and handling procedures and were performed in accordance with the UK Animals (Scientific Procedures) Act, 1986 currently under Home Office licence 30-3379 and Natural England licence 2017-27589-SCI-SCI.

2.2 Scent-Playback Experiments and Behavioural Observations

Scent-playback experiments using contemporaneous urine samples were conducted during the spring cub-rearing season from 24 May 2017 to 12 June 2017.

Urine samples were defrosted within 1–2 h prior to scent playback and the lids of the Eppendorf tubes removed. A new pair of latex gloves was used for handling each sample to prevent cross-contamination. Around 1–2 h before badger emergence, 1 urine sample and 1 control sample (an Eppendorf tube filled with 0.5 ml of water) were positioned 50 cm—about one head–body length of an average badger—apart on either side along badger paths near each of the 3 focal badger setts (between 2 and 8 m from the sett). The Eppendorf tubes were pushed into the ground to ensure that the rim was level with the soil. Eppendorf tubes were removed in the mornings and replaced in the afternoon with fresh urine and water samples. Urine and water placements were moved about 50 cm after each trial, and relative positions of urine and water (i.e. urine placement on left or right side of badger path) were alternated every day to prevent familiarisation. Urine samples were presented randomly each night regarding donor sex, familiarity, and age to avoid scent habituation.

Badger reactions to urine and water were observed remotely using Infrared Crenova 2.4 LCD Display 12MP 1080P HD camera traps activated by three motion sensors. Camera traps were set up on a nearby tree at a height of 1.5 m and about 1.5–3.0 m away from the sample, and the video length was set to 60 s with 15 frames per second. The previous night's footage was downloaded each morning.

Badger proximity to the samples was estimated by sight using body size as an indicator. No response was recorded if a badger did not approach a sample within a 50 cm (i.e. 1 body length) radius. Responses were recorded as 'potential' if a badger approached the samples within a 50 cm radius due to the likelihood that the tube content could be smelled (Tinnesand et al. 2015). Responses were recorded as 'actual' if a badger investigated the tubes more closely (≤ 10 cm) by sniffing or licking/direct contact.

For each potential and actual response, we recorded:

- Sniff duration (in seconds)
- Subcaudal over-marking (within 10 cm of tube) and proximity marking (between 10–50 cm of a vial)

2.3 Enzyme Immunoassay (EIA) Analyses of Urinary Oestradiol and Testosterone Metabolites

EIA analyses were performed at Leibniz Institute for Zoo and Wildlife Research (IZW) in Berlin, Germany. Sex-steroid levels were analysed by mixing 0.1 ml of urine with 0.5 ml of 0.05 M acetate buffer (pH = 4.8). ß-glucuronidase/arylsulfatase from Helix pomatia was added to the mixture, and samples were incubated for 2 h at 37 °C to liberate the steroids from their conjugates. For extraction, 2.5 ml of a mixture of *tert*-butyl methyl ether and petroleum ether 30/70% v/v was added. After being shaken for 20 min, the samples were frozen at −80 °C. The organic phase was then decanted, dried at 50°C under a stream of nitrogen, and dissolved in 1 mL 40% methanol. For both oestradiol and testosterone analysis, 20 μl aliquots were used.

2.3.1 Oestrogen EIA (Ng/Ml)

Total oestrogen analyses were carried out using a polyclonal antibody raised in rabbits immunised against 1,3,5(10)-oestratrien-3,17ß-diol-17-HS-BSA and 1,3,5(10)-oestratrien-3,17ß-diol-17-HS-peroxidase. Cross-reactivities to oestrogens were as follows: 1,3,5(10)-oestratrien-3,17b-diol (*17β-oestradiol*), 100%; 1,3,5(10)-oestratrien-3,17-one (*oestrone*), 100%; 1,3,5(10)-oestratrien-3,17α-diol (*17α-oestradiol*), 66%; 1,3,5(10)-oestratrien-3,16a,17b-triol (*oestriol*), 1.5%; and 0.1% for 19-nortestosterone, progesterone (P4), and testosterone.Intra- and inter-assay coefficient of variations for biological samples with low and high concentrations were 2.4% and 6.9% (nZ12) and 1.3 and 4.7% (n = 18), respectively.

2.3.2 Testosterone EIA (Ng/Ml)

Antibodies (provided by Prof. Meyer, Weihenstephan, Germany) were raised in rabbits immunised against 17a-OH-testosterone-HS-BSA. Testosterone-3-CMO-peroxidase was used as an enzyme conjugate. Cross-reactivity of the antibody with other steroids was as follows: 4-androsten-17ß-ol-3-one (*testosterone*), 100%; 5a-androstan-17ß-ol-3-one (*dihydrotestosterone [DHT]*), 13.6%; 5a-androst-2-en-17ß-ol (1.5%), and <0.1% for 5a-androstan-3ß-ol-17-one (*epi-androsterone*), 5a-androstan-3a-ol-17-one (*androsterone*) and 5a-androst-2-en-17-one.

2.3.3 Normalisation of Urinary Hormone Concentration with Specific Gravity

Specific gravity (SG) measurements for all urine samples were collected with a urine SG refractometer (Atago Uricon-PN; NSA Precision Cells, Inc.) after the first freeze-thaw cycle. The SG population mean was calculated to establish the SG_{target}. Following the methodology detailed in Miller et al. (2004), all sex-steroid levels were corrected using the equation:

$$Raw\, hormone\, concentration\, x \frac{(SGtarget - 1.0)}{(SGsample - 1.0)}$$

2.4 Statistical Analysis

2.4.1 Behavioural Observations with Scent-Playback Experiments

To analyse if urine encodes any biologically relevant information that influences responder behaviour, we conducted contingency chi-square tests comparing responses to urine versus water as a control by using response likelihood (proportion of actual responses to total [actual + potential] responses) and subcaudal countermarking responses (proportion of proximity marking and over-marking to total responses). No urine marking was observed in any of the recordings.

To improve the fit of the chi-square tests for countermarking analysis, categories were reclassified as 'marking' (combining both over-marking and proximity marking) and 'no marking'. Highly skewed sniff durations prompted the use of permutation tests (*lmPerm* package, Wheeler 2010) to analyse any variation between urine and water responses. Variation in sniff duration was quantified as ± standard error (SE). To analyse if the duration samples were frozen for affected the sniff duration of urine, we used permutation tests due to highly skewed data.

Additional contingency chi-square tests were conducted to relate donor and responder traits in the specific context of reproductive advertisement (i.e. age [adult or cub], sex, familiarity [resident or non-resident]) to response likelihood and countermarking responses. Sniff duration was analysed using permutation tests. A low number of observations of cubs as responders ($n = 9$) prevented any statistical analysis on responder age. Thus, all subsequent analyses focused solely on adult responders.

Analyses involving interactions between traits were conducted using general linear models (for response likelihood and countermarking responses) and a permutated two-way ANOVA using the *aovp* function (for sniff duration; *lmPerm* package,

Wheeler 2010). Significant interactions were analysed with a post-hoc pairwise test using the *lsmeans* package (Lenth 2016).

2.4.2 Correlating Behavioural Observations with EIA Analyses

To determine if sex-steroid metabolites affect behavioural responses, we ran three generalised linear models (GLMs) and linear models (only for sniff duration of opposite sex urine) with donor identity and responder identity as random effects:

(1) *Response likelihood—Oestradiol/Testosterone levels (ng/ml)*
(2) *Urine sniff duration—Oestradiol/Testosterone levels (ng/ml)*
(3) *Marking response— Oestradiol/Testosterone levels (ng/ml)*

Male (n = 9) and female donors (n = 14) were analysed separately, as were male (n = 7) and female responders (n = 7). GLMs were modified to incorporate responder age in months or responder reproductive stage. Hormone levels and age in months were re-scaled using the *scale* function in base R when analysed together. Reproductive life stages of both donors and responders were based on age following Sugianto et al. submitted a, b (*sexually immature*: ≤ 2 years old [n = 9; 1 male donor, 6 female donors, 1 male responder, 1 female responder]; *reproductively active*: 2–8 years old [n = 24; 8 male donors, 6 female donors, 5 male responders, 5 female responders]; *reproductively senescent*: ≥ 9 years old [n = 2; 1 female donor, 1 male responder].

All analyses were performed in R (version 3.5.1) and R Studio (version 1.1.453).

3 Results

3.1 Behavioural Observations with Scent-Playback Experiments

143 of the 197 videos analysed contained badger appearances involving 26 different individually identifiable adults and two unmarked cubs. In 22 badger appearances, the adult individuals could not be identified unambiguously. However, in 33 appearances (23.0% of appearances), badgers did not come within 50 cm of either water or urine sample (i.e. no potential reaction). Potential, but not actual responses were recorded in a further 74 appearances (51.7%; 15 within urine only and 8 within water only), while 63 appearances (44.0%) included actual responses where badgers sniffed in 24 appearances (38.1%) only the urine, in 23 (36.5%) only the water sample, and in 16 appearances both (25.4%) urine and water. Because only two cubs responded to any of the samples, we analysed only adult responses in the following.

Badgers exhibited similar response likelihoods ($\chi^2 = 0.34$, $df = 1$, $p = 0.56$) for both urine and water. Regardless, badgers sniffed urine (n = 40, 5.02 ± 1.58 s) significantly longer than water (n = 23, 3.05 ± 0.49 s; estimate = 0.95, $F_{1,77} = 4.71$, Pr (Prob) = 0.02). Individuals that sniffed both urine and water (n = 16) were removed from countermarking analysis due to the uncertainty about which sample they countermarked against. Removing these individuals showed that badgers significantly countermarked urine more than water ($\chi^2 = 14.08$, $df = 1$, $p = <0.001$). Badgers demonstrated no significant variation in sniff duration regardless of sample freezing duration (estimate = −0.19, $F_{1,\ 38} = 1.09$, Pr(Prob) = 0.46).

3.2 Effects of Age-Related Information on Behavioural Responses

Cub urine elicited slightly more responses overall than did adult urine amongst adult responders, although this effect was not significant ($\chi^2 = 1$, $df = 1$, $p = 0.32$). Cub and adult urine were countermarked equally often, but only one single female and one male were observed countermarking cub urine (albeit at high frequency).

Both cub urine and adult urine elicited more responses from adult females than from adult males (Table 2, Fig. 1), with adult females generally sniffing urine longer than males (males: $n = 6$, 2.35 ± 0.86 s; females: $n = 4$, 5.37 ± 2.39 s; Fig. 2).

Because all cub donors were non-residents at the setts provisioned, we repeated this analysis including only unfamiliar donors of either age class. As before, adult badgers

Table 2. Pairwise tests comparing behavioural responses based on donor sex/age and responder sex/age

Response likelihood (Donor sex*Responder sex)				
Contrast	Estimate	SE	Z-ratio	P
Female, female—Male, female	−0.436363636	3.532708e−16	−1.235210e+15	<0.0001[a]
Female, female—Female, male	0.006493506	3.532708e−16	1.838110e+13	<0.0001[a]
Female, female—Male, male	0.163636364	3.532708e−16	4.632038e+14	<0.0001[a]
Male, female—Female, male	0.442857143	3.532708e−16	1.253591e+15	<0.0001[a]
Male, female—Male, male	0.600000000	3.532708e−16	1.698414e+15	<0.0001[a]
Female, male—Male, male	0.157142857	3.532708e−16	4.448227e+14	<0.0001[a]
Countermarking responses (Donor sex*Responder sex)				
Female, female—Male, female	−0.21818182	6.798700e−17	−3.209170e+15	<0.0001[a]
Female, female—Female, male	0.03896104	7.447602e−17	5.231353e+14	<0.0001[a]
Female, female—Male, male	−0.01818182	7.447602e−17	−2.441298e+14	<0.0001[a]
Male, female—Female, male	0.25714286	6.798700e−17	3.782236e+15	<0.0001[a]
Male, female—Male, male	0.20000000	6.798700e−17	2.941739e+15	<0.0001[a]
Female, male—Male, male	−0.05714286	7.447602e−17	−7.672651e+14	<0.0001[a]
*Countermarking responses (Responder sex*Donor age)*				
Female, adult—Male, adult	0.1190476	2.019047e−16	5.896228e+14	<0.0001[a]
Female, adult—Female, cub	−0.7142857	2.553915e−16	−2.796826e+15	<0.0001[a]
Female, adult—Male, cub	0.1190476	2.019047e−16	5.896228e+14	<0.0001[a]
Male, adult—Female, cub	−0.8333333	2.708836e−16	−3.076352e+15	<0.0001[a]
Male, adult—Male, cub	0.0000000	2.211755e−16	0.000000e+00	1.0000
Female, cub—Male, cub	0.8333333	2.708836e−16	3.076352e+15	<0.0001[a]
*Response likelihood (Donor group * Donor sex * Responder sex)*				
Familiar, female, female—Unfamiliar, female, female	−4.166667e−02	0.04014586	−1.038	0.9687
Familiar, female, female—Familiar, male, female	−3.347222e−01	0.05051576	−6.626	<0.0001[a]
Familiar, female, female—Unfamiliar, male, female	−6.263889e−01	0.05794555	−10.810	<0.0001[a]
Familiar, female, female—Familiar, female, male	4.166667e−02	0.04014586	1.038	0.9687
Familiar, female, female—Unfamiliar, female, male	1.526557e−16	0.05677482	0.000	1.0000

(continued)

Table 2. *(continued)*

Response likelihood (Donor sex*Responder sex)				
Contrast	Estimate	SE	Z-ratio	P
Familiar, female, female—Familiar, male, male	3.736111e−01	0.05794555	6.448	<0.0001[a]
Familiar, female, female—Unfamiliar, male, male	8.194444e−02	0.05051576	1.622	0.7370
Unfamiliar, female, female—Familiar, male, female	−2.930556e−01	0.05051576	−5.801	<0.0001[a]
Unfamiliar, female, female—Unfamiliar, male, female	−5.847222e−01	0.05794555	−10.091	<0.0001[a]
Unfamiliar, female, female—Familiar, female, male	8.333333e−02	0.05677482	1.468	0.8247
Unfamiliar, female, female—Unfamiliar, female, male	4.166667e−02	0.04014586	1.038	0.9687
Unfamiliar, female, female—Familiar, male, male	4.152778e−01	0.05794555	7.167	<0.0001[a]
Unfamiliar, female, female—Unfamiliar, male, male	1.236111e−01	0.05051576	2.447	0.2188
Familiar, male, female—Unfamiliar, male, female	−2.916667e−01	0.04916843	−5.932	<0.0001[a]
Familiar, male, female—Familiar, female, male	3.763889e−01	0.05051576	7.451	<0.0001[a]
Familiar, male, female—Unfamiliar, female, male	3.347222e−01	0.05051576	6.626	<0.0001[a]
Familiar, male, female—Familiar, male, male	7.083333e−01	0.04916843	14.406	<0.0001[a]
Familiar, male, female—Unfamiliar, male, male	4.166667e−01	0.05677482	7.339	<0.0001[a]
Unfamiliar, male, female—Familiar, female, male	6.680556e−01	0.05794555	11.529	<0.0001[a]
Unfamiliar, male, female—Unfamiliar, female, male	6.263889e−01	0.05794555	10.810	<0.0001[a]
Unfamiliar, male, female—Familiar, male, male	1.000000e+00	0.08029171	12.455	<0.0001[a]
Unfamiliar, male, female—Unfamiliar, male, male	7.083333e−01	0.04916843	14.406	<0.0001[a]
Familiar, female, male—Unfamiliar, female, male	−4.166667e−02	0.04014586	−1.038	0.9687
Familiar, female, male—Familiar, male, male	3.319444e−01	0.05794555	5.729	<0.0001[a]
Familiar, female, male—Unfamiliar, male, male	4.027778e−02	0.05051576	0.797	0.9933
Unfamiliar, female, male—Familiar, male, male	3.736111e−01	0.05794555	6.448	<0.0001[a]
Unfamiliar, female, male—Unfamiliar, male, male	8.194444e−02	0.05051576	1.622	0.7370
Familiar, male, male—Unfamiliar, male, male	−2.916667e−01	0.04916843	−5.932	<0.0001[a]

[a]denote statistical significance.

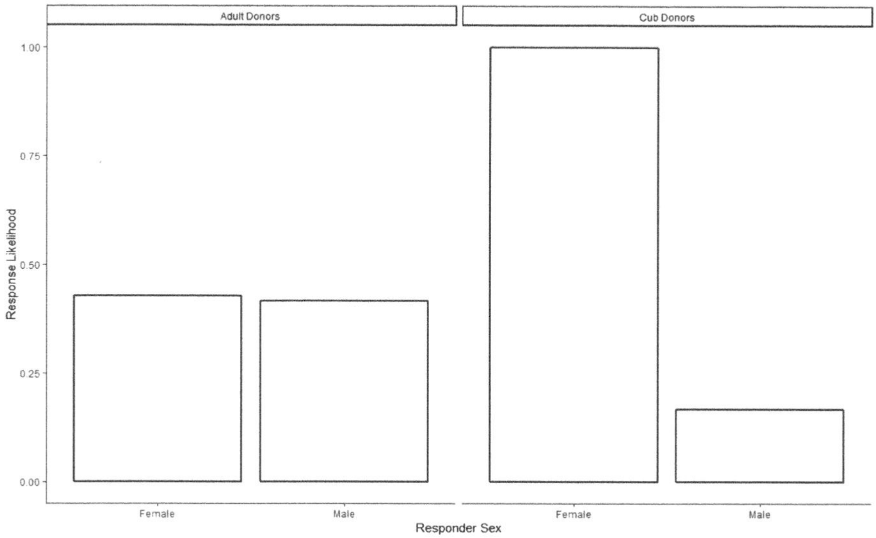

Fig. 1. Interactions between donor age and responder sex for response likelihood. When controlling for donor familiarity, female adults responded more often towards cub urine than adult urine (Table 3)

demonstrated similar response likelihoods ($\chi^2 = 1$, $df = 1$, $p = 0.32$) and countermarking responses (Table 2) towards adult and cub urine. Sniff duration, however, differed significantly between cub and adult urine samples (*all donors*: estimate = −1.78, $F_{1,\ 38} = 4.32$, Pr(Prob) = 0.03; *unfamiliar donors only*: estimate: −1.78, $F_{1,\ 38} = 4.42$, Pr(Prob) = 0.02) as responders sniffed cub urine (*all cub donors*: $n = 13$, 6.46 ± 1.86 s) longer than adult urine (*all adult donors*: $n = 27$, 2.90 ± 0.77 s; *unfamiliar adult donors*: $n = 17$, 3.05 ± 1.11 s; Fig. 3).

3.3 Effects of Sex-Related Information on Behavioural Responses

Overall, males and females were equally likely to respond to conspecific urine (*response likelihood*: $\chi^2 = 1$, $df = 1$, $p = 0.32$) and did not differ in their propensity to countermark ($\chi^2 = 1.70$, $df = 1$, $p = 0.19$) or sniff duration (estimate = 0.86, $F_{1,13} = 0.96$, Pr(Prob) = 0.64). Similarly, overall, the sex of the donor urine sample donor did not elicit significantly different responses among responders (*response likelihood*: $\chi^2 = 1$, $df = 1$, $p = 0.32$; *countermarking responses*: $\chi^2 = 2.67$, $df = 1$, $p = 0.10$; *sniff duration*: estimate = 1.62, $F_{1,\ 29} = 2.34$, Pr(Prob) = 0.19).

However, when responder sex was considered, adult females responded more to male donors (response likelihood = 0.80; Table 3; Fig. 4) than to female donors (response likelihood = 0.36), whereas adult males responded more to female donors (response likelihood = 0.36; Fig. 4) than to male donors (response likelihood = 0.20), although sniff duration was not affected by the interaction between donor–responder sex dyads (Table 3).

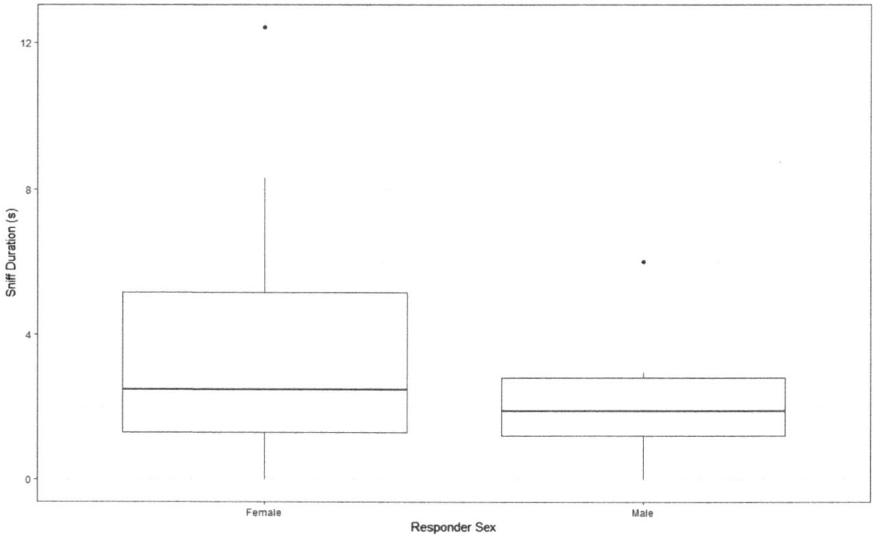

Fig. 2. Urine sniff duration in relation to donor age and responder sex. When analysed alongside donor age, adult females (n = 4; 5.37 ± 2.39 s) generally sniffed urine longer than adult males (n = 6; 2.35 ± 0.86 s)

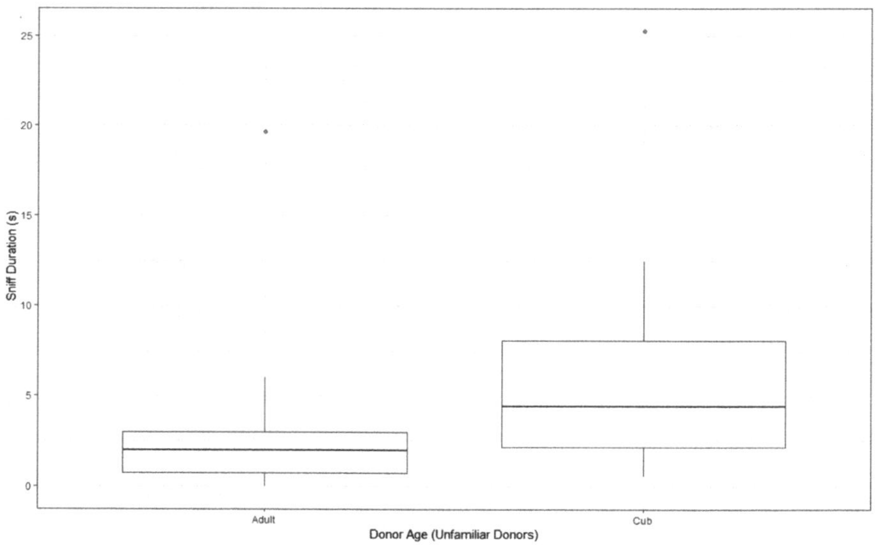

Fig. 3. Sniff duration of unfamiliar adult urine and cub urine by adult responders. When accounting for familiarity, adult badgers sniffed unfamiliar cub urine significantly longer (n = 13, 6.46 ± 1.86 s) than they did unfamiliar adult urine (n = 17, 3.05 ± 1.11 s; estimate: −1.78, $F_{1, 38}$ = 4.42, Pr(Prob) = 0.02)

Table 3. Behavioural responses based on the interaction between donor-related traits and responder-related sex/age

Response likelihood					
		Estimate	Std. error	t-value	Pr(>|t|)
Donor sex		4.364e−01	3.533e−16	1.235e+15	<2e−16[a]
Responder sex		−6.494e−03	3.533e−16	−1.838e+13	<2e−16[a]
Donor sex*Responder sex		−5.935e−01	4.996e−16	−1.188e+15	<2e−16[a]
Countermarking responses					
		Estimate	*Std. error*	*t-value*	*Pr(>|t|)*
Donor sex		2.182e−01	6.799e−17	3.209e+15	<2e−16[a]
Responder sex		−3.896e−02	7.448e−17	−5.231e+14	<2e−16[a]
Donor sex*Responder sex		−1.610e−01	1.008e−16	−1.597e+15	<2e−16[a]
Sniff duration					
	Df	*R sum sq*	*R mean sq*	*Iterations*	*Pr(Prob)*
Donor sex	1	28.881	28.8806	807	0.1103
Responder sex	1	12.890	12.8904	240	0.2958
Donor sex*Responder sex	1	11.060	11.0595	58	0.6379
Response likelihood					
		Estimate	*Std. error*	*t-value*	*Pr(>|t|)*
Responder sex		−1.190e−02	2.691e−16	−4.424e+13	<2e−16[a]
Donor age		5.714e−01	3.296e−16	1.734e+15	<2e−16[a]
Responder sex* Donor age		−8.214e−01	4.255e−16	−1.931e+15	<2e−16[a]
Countermarking responses					
		Estimate	*Std. error*	*t-value*	*Pr(>|t|)*
Responder sex		−1.190e−01	2.019e−16	−5.896e+14	<2e−16[a]
Donor age		7.143e−01	2.554e−16	2.797e+15	<2e−16[a]
Responder sex*Donor age		−7.143e−01	3.379e−16	−2.114e+15	<2e−16[a]
Sniff duration					
		Estimate	*Iterations*		*Pr(Prob)*
Responder sex		2.593	2028		0.0473[a]
Donor age		−2.523	3444		0.0285[a]
Responder sex*Donor age		−2.302	1369		0.0687
Response likelihood					
		Estimate	*Std. error*	*t-value*	*Pr(>|t|)*
Donor familiarity		0.04167	0.04015	1.038	0.329675
Donor sex		0.33472	0.05052	6.626	0.000165[a]
Responder sex		−0.04167	0.04015	−1.038	0.329675
Donor sex*responder sex		−0.66667	0.06348	−10.503	5.88e−06[a]
Donor familiarity* Donor sex		0.25000	0.06348	3.938	0.004305[a]

(*continued*)

Table 3. (*continued*)

Response likelihood				
	Estimate	Std. error	t-value	Pr(>|t|)
Countermarking responses				
	Estimate	*Std. error*	*t-value*	*Pr(>|t|)*
Donor familiarity	0.04333	0.08495	0.510	0.6238
Donor sex	0.12333	0.10967	1.125	0.2934
Responder sex	−0.04000	0.08495	−0.471	0.6503
Donor sex*Responder sex	−0.24333	0.12014	−2.025	0.0774
Donor familiarity* Donor sex	0.15667	0.12014	1.304	0.2285

[a]denote statistical significance.

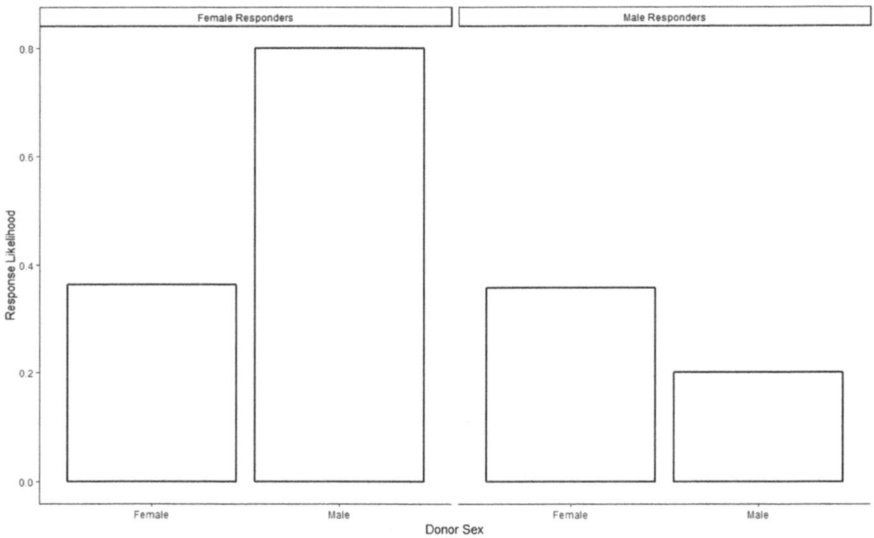

Fig. 4. Interactions between donor sex and responder sex for response likelihood. Adult females showed more actual responses towards male urine than female urine, while adult males showed more actual responses towards female urine than male urine (Tables 2 and 3)

Both sexes countermarked male urine (*female responders*: countermarking proportion = 0.4; *male responders*: countermarking proportion = 0.2) significantly more than female urine (*female responders*: countermarking proportion = 0.18; *male responders*: countermarking proportion = 0.14; Fig. 5), with all donor-responder combinations differing significantly from each other (Table 3).

Responder sex, donor age and their interaction all significantly affected response likelihood (Table 3). Females responded more often than males to adult and to cub donors (Fig. 4). Responder sex and donor age, but not their interaction, also significantly affected sniff duration, with females (n = 4; 5.37 ± 2.39 s) investigating urine longer than males (n = 6; 2.35 ± 0.86 s; Fig. 2) and all responders investigating cub urine (n = 2; 7.54 ± 4.89 s) for longer durations than adult urine (n = 8; 2.56 ± 0.67 s; Fig. 6).

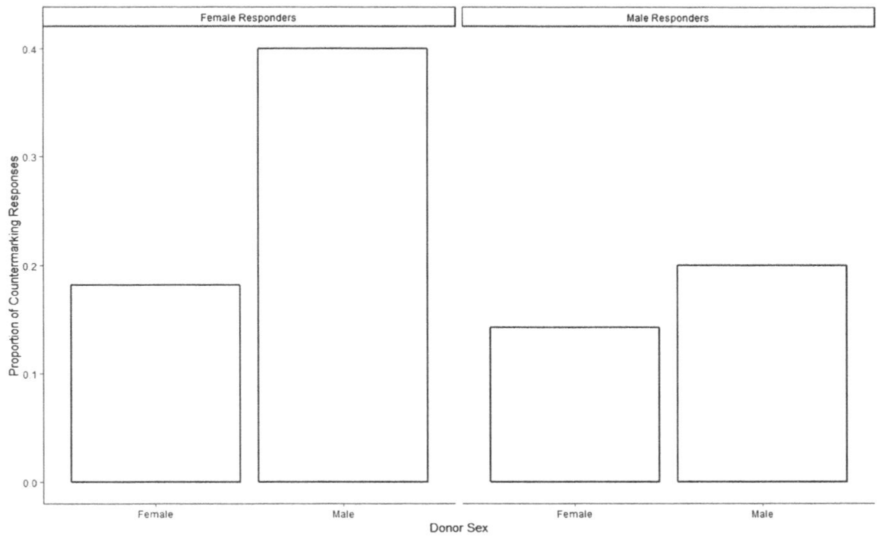

Fig. 5. Interactions between donor sex and responder sex for countermarking responses. Urine from male donors elicited more countermarking responses than urine from female donors, particularly amongst adult female responders (Tables 2 and 3)

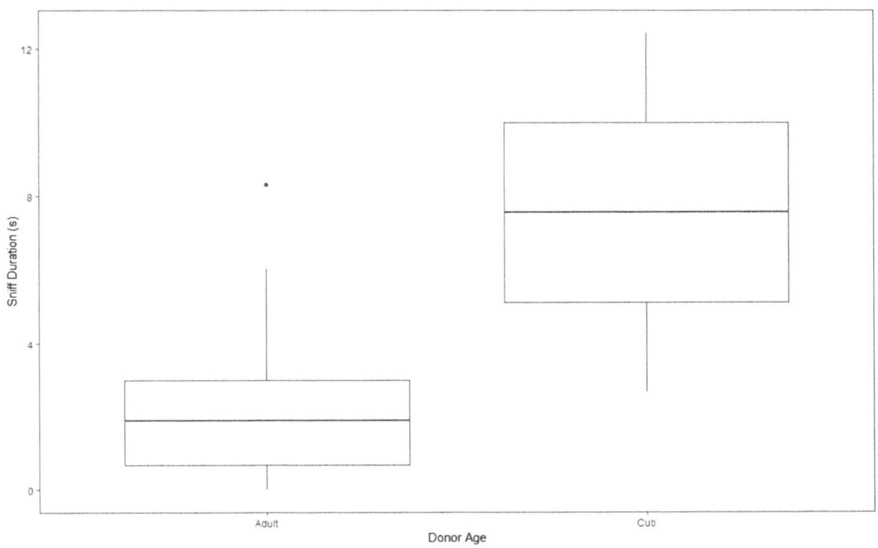

Fig. 6. Urine sniff duration in relation to donor age and responder sex. When analysed alongside responder sex, cub urine (n = 2; 7.54 ± 4.89 s) was sniffed longer than adult urine (n = 8; 2.56 ± 0.67 s). However, only two individuals sniffed cub urine

Countermarking responses were affected by responder sex, donor age, and their interaction (Table 3).

3.4 Effects of Familiarity on Behavioural Responses

Overall, donor familiarity did not affect behavioural responses (*response likelihood*: $\chi^2 = 1$, $df = 1$, $p = 0.32$; *sniff duration*: estimate = −0.55, $F_{1,13} = 0.33$, Pr(Prob) = 1; *countermarking responses*: $\chi^2 = 1.70$, $df = 1$, $p = 0.19$). All adults responded more to unfamiliar male donors than to familiar females, unfamiliar females, and familiar males (Table 2).

3.5 Correlating Behavioural Observations with Sex-Steroid Levels

3.5.1 Effects of Female Donor Oestradiol Levels

Twenty-five of 54 videos (for which we had oestradiol levels for the provisioned donor sample) involved identifiable responders (adult females = 8, adult males = 17). For female donors, female response likelihood was not influenced significantly by donor oestradiol levels, responder age and responder reproductive stage ($n = 6$). No female responders were observed countermarking urine, while only two female responders sniffed urine, thereby preventing further statistical analysis.

Male response likelihood to female donors was also not influenced significantly by donor oestradiol levels, responder age, and responder reproductive stage. However, male responders spent significantly more time sniffing urine containing lower oestradiol levels ($n = 5$, estimate = −2.05, standard error = 0.71, t-value = −2.88; Fig. 7). There were also no effects on countermarking.

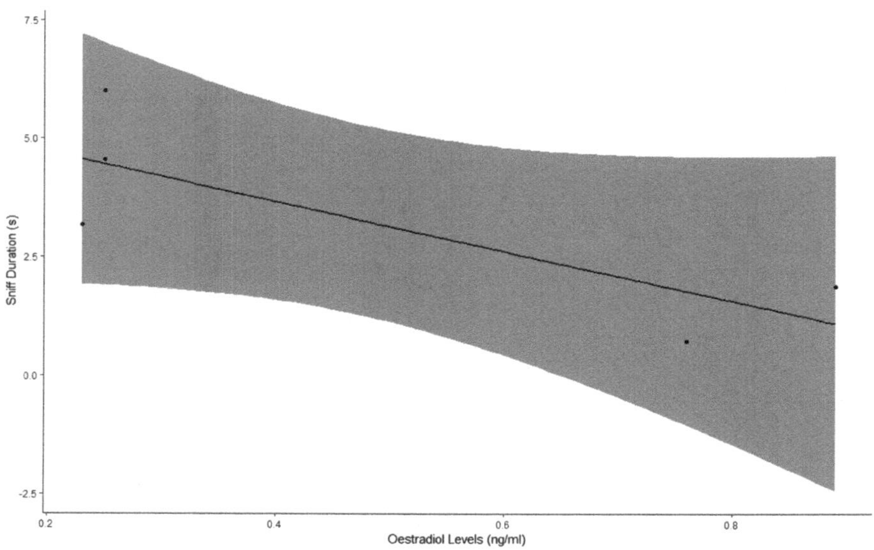

Fig. 7. Sniff duration of adult male responders towards urine with oestradiol. Male responders spent significantly longer time investigating urine with lower oestradiol levels (Table 3)

3.5.2 Effects of Male Donor Testosterone Levels

Fifteen of 20 videos (for which we had testosterone levels for the provisioned donor sample) involved identifiable responders (adult females = 6, adult males = 9).

There were no effects on response likelihood, and sniff duration by female responders to male urine was not influenced by testosterone levels, responder age, or responder reproductive stage. Limited responses ($n = 6$) and observed countermarking ($n = 1$) prevented further analysis. Response likelihoods and countermarking by male responders were not influenced by testosterone levels, responder age or responder reproductive stage. Only two males sniffed urine, thereby preventing any further analysis.

4 Discussion

Adult badgers were demonstrably able to decipher relevant information concerning sex, age and group familiarity encoded in the urine of conspecifics. In response, they modified their behaviour according to the signals' pertinence to their own traits/condition, investigating and countermarking urine from opposite-sex donors more often than urine from same-sex donors. Heightened interest towards opposite-sex donors occurred regardless of donor familiarity. Additionally, cub urine stimulated more responses and countermarking as well as longer sniff durations than to adults (but limited by sample size).

Adult badgers responded more often to and countermarked urine from opposite-sex donors than they did same-sex donors. Badgers were likely responding to the potentially volatile sex-specific information encoded within urine, reflected by GC-MS profile variations between reproductively active males and oestrous females (Tinnesand et al. unpublished data). This aligns with previous sex-related differences in response behaviour amongst badgers (Service et al. 2001). Other species have been shown to exhibit considerable sex-related variation in urinary scent profiles (Barman et al. 2013), and these variations may augment olfactory sex discrimination abilities (Charlton 2014; Gilfillan et al. 2017; Leonard et al. 2001; Linklater et al. 2013). Surprisingly, however, the excreted sex-steroid metabolite (oestradiol and testosterone) levels in urine did not elicit clear response patterns, aside from male (testosterone) samples receiving more attention than female (oestradiol) samples, conforming with Service and Harris (2001). Interestingly, in this context, our pilot study indicated that male responders sniffed female urine with low oestradiol content significantly longer than samples with high oestradiol content.

Oestradiol fluctuations during oestrus, or elevated testosterone levels during the mating season, are likely reflected in altered urinary odours detectable by conspecifics. Surprisingly, however, we found that the excreted sex-steroid metabolite (oestradiol and testosterone) levels in urine did not elicit clear response patterns, except that male responders sniffed urine provisioned from females with low oestradiol for significantly longer durations. This is contrary to typical behaviour in other mammals where males typically show heightened interest in urine from oestrous females and engage in flehmen behaviour to transfer sex-steroid metabolites to their vomeronasal organ (Crump et al. 1984; Rajanarayanan and Archunan 2011).

Our findings here, however, imply that either (i) a donor's reproductive quality may take longer to discern when sex-steroid metabolite levels are lowered or (ii) any potential olfactory cue information that sex-steroid metabolites might imbue to urine is not being used as any discernible quantitative signal for individual advertisement. This is consistent with the theory that urinary proteins mediate individual recognition (mice: Hurst et al. 2001), superseding former ideas that the Major Histocompatibility Complex (MHC) played a role (Hurst et al. 2005); thus, this mediator's signal quality might exceed the cue potential of sex metabolites, although we re-emphasise here that we analysed only small sample sizes in this pilot study.

Although we found no gross effect of familiarity, when re-examined by the sex of the donor/responder, unfamiliar male scent marks elicited the greatest responses and familiar females the least. Unfamiliar cubs also elicited a strong response, particularly from adult females since such signals would be very out of context, given that cubs would not be expected to visit other social groups (Macdonald et al. 2008). Within other mammalian species (e.g. Rostain et al. 2004), individuals—particularly males—often investigate unfamiliar scents more often than familiar scents. This may serve to either locate new females as part of their philandering search for mates or enable them to detect the risk of competing males intruding into their territory (Gosling and Roberts 2001; Tinnesand et al. 2013). This has been formalised as the 'Dear Enemy Phenomenon' (DEP) which theorises that animals will respond less aggressively towards neighbours than strangers (Fisher 1954). This DEP becomes evident in how this same badger population reacts to anal gland secretions from neighbouring and unfamiliar individuals (Tinnesand et al. 2013). Insofar as successive studies of the socio-spatial system of these same badgers have provided little corroboration for the classical model of active territorial defence (Buesching and Macdonald 2001; Kilshaw et al. 2009; Macdonald et al. 2015; Stewart et al. 1997), it is progressively less parsimonious to interpret urine (or other) marking in this context. Consequently, this adds weight to the possibility that individuals investigate unfamiliar scents longer to discern potential mates that have yet to be encountered.

These sex and age effects, but only weak effects of familiarity within this social system, and no strong evidence for any discrimination based on sex-steroid metabolite signalling, illustrate the species- and context- specific nature of scent communication and the active use of urinary scents in advertisement. Moreover, the deficits in what badgers glean from conspecific scent marks are perhaps more informative to the olfactory communication paradigm than what we demonstrate they may be detecting, where there is a general trend for signalling individuality to increase with social group size and the extent of social behaviour within a clade, best established thus far in studies of acoustic communication (Pollard and Blumstein 2011). Given that European badgers living at high-density in the UK display only rudimentary social behaviours, and given the high rate of trespassing and thus integration at the population level in our study population (Macdonald et al. 2015), our pilot study may be exposing a stage in the evolution of social scent-marking. The relatively recent adaptation of badgers to the higher carrying capacity supported by the agricultural mosaic of land use in parts of the UK (and Ireland; Macdonald et al. 2015) may have not yet led them to fully exploit urinary scent discrimination. Alternatively, given that we have established that badgers are discriminating a lot of information from the anal gland secretion coating faeces at latrines (Tinnesand et al. 2015), and given that badgers uniquely engage in quite subtle social signalling of individual qualities using their subcaudal gland secretion (Buesching et al. 2002b, c), then urinary marking may play a more subsidiary role in this macrosmatic mammal species.

References

Albone ES (1984) Mammalian semiochemistry: the investigation of chemical signals between mammals. Wiley, Chichester, United Kingdom

Amendolagine L, Schoffne T, Koscielny L, Schook M, Copeland D, Casteel J, Duff B, Koester D (2018) In-house monitoring of steroid hormone metabolites in urine informs breeding management of a giant anteater (*Myrmechophaga tridactyla*). Zoo Biol 37:40–45

Annavi G, Newman C, Dugdale HL, Buesching CD, Sin Y-W, Burke T, Macdonald DW (2014) Neighbouring-group composition and within-group relatedness drive extra-group paternity rate in the European badger (*Meles meles*). J Evol Biol 27(10):2191–2203

Barman P, Yadav MC, Kuma H, Meur SK, Ghosh SK (2013) Gas chromatographic-mass spectrometric analysis of chemical volatiles in buffalo (*Bubalus bubalis*) urine. Theriogenology 80(6):654–658

Behringer V, Deschner T (2017) Non-invasive monitoring of physiological markers in primates. Horm Behav 91:3–18

Brown JL, Wasser SK, Wildt DE, Graham LH (1994) Comparative aspects of steroid hormone metabolism and ovarian activity in felids, measured noninvasively in feces. Biol Reprod 51:776–786

Buesching CD, Macdonald DW (2001) Scent-marking behaviour of the European badger (*Meles meles*): resource defence or individual advertisement? In: Marchlewska-Koj A, Lepri JJ, Müller- Schwarze D (eds) Chemical signals in vertebrates 9. Kluwer Academic/Plenum Publishers, New York, pp 321–327

Buesching CD, Stankowich T (2017) Communication amongst the musteloids: signs, signals, and cues. In: Macdonald DW, Newman C, Harrington LA (eds) Biology and Conservation of Musteloids. Oxford University Press, Oxford, UK, pp 149–166

Buesching CD, Newman C, Macdonald DW (2002a) Variations in colour and volume of the subcaudal gland secretion of badgers (*Meles meles*) in relation to sex, season and individual-specific parameters. Mamm Biol 67:147–156

Buesching CD, Waterhouse JS, Macdonald DW (2002b) Gas-chromatographic analyses of the subcaudal gland secretion of the European badger (*Meles meles*) part I: chemical differences related to individual parameters. J Chem Ecol 28(1):57–69

Buesching CD, Waterhouse JS, Macdonald DW (2002c) Gas-chromatographic analyses of the subcaudal gland secretion of the European badger (*Meles meles*) part II: time-related variation in the individual-specific composition. J Chem Ecol 28(1):41–56

Buesching CD, Stopka P, Macdonald DW (2003). The social function of allo-marking in the European badger (*Meles meles*). Wildl Conserv 965–980

Buesching CD, Heistermann M, Macdonald DW (2009) Seasonal and inter-individual variation in testosterone levels in badgers *Meles meles*: evidence for the existence of two endocrinological phenotypes. J Comp Physiol A: Neuroethol, Sens, Neural, Behav Physiol 195(9):865–871

Campbell CJ, Shideler SE, Todd HE, Lasley BL (2001) Fecal analysis of the ovarian cycle of female black-handed spider monkeys (*Ateles geoffroyi*). Am J Primatol 54:79–89

Charlton BD (2014) Discrimination of sex and reproductive state in koalas, *Phascolarctos cinereus*, using chemical cues in urine. Anim Behav 91:118–124

Converse LJ, Carlson AA, Ziegler TE, Snowdon CT (1995) Communication of ovulatory state to mates by female pygmy marmosets, *Cebuella pygmaea*. Anim Behav 49(3):615–621

Courant F, Aksglaede L, Antignac JP, Monteau F, Sorensen K, Andersson AM, Skakkebaek NE, Juul A, Bizec BL (2010) Assessment of circulating sex steroid levels in prepubertal and pubertal boys and girls by a novel ultrasensitive gas chromatography-tandem mass spectrometry method. J Clin Endocrinol Metab 95(1):82–92

Crump D, Swigar AA, West JR, Silverstein RM, Müller-Schwarze D, Altieri R (1984) Urine fractions that release flehmen in black-tailed deer, *Odocoileus hemionus columbianus*. J Chem Ecol 10(2):203–215

da Silva J, Macdonald DW (1989) Limitations to the use of tooth wear as a means of ageing Eurasian badgers, *Meles meles*. Rev D Ecol 44:275–278

Findlay JK, Liew SH, Simpson ER, Korach KS (2010) Estrogen signaling in the regulation of female reproductive functions. Fertility control, handbook of experimental pharmacology, vol 19. Springer, Heidelberg, pp 29–36

Fisher J (1954) Evolution and bird sociality. Evolution as a process, vol 71., p 83

Fobes JL, King JE (1977) Prosimian sensory capacities. Primates 18(3):713–730

Gilfillan GD, Vitale JDT, McNutt JW, McComb K (2017) Spontaneous discrimination of urine odours in wild African lions, *Panthera leo*. Anim Behav 126:177–185

Gosling LM, Roberts SC (2001) Scent-marking by male mammals: cheat-proof signals to competitors and mates. Adv Study Behav 30:169–217

Goymann W (2012) On the use of non-invasive hormone research in uncontrolled, natural environments: the problem with sex, diet, metabolic rate and the individual. Methods in Ecol Evol 3(4):757–765

Henkel S, Lambides A, Berger R, Thomsen A, Widdig R (2015) Rhesus macaques (*Macaca mulatta*) recognize group membership via olfactory cues alone. Behavl Ecol Sociobiol 69 (12):2019–2034

Hradecký P (1986) Volatile fatty acids in urine and vaginal secretions of cows during reproductive cycle. J Chem Ecol 12(1):187–196

Hurst JL, Payne CE, Nevison CM, Marie AD, Humphries RE, Robertson DH, Cavaggioni A, Beynon RJ (2001) Individual recognition in mice mediated by major urinary proteins. Nature 414(6864):631

Hurst JL, Thom MD, Nevison CM, Humphries RE, Beynon RJ (2005) MHC odours are not required or sufficient for recognition of individual scent owners. Proc R Soc Lond B Biol Sci 272(1564):715–724

Hutchings MR, Service KM, Harris S (2002) Is population density correlated with faecal and urine scent marking in European badgers (*Meles meles*) in the UK? Mamm Biol 67:286–293

James MO, Smith R, Williams RT, Reidenberg M (1972) The conjugation of phenylacetic acid in man, sub-human primates and some non-primate species. Proc R Soc Lond B Biol Sci 182 (1066):25–35

Kilshaw K, Newman C, Buesching CD, Bunyan J, Macdonald DW (2009) Coordinated latrine use by European badgers, *Meles meles*: potential consequences for territory defense. J Mammal 90(5):1188–1198

Lenth RV (2016) Least-squares means: the R package *lsmeans*. J Stat Soft 69(1):1–33

Leonard ST, Ferkin MH, Johnson MM (2001) The response of meadow voles to an over-mark in which the two donors differ in gonadal hormone status. Anim Beh 62(6):1171–1177

Linklater WL, Mayer K, Swaisgood RR (2013) Chemical signals of age, sex and identity in black rhinoceros. Anim Beh 85(3):671–677

Lu KH, Hopper BR, Vargo TM, Yen SSC (1979) Chronological changes in sex steroid, gonadotropin, and prolactin secretion in aging female rats displaying different reproductive states. Biol Reprod 21:193–203

Macdonald DW (1979) Some observations and field experiments on the urine marking behaviour of the red fox, *Vulpes vulpes L.* Z Tierpsychol 51(1):1–22

Macdonald DW, Newman C, Buesching CD (2015) Badgers in the rural landscape—conservation paragon or farmland pariah? Lessons from the wytham badger project. In: Macdonald DW, Willis KJ (eds) Wildlife conservation on farmland, vol 2. Oxford University Press, Oxford UK, pp 65–95

Macdonald DW, Newman C, Buesching CD, Johnson PJ (2008) Male-biased movement in a high-density population of the Eurasian badger (*Meles meles*). J Mammal 89(5):1077–1086

Macdonald DW, Newman C, Nouvellet PM, Buesching CD (2009) An analysis of Eurasian badger (*Meles meles*) population dynamics: implications for regulatory mechanisms. J Mammal 90(6):1392–1403

McLaren GW, Thornton PD, Newman C, Buesching CD, Baker SE, Mathews F, Macdonald DW (2005) High rectal temperature indicates an increased risk of unexpected recovery in anaesthetized badgers. Vet Anaesth Analg 32(1):48–52

Miller RC, Brindle E, Holman DJ, Shofer J, Klein NA, Soules MR, O'Connor KA (2004) Comparison of specific gravity and creatinine for normalizing urinary reproductive hormone concentrations. Clin Chem 50(5):924–932

Penn D, Potts WK (1998) Chemical signals and parasite-mediated sexual selection. Trends Ecol Evol 13(10):391–396

Pollard KA, Blumstein DT (2011) Social group size predicts the evolution of individuality. Curr Biol 21(5):413–417

Preston BT, Stevenson IR, Lincoln GA, Monfort SL, Jill G, Wilson K (2012) Testes size, testosterone production and reproductive behaviour in a natural mammalian mating system. J Anim Ecol 81(1):296–305

Rajanarayanan S, Archunan G (2011) Identification of urinary sex pheromones in female buffaloes and their influence on bull reproductive behaviour. Res Vet Sci 91(2):301–305

Rostain RR, Ben-David M, Groves P, Randall JA (2004) Why do river otters scent-mark? An experimental test of several hypotheses. Anim Behav 68(4):703–711

SankarGanesh D, Ramachandran R, Muniasamy S, Saravanakumar VR, Suriyakalaa U, Kannan S, Archunan G, Achiraman S (2014) A correlation of fecal volatiles and steroid hormone profiles with behavioral expression during estrous cycle of goat, *Capra hircus*. Gen Comp Endocrinol 206:178–183

Savill P, Perrins C, Fisher N, Kirby K (2011) Wytham woods: Oxford's ecological laboratory. Oxford University Press

Service KM, Brereton RG & Harris S (2001) Analysis of badger urine volatiles using gas chromatography-mass spectrometry and pattern recognition techniques. The Analyst 126:615–623

Service KM, Harris S (2001) Remote monitoring of badgers for testing discrimination between urine samples from donors of different age and sex categories. In: Marchlewska-Koj A, Lepri JJ, Müller-Schwarze D (eds) Chemical signals in vertebrates 9. Springer, Boston, M.A, pp 451–457

Sharpe LL (2015) Handstand scent marking: height matters to dwarf mongooses. Anim Behav 105:173–179

Snoj T, Dolenc J, Kobal S (2014) Sex steroid levels in urine of cattle of different ages: evaluation of abuse control procedures. Food Addit Contam Part A Chem Anal Control Exposure Risk Assess 31(4):614–620

Stewart PD, Anderson C, Macdonald DW (1997) A mechanism for passive range exclusion: evidence from the European badger (*Meles meles*). J Theor Biol 184(3):279–289

Stewart PD, Macdonald DW (1997) Age, sex, and condition as predictors of moult and the efficacy of a novel fur-clip technique for individual marking of the European badger (*Meles meles*). J Zool 241(3):543–550

Stewart PD, Macdonald DW, Newman C, Tattersall FH (2002) Behavioural mechanisms of information transmission and reception by badgers, *Meles meles*, at latrines. Anim Behav 63 (5):999–1007

Stubbe M (1971) Die analen Markierungsorgane des Dachses, *Meles meles (L.)*. Zool Garten N. F. Leipzig 40:125–135

Sugianto NA, Newman C, Macdonald DW, Buesching CD (in press, a) Reproductive and somatic senescence of the European badger (*Meles meles*): evidence through lifetime sex-steroid profiles

Sugianto NA, Newman C, Macdonald DW, Buesching CD (in press, b) Heterochrony of puberty in the European badger (*Meles meles*) can be explained by growth rate and group size: evidence for two endocrinological phenotypes. Zoology

Thomas C, Van Den Berg R, Segers M, Geelen L, Hollanders J, Doesburg W, Houx P (1990) Time-resolved fluoroimmunoassay for unconjugated estrogen in urine: comparison with a fluorometric assay for total estrogen and application in an in vitro fertilization program. Clin Chem 36(10):1774–1778

Thonhauser KE, Raveh S, Hettyey A, Beissmann H, Penn DJ (2013) Scent marking increases male reproductive success in wild house mice. Anim Behav 86(5):1013–1021

Tinnesand HV, Buesching CD, Noonan MJ Newman C, Zedrosser A, Rosell F, Macdonal DW (2015) Will trespassers be prosecuted or assessed according to their merits? A consilient interpretation of territoriality in a group-living carnivore, the European badger (*Meles meles*). PLoS One 10(7)

Tinnesand HV, Jojola S, Zedrosser A, Rosell F (2013) The smell of desperadoes? Beavers distinguish between dominant and subordinate intruders. Behav Ecol Sociobiol 67(6):895–904

Touma C, Palme R (2005) Measuring fecal glucocorticoid metabolites in mammals and birds: the importance of validation. Ann NY Acad Sci 1046(1):54–74

Vicente J, Delahay RJ, Walker NJ, Cheeseman CL (2007) Social organization and movement influence the incidence of bovine tuberculosis in an undisturbed high-density badger *Meles meles* population. J Anim Ecol 76:348–360

Vogt K, Zimmermann F, Kölliker M, Breitenmoser U (2014) Scent-marking behaviour and social dynamics in a wild population of Eurasian lynx *Lynx lynx*. Behav Proc 106:98–106

Wasser SK, Risler L, Steiner RA (1988) Excreted steroids in primate feces over the menstrual cycle and pregnancy. Biol Reprod 39(4):862–872

Wheeler RE (2010) multResp() *lmPerm*. The R project for statistical computing http://www.r-project.org/

White PCL, Brown JA, Harris S (1993) Badgers (*Meles meles*), cattle and bovine tuberculosis (*Mycobacterium bovis*): a hypothesis to explain the influence of habitat on the risk of disease transmission in southwest England. Proc R Soc Lond B 253(1338):277–284

White AM, Swaisgood RR, Zhang H (2002) The highs and lows of chemical communication in giant pandas (*Ailuropoda melanoleuca*): effect of scent deposition height on signal discrimination. Behavl Ecol Sociobiol 51(6):519–529

Wira CR, Rodriguez-Garcia M, Patel MV (2015) The role of sex hormones in immune protection of the female reproductive tract. Nat Rev Immunol 15(4):217–230

Wong P, Collins CE, Kaas JH (2010) Overview of sensory systems of *Tarsius*. Int J Primatol 31 (6):1002–1031
Wyatt TD (2014) Pheromones and animal behavior: chemical scents and signatures, 2nd edn. Cambridge University Press, New York City, N.Y
Zala SM, Potts WK, Penn DJ (2004) Scent-marking displays provide honest signals of health and infection. Behav Ecol 15(2):338–344

LPS-Induced Immune System Stimulation Alters Urinary Volatiles and Behaviour in Growing Pigs

Sankarganesh Devaraj[1], Anoosh Rakhshandeh[2], Edgar Aviles-Rosa[1], and John J. McGlone[1](✉)

[1] Laboratory of Animal Behavior, Physiology and Welfare, Department of Animal and Food Sciences, Texas Tech University, Lubbock, TX, USA
sankarganesh.eco@gmail.com, john.mcglone@ttu.edu
[2] Department of Animal and Food Sciences, Texas Tech University, Lubbock, TX, USA

Abstract. Volatile compounds from body fluids and secretions may indicate the health status of individuals to other conspecifics. We sought to identify urinary volatiles that might act as biomarkers indicative of the health status of growing pigs. Immune system stimulation (ISS) was induced in pigs (N = 5) by intramuscular (IM) injection of bacterial lipopolysaccharide (LPS; 25 μg/kg BW). Urine samples were collected before (baseline) and after LPS treatment. The urinary headspace volatile compounds were identified using Solid-phase microextraction (SPME) coupled with GC-MS. Pig eye temperatures was measured using an infrared imaging technique, and behaviours (i.e. standing, feeding, drinking, moving, lying, sitting and active) were video-recorded and quantified 24 h before and after ISS. Data were analysed using mixed model ANOVA examining periods before and after LPS administration with pigs serving as their own control. Pig eye/body temperature increased post-ISS (at 6 h, $P = 0.03$) and correlated negatively with active behaviours (r = −0.64, $P = 0.048$). Relative to pre-ISS, pigs spent less time standing ($P = 0.018$), moving ($P = 0.061$) and active ($P < 0.0001$) during the post-ISS period. SPME-GC-MS revealed 7.2- and 4.3-fold increases in 3-Methylphenol ($P < 0.0001$) and 4-Ethylphenol ($P = 0.005$), respectively, in post-ISS urine compared with urine collected during the pre-ISS period. In conclusion, ISS reduced pig activity, and elevated ($P < 0.01$) urinary levels of 3-Methylphenol and 4-Ethylphenol. Behavioural changes and volatile molecules may be used as a non-invasive method for detection of fever, illness and health in farm and possibly other animals. Furthermore, the effects of 3-Methylphenol and 4-Ethylphenol on the behaviour of other pigs remain to be determined.

1 Introduction

Olfaction is the primordial form of communication in animals (Kimball et al. 2016b). Olfactory cues are often volatile compounds present in animal secretions/excretions. These volatiles convey information about species, sex, identity, age and health of the

© Springer Nature Switzerland AG 2019
C. D. Buesching (Ed.): *Chemical Signals in Vertebrates 14*, pp. 60–70, 2019.
https://doi.org/10.1007/978-3-030-17616-7_5

animal to other animals within or between species (Poddar-Sarkar and Brahmachary 2014). A number of studies have investigated the effects of disease on volatiles in the urine, faeces, saliva and exhaled breath of various species (Shirasu and Touhara 2011; Kimball et al. 2013, 2016a; Pijls et al. 2016; Kwak and Preti 2011). Urine of animals contains volatile compounds, which can be biomarkers indicative of physiological and health conditions (Tirindelli et al. 2009; Khalid et al. 2015). For instance, the urine of cancer patients revealed differential volatile compounds that act as biomarkers (Silva et al. 2011). The urinary volatile metabolome of diseased/infected mice shows noticeable changes in volatile compounds (Kimball et al. 2016a). The volatilome of body exudates is thus an emerging area of research to detect an individual's health and develop disease biomarkers (Angle et al. 2016).

Urinary volatiles altered in response to infection/disease might provide a distinct pattern of signals to their conspecifics that help them avoid a diseased animal for social/sexual behavioural interactions. It may also reduce the chance of spreading disease/harmful pathogenic agents to other animals in a colony, thereby increasing the survival rates of the other animals in the same colony. These contextually excreted volatile molecules of infected/diseased animals can act as alarm signals. The production of alarm pheromones is conditional in animals. For instance, mice produce alarm pheromones in their urine when encountering a predator (Brechbuhl et al. 2013).

Current practices in pig farming systems, such as dense housing conditions or air quality issues, may influence the production of alarm pheromones in pigs and have negative impacts on their growth and welfare (Vieuille-Thomas and Signoret 1992). Amory and Pearce (2000) collected restrained pigs' urine and sprayed it over the weaner pigs and observed significant behavioural changes. They conclude that restrained pigs produce different urinary molecules, which possibly changed the behaviour in weaner pigs. Olsson et al. (2014) evidenced that lipopolysaccharide (LPS) injection activated inflammatory pathways and produced a characteristic body odour in humans. LPS is a component of the membrane of bacteria that causes immune system stimulation (ISS) in animals and humans. LPS injection induces fever and sickness-associated behaviours in pigs (Rakhshandeh and de Lange 2012). ISS in pigs may also result in the production of specific volatiles as alarm signals. However, the effects of ISS on urinary volatile changes have not been evaluated. Therefore, we postulate that urinary headspace volatiles would change in pigs after stimulating their immune system by injecting LPS.

2 Materials and Methods

2.1 Pigs and Urine Collection

Five prepubertal female pigs (25 ± 5 kg BW) were housed individually in metabolism crates. To induce ISS in pigs, LPS (05:55B5) was injected intramuscularly at a dose of 25 μg/kg BW. Urine samples were collected 24 h before (baseline; pre-ISS) and after LPS injection (post-ISS; 6 h later) in clean glass beakers. The samples were aliquoted into glass vials, transported to the laboratory on ice, and kept frozen (–20°C) until analysis. All animal procedures were approved by the Institutional Animal Care and Use Committee, Texas Tech University, TX and followed the recommendations in the Guide for the Care and Use of Agricultural Animals (American Society of Animal Science, 2010).

2.2 Eye Temperature Measurement and Recording of Pig Behaviour

Pig eye temperatures were measured every 2 h up to 10 h and at 24 h, including pre- and post-ISS conditions, using an infrared imaging technique (Petry et al. 2017). LPS injection increased eye temperature confirming effective ISS. Pig behaviours were recorded during baseline and post-ISS conditions (24 h each) using digital video cameras connected to a Digital Video Recorder. Standing, feeding, drinking, moving, lying, sitting and active behaviours were quantified by validated observers by each hour.

2.3 *Headspace Volatile Extraction and Analysis Using SPME-GC-MS*

Solid-phase microextraction (SPME) (65 μm PDMS/DVB fiber, Supelco, Sigma-Aldrich, Belafonte, PA) -GC-MS (Thermo Ultra GC) was employed to extract and analyse the headspace volatiles emanated from the urine samples kept in a 20-mL glass vial. The septum in the screw cap (Supelco, w/PTFE/Silicone Septa 18 mm) was pierced, through which a preconditioned SPME fiber was inserted. The urine sample was heated (60 °C) and stirred (800 rpm) for 30 min using a hot plate with magnetic stirrer, and the fiber was allowed to absorb the emanating volatiles. The fiber was then removed from the vial and injected into the GC-MS to desorb the volatiles at 220 °C (30 s in split mode). The column used in GC-MS was Rtx-5 (30 m length × 0.25-mm i.d.; film thickness 0.25 μm) (Restek, Belafonte, PA, USA). The initial oven temperature was 40 °C for 5 min, raised to 160 °C (5 °C/min) and held constant for 5 min, further raised to 260 °C (10 °C/min) and held constant for 10 min. The ion source and transfer line temperatures were 200°C and 230 °C, respectively. Electron ionization was performed at 70 eV and helium was used as carrier gas (1.2 mL/min). The compounds were identified by probability matching to the NIST library. A few compounds were purchased as analytical standards (3-Methylphenol, Nonanal, 4-Ethylphenol, Ethanone (2'-Aminoacetophenone), Indole, and 2-Methoxy-4-vinylphenol) from Sigma-Aldrich (St. Louis, MO) and the resulting characteristic major ions in the mass spectrum were used to confirm the compounds (Table 1).

Table 1. List of characteristic major ions in the spectrum of each compound that was used to identify and confirm the specific compounds

Name of the compound	Characteristic major ions	Molecular formula	Molecular weight (Da)
3-Methylphenol	39, 51, 77, 79, 90, 108, 110	C_7H_8O	108
Nonanal	41, 57, 70, 82, 98, 114, 124, 125	$C_9H_{18}O$	142
4-Ethylphenol	39, 65, 77, 91, 107, 122, 123	$C_8H_{10}O$	122
Indole	39, 50, 63, 64, 89, 90, 117, 118	C_8H_7N	117
Ethanone (2'-Aminoacetophenone)	37, 39, 65, 92, 120, 135, 136	C_8H_9NO	135
2-Methoxy-4-vinylphenol	15, 51, 53, 77, 107, 135, 150, 151	$C_9H_{10}O_2$	150

2.4 Peak Area Measurement

Xcalibur™ software (Thermo Fisher Scientific) was used to measure the area under each peak in the chromatograms, and the measurements were copied to a spreadsheet (Microsoft Excel). The measurements for each group (pre- and post-ISS) were then averaged into one value and compared between groups to calculate the relative variation in the area under the curve for each peak.

2.5 Statistical Analysis

Statistical evaluations were performed using SAS software version 9.4 (SAS Inst. Inc., Cary, NC) and a completely randomized design. Original peak area data were transformed using Empirical Transformation Method to approach normality (Kuehl 2000). For eye temperature and peak area of volatile compounds, MIXED procedure was used with time (pre- vs. post-ISS) and pig as fixed and random effects, respectively. An appropriate covariance structure was selected by fitting the model with the structure, which provided the 'best' fit based on the Akaike information criterion and the Schwarz Bayesian criterion. For behaviour data, general linear model was used with treatment (pre- vs. post-ISS), time (hour) and, the interaction of treatment by time as fixed effects. Correlation procedures (PROC CORR) were used to compute the correlation coefficient (r) between eye temperature and measures of behaviour. Treatment effects were considered significant at $P < 0.05$. A tendency towards a significant difference between treatment means was also considered at $P < 0.10$. The 95% confidence interval was determined and reported for transformed data.

3 Results

3.1 LPS Treatment Increases Body Temperature and Induces Sickness Behaviours in Pigs

The eye temperature of the pigs increased into the fever range during post-ISS and varied at six hours ($P = 0.03$) from baseline and other time points (Fig. 1). The Pearson correlation coefficient (PCC) showed that temperature and pig active behaviours were negatively correlated ($r = -0.63$, $P = 0.04$). ISS also reduced pig behaviours such as standing, moving, and active compared with their baseline ($P < 0.05$), whereas lying down was not different (Fig. 2). Eating and drinking behaviours also did not differ between baseline and post-ISS. The treatment by hour interaction was significant for standing ($P = 0.0002$), moving ($P = 0.001$) and active behaviours ($P < 0.0001$) (Figs. 3, 4 and 5). All gilts administered with LPS thus showed typical symptoms of fever and sickness behaviours (especially decreased activity).

3.2 LPS Treatment Quantitatively Changes the Urinary Headspace Volatile Compounds

We identified 14 volatile compounds in the headspace of the pig urine samples. The peak area of each compound was quantitatively different between baseline and post-ISS

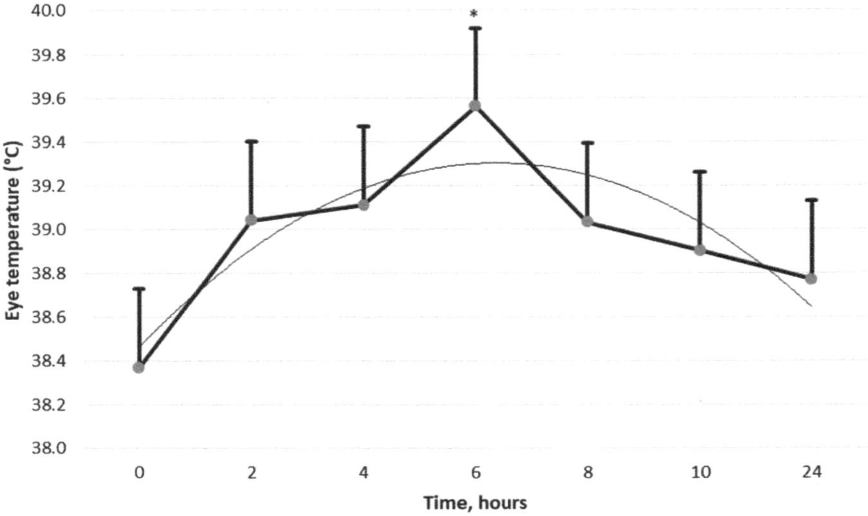

Fig. 1. Pig eye temperatures at different time points during baseline (0) and post-LPS administration (2, 4, 6, 8, 10 and 24 h) (N = 5). The figure shows Least Squares means ± s-tandard error comparison of eye temperatures between baseline (0) and post-LPS conditions. *indicates significant difference to other time points

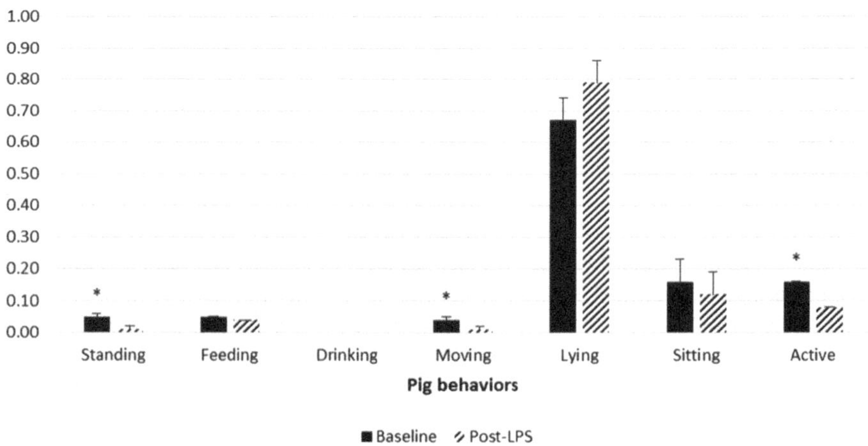

Fig. 2. The percent of the time of pig behaviours during baseline and post-LPS treatment conditions (N = 5). The figure shows Least Squares means ± standard error comparison of behaviours between baseline and post-LPS conditions during 24 h. *indicates a significant difference between baseline and post-LPS conditions

conditions. Specifically, the peak area was increased for 3-Methylphenol and 4-Ethylphenol (Table 2). The peak areas of 3-Methylphenol and 4-Ethylphenol were increased 7.2- and 4.3-fold, respectively in post-ISS than baseline on average. The 3-

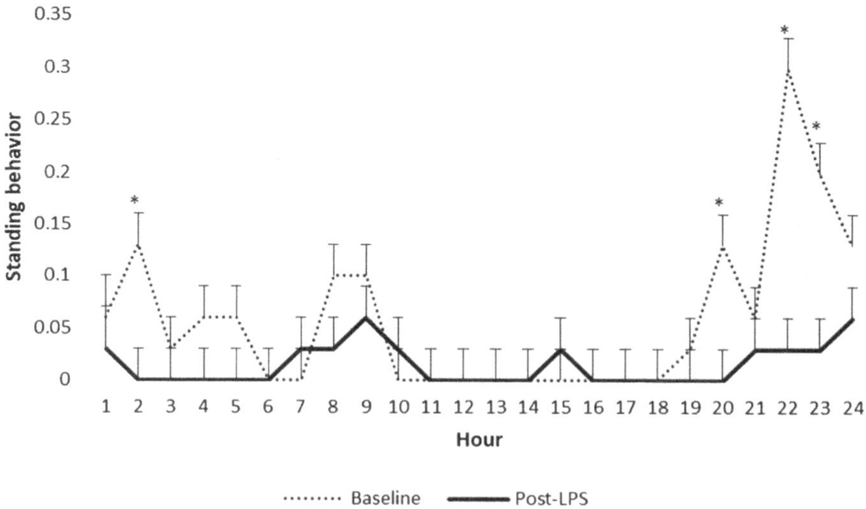

Fig. 3. Pig standing behaviour (percent of the time) (treatment by hour) during baseline and post-LPS treatment conditions (N = 5). The figure shows Least Squares means ± standard error comparison of standing behaviour between baseline and post-LPS conditions during 24 h. *indicates significant difference between baseline and post-LPS conditions

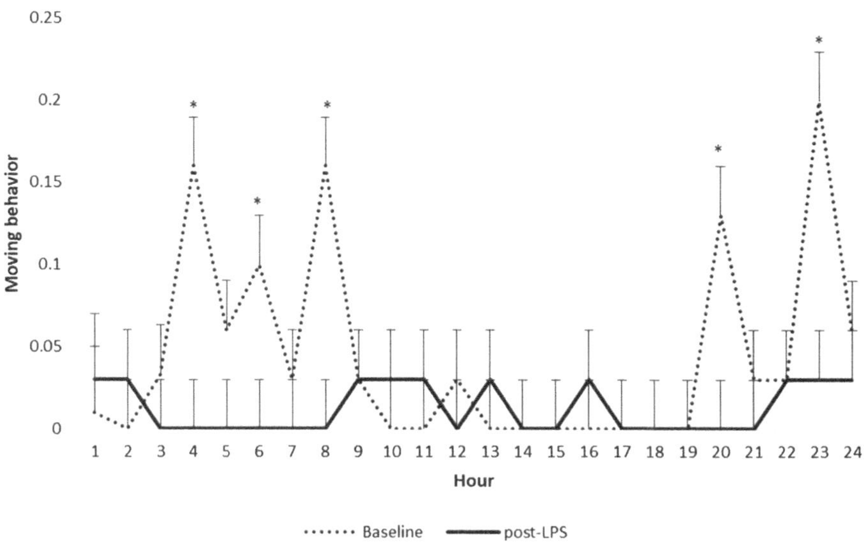

Fig. 4. Pig moving behaviour (percent of the time) (treatment by hour) during baseline and post-LPS treatment conditions (N = 5). The figure shows Least Squares means ± standard error comparison of moving behaviour between baseline and post-LPS conditions during 24 h. *indicates significant difference between baseline and post-LPS conditions

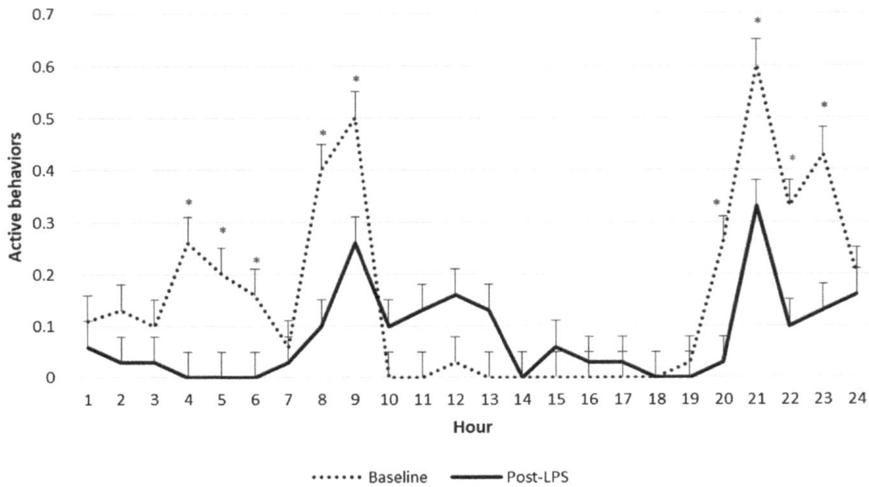

Fig. 5. Pig active behaviour (percent of the time) (treatment by hour) during baseline and post-LPS treatment conditions (N = 5). The figure shows Least Squares means ± standard error comparison of active behaviour between baseline and post-LPS conditions during 24 h. *indicates a significant difference between baseline and post-LPS conditions

Table 2. Peak area variation of the compounds before- and after-ISS. The table shows Least Squares means of the level of compounds at 95% confidence interval. Compounds with an *indicate significant difference ($P < 0.05$) after ISS

Compound name	Pre-ISS	Post-ISS	95% confidence interval
3-Methylphenol	15.16	110.3*	46.63
Nonanal	5.13	1.97	1.548
4-Ethylphenol	4.14	18.06*	6.821
1-Nonen-3-one	0.86	0.92	0.031
3-Decen-2-one	1.22	1.15	0.037
Indole	0.87	1.23	0.174
Ethanone (2-Aminoacetophenone)	7.33	5.19	1.047
2-Methoxy-4-vinylphenol	5.39	2.32	1.503
2,4,4-Trimethylhexa-5-En-1-ol	5.05	2.23	1.380
Cis-Jasmone	4.25	1.56	1.319
Spiro-decane	3.76	1.09	1.306
3-Octen-2-one	1.88	0.86	0.501
Propionic acid	5.80	4.72	0.530
Cyclopentanone2-(2-Octenyl)	0.77	0.27	0.247

Methylphenol was the most abundant compound in pig urine of both baseline and post-ISS conditions. The molecular weight of the identified volatile compounds was less than 300 Da (the highest of 284 Da for propionic acid), which suggests the compounds are highly volatile. Further, the characteristic ions of the analytical standards of

3-Methylphenol, Nonanal, 4-Ethylphenol, Indole, Ethanone (2-Aminoacetophenone), and 2-Methoxy-4-vinylphenol were synonymous to that of identified compounds by mass spectrum.

4 Discussion

Behavioural expressions are crucial in animal communication. Diseases or microbial infections may result in profound changes of animal behaviour. The altered behavioural pattern denotes the health status of diseased animals that may result in changes in their body odour (Gervasi et al. 2018). For instance, in rats, administration of LPS has been shown to reduce movement behaviours and LPS-injected rats were immobile in an elevated plus maze. They often laid down or slept and were behaviourally inactive, which showed that LPS-induced immunization altered the behavioural patterns (Gaykema and Goehler 2009). Similarly, immune-challenged mice also show reduced moving behaviour, and avoid social contact with their conspecifics (Lopes et al. 2016). In a previous study, LPS-administered pigs showed elevated body temperature, a clear indication of ISS. They were also behaviourally less active and ate less than healthy/not injected control animals (Johnson and von Borell 1994). Similarly, in our study, the measured eye temperature of LPS-administered pigs increased to the fever range and was negatively correlated with active behaviours. Standing, moving and active behaviours were significantly reduced after LPS treatment, as shown by significant treatment by time interactions. However, we did not observe significant differences in feeding or drinking behaviours.

Secretions from infected animals can contain discrete olfactory cues that might denote animal health status. Odorous olfactory molecules (volatile compounds) can be found in fluids such as urine, faeces, saliva, sweat, and sputum (Kimball 2016) that can convey social, sexual and health statuses and such volatile secretions can be used as biomarkers. Odour signals in excretions are useful because they can often be collected by non-invasive means (e.g. urine: Bouatra et al. 2013; Khamis et al. 2017), and can help to distinguish between diseased and healthy individuals, potentially providing a useful tool for the development of disease biomarkers (Buljubasic and Buchbauer 2015) in clinical settings.

In our study, there was considerable inter-individual variation in the urinary volatiles of pigs, where the urinary compounds were found to originate mostly from amino acids. Two of these volatile compounds were increased significantly (3-Methylphenol 7.2-fold, 4-Ethylphenol 4.3-fold) in the urine of post-ISS pigs compared with baseline urine samples.

Variation in volatiles in response to disease may include metabolites of the infectious agent or molecules produced by the host, or both (Sethi et al. 2013). For example, previous studies have shown that influenza virus infection in male mice profoundly changed the odour of their urine, which made it less attractive to females. This change in the odour profile of their urinary volatiles was likely a consequence of infection-dependent changes in their immune system and hormone milieu, possibly removing essential attractive pheromones (Penn et al. 1998). Similarly, administration of LPS produced a distinct urine odour in mice that was differentiated by biosensor mice (Kimball et al. 2016b). Also, in another study, the bedding material of LPS-treated mice was avoided by their conspecifics (Arakawa et al. 2010). We suggest here that, in pigs, the observed changes in urinary volatiles were due to the induced host response because no new volatiles were present after administration of LPS; they varied only in concentration before and after ISS. Haick et al. (2014) reported VOCs could be altered by two ways during ill conditions, viz., altering the existing VOCs quantitatively or producing new VOCs. Avian Influenza Virus (AIV) infection in mallard ducks changed the concentration of faecal volatiles significantly (acetoin was elevated, and 1-octen-3-ol was reduced; Kimball et al. 2013). In humans, certain diseases (ex., some cancers, ulcerative colitis, neurological, cardiovascular and kidney disease) can be associated with a unique volatile molecular print in breath compared to both healthy and control (Nakhleh et al. 2017). In this study, we propose that quantitative differences solely denote the status of sickness by producing a distinctive odour in the urine that represents ISS. We also hypothesize that reduced tissue catabolism may cause the increase in particular volatiles in sick pigs that may be the products of amino acid catabolism rather than the formation of new volatiles.

The dispersal nature of any odour depends on the molecular weight of the volatile compounds, where molecules weighing less than 300 Da are likely highly volatile and thus predominantly affect the characteristics of the odour source (Apfelbach et al. 2015). Here, we found as many as 14 volatile compounds below 300 Da in the urine of pigs during baseline and post-ISS conditions, signifying that pig urine odour is highly volatile and thus detectable to conspecifics from a distance (i.e. without direct physical contact/licking to engage the VNO).

The pig urinary compounds identified here have previously been reported to serve as metabolic by-products and/or odour cues in other species: For example, Decanal is present in human urine as a metabolite (Human Metabolome Database), although it is not yet known if people can detect and/or react to its odour. 4-Ethylphenol is present in the urine of male rats and may function as a pheromone (Zhang et al. 2008). Excretion of indole, propionic acid, and phenolic compounds occur in high concentration in air from swine operations (Ni et al. 2012). Most of the compounds identified were also found in pig manure (Lo et al. 2008). Indole and Ethanone (a metabolite of skatole) were reported in the urine of pigs (Brunius et al. 2016). Spiro decane was identified as a metabolite in human urine (Bouatra et al. 2013). 3-Methylphenol was identified in the urine of mares as an oestrus indicator and was shown behaviourally to be a potential olfactory stimulus in stallions (Mozuraitis et al. 2012; Būda et al. 2012). In the present study, 3-Methylphenol and 4-Ethylphenol increased in concentration in pig urine, together with other compounds, represents the condition of ISS, implying that the same volatile compounds can show functional diversity in multiple species. Kimball et al. (2016a) reported that disruption of amino acid (phenylalanine) metabolism increased the concentration phenylacetone in Alzheimer's disease mice model. We suggest that a similar mechanism may occur in pigs with a fever that produce volatile molecules, which are derived from amino acid catabolism.

In conclusion, both behaviour and volatile secretions are important cues in differentiating healthy versus diseased status of pigs. We report here that ISS of pigs caused both behavioural changes and quantitative changes in at least 2 volatile urinary molecules. The volatile compounds (3-Methylphenol and 4-Ethylphenol) were significantly increased in post-ISS pigs and may have behavioural significance to pigs, perhaps by acting as warning cues. Testing the behavioural effects of these volatile molecules is in progress.

Acknowledgements. This work was supported by funding from Animal Biotech LLC., to Texas Tech University. JJM declares a conflict of interest in that he is an equity partner in Animal Biotech LLC.

References

American Society of Animal Science (2010) 3rd edn. Champaign, IL, USA
Amory JR, Pearce GP (2000) Alarm pheromones in urine modify the behavior of weaner pigs. Anim Welf 9:167–175
Angle C, Waggoner LP, Ferrando A et al (2016) Canine detection of the volatilome: a review of implications for pathogen and disease detection. Front Vet Sci 3:47
Apfelbach R, Parsons MH, Soini HA et al (2015) Are single odorous components of a predator to elicit defensive behaviors in prey species? Front Neurosci 9:263
Arakawa H, Arakawa K, Deak T (2010) Acute illness induces the release of aversive odor cues from adult, but not prepubertal, male rats and suppresses social investigation by conspecifics. Behav Neurosci 123:964–978
Bouatra S, Aziat F, Mandal R et al (2013) The human urine metabolome. PLoS One 9:e73076
Brechbuhl J, Moine F, Klaey M et al (2013) Mouse alarm pheromone shares structural similarity with predator scents. Proc Natl Acad Sci USA 110:4762–4767
Brunius C, Vidanarachchi JK, Tomankova J et al (2016) Skatole metabolites in urine as a biological marker of pigs with enhanced hepatic metabolism. Animal 10:1734–1740
Būda V, Mozūraitis R, Kutra J et al (2012) p-Cresol: a sex pheromone component identified from the estrous urine of mares. J Chem Ecol 38:811–813
Buljubasic F, Buchbauer G (2015) The scent of human diseases: a review on specific volatile organic compounds as diagnostic biomarkers. Flavour Frag J 30:5–25
Gaykema RPA, Goehler LE (2009) Lipopolysaccharide challenge-induced suppression of Fos in hypothalamic orexin neurons: their potential role in sickness behavior. Brain Behav Immun 23:926–930
Gervasi SK, Opiekun M, Martin T, Beauchamp GK, Kimball BA (2018) Sharing an environment with sick conspecifics alters odor of health animals. Sci Rep 8:14255
Haick H, Broza YY, Pawel M et al (2014) Assessment, origin, and implementation of breath volatile cancer biomarkers. Chem Soc Rev 43:1423–1449
Human Metabolome Database. www.hmdb.ca/metabolites/HMDB11623. Accessed on 3rd Mar 2017
Johnson RW, von Borell E (1994) Lipopolysaccharide-induced sickness behavior in pigs is inhibited by pretreatment with indomethacin. J Anim Sci 72:309–314
Khalid T, Aggio R, White P et al (2015) Urinary volatile organic compounds for the detection of prostrate cancer. PLoS One 10:e0143283
Khamis MM, Adamko DJ, El-Aneed A (2017) Mass spectrometric based approaches in urine metabolomics and biomarker discovery. Mass Spectrom Rev 36:115–134
Kimball BA (2016) Volatile metabolome: problems and prospects. Bioanalysis 8:1987–1991
Kimball BA, Yamazaki K, Kohler D et al (2013) Avian influenza infection alters fecal odor in mallards. PLoS One 8:e75411

Kimball BA, Wilson DA, Daniel WW (2016a) Alterations of the volatile metabolome in mouse models of Alzheimer's disease. Sci Rep 6:19495

Kimball BA, Cohen AS, Gordon AR et al (2016b) Brain injury alters volatile metabolome. Chem Senses 41:407–414

Kuehl RO (2000). Design of Experiments: Statistical principles of research design and analysis. 2nd edn, p 136

Kwak J, Preti G (2011) Volatile disease biomarkers in breath: a critique. Curr Pharma Biotech 12:1067–1074

Lo YC, Koziel JA, Cai L et al (2008) Simultaneous chemical and sensory characterization of volatile organic compounds and semi-volatile organic compounds emitted from swine manure using solid phase microextraction and multidimensional gas chromatography-mass spectrometry-olfactometry. J Environ Qual 37:521–534

Lopes PC, Block P, Konig B (2016) Infection-induced behavioural changes reduce connectivity and the potential for disease spread in wild mice contact networks. Sci Rep 6:31790

Mozuraitis R, Buda V, Kutra J et al (2012) p- and m-Cresols emitted from estrous urine are reliable volatile chemical markers of ovulation in mares. Anim Reprod Sci 130:51–56

Nakhleh MK, Amal H, Jeries R et al (2017) Diagnosis and classification of 17 diseases from 1404 subjects via pattern analysis of exhaled molecules. ACS Nano 11:112–125

Ni JQ, Robarge WP, Xiao C et al (2012) Volatile organic compounds at swine facilities: a critical review. Chemosphere 89:769–788

Olsson MJ, Lundstrom JN, Kimball BA et al (2014) The scent of disease: human body odor contains an early chemosensory cues of sickness. Psychol Sci 25:817–823

Penn D, Schneider G, White K et al (1998) Influenza infection neutralizes the attractiveness of male odour to female mice (*Mus musculus*). Ethology 104:685–694

Petry A, McGilvray W, Rakhshandeh AR et al (2017) Technical note: assessment of an alternative technique for measuring body temperature in pigs. J Anim Sci 95:3270–3274

Pijls KE, Smolinska A, Jonkers DM et al (2016) A profile of volatile organic compounds in exhaled air as a potential non-invasive biomarker for liver cirrhosis. Sci Rep 6:19903

Poddar-Sarkar M, Brahmachary RL (2014) Pheromones of tiger and other big cats. In: Mucignat-Caretta C (ed) Neurobiology of chemical communication. CRC Press/Taylor & Francis, Boca Raton (FL); Chapter 15

Rakhshandeh A, de Lange CF (2012) Evaluation of chronic immune system stimulation models in growing pigs. Animal 6:305–310

SAS Institute 2012. SAS statistics, Cary, NC. Version 9.4

Sethi S, Nanda R, Chakraborty T (2013) Clinical applications of volatile organic compound analysis in detecting infectious diseases. Curr Microbiol Rev 26:462–475

Shirasu M, Touhara K (2011) The scent of disease: volatile organic compounds of the human body related to disease and disorder. J Biochem 150:257–266

Silva CL, Passos M, Camara JS (2011) Investigation of urinary volatile organic metabolites as potential cancer biomarkers by solid-phase microextraction in combination with gas chromatography-mass spectrometry. Br J Cancer 105:1894–1904

Tirindelli R, Dibattista M, Pifferi S et al (2009) From pheromones to behavior. Physiol Rev 89:921–956

Vieuille-Thomas C, Signoret JP (1992) Pheromonal transmission of an aversive experience in domestic pig. J Chem Ecol 18:1551–1557

Zhang JX, Sun L, Zhang JH et al (2008) Sex- and gonad-affecting scent compounds and 3 male pheromones in the rat. Chem Senses 33:611–621

A Field Study of Wild Echidna Responses to Conspecific Odour

Rachel L. Harris[1,2(✉)], Elissa Z. Cameron[1,3], and Stewart C. Nicol[1]

[1] School of Biological Sciences, University of Tasmania, Private Bag 55, Hobart, TAS 7001, Australia
rachelharris138@outlook.com
[2] Department of Evolutionary Anthropology, Duke University, Box 90383, Durham, NC 27708, USA
[3] School of Biological Sciences, University of Canterbury, Private Bag 4800, Christchurch 8140, New Zealand

Abstract. Mammalian olfactory cues play a crucial role in coordinating mating behaviour as they contain pertinent information on sex, reproductive status and receptivity to mating. Males of many species are particularly attentive to changes in female odour cues, as these can be used to find potential mates and monitor changes in reproductive condition. Such sensitivity is likely to be especially important for usually solitary or seasonally breeding individuals. We used remote scent trials in the field to examine if wild short-beaked echidnas are attracted to conspecific female odour. Our results are consistent with the hypothesis that male echidnas use olfactory cues during the breeding season to locate females: Echidnas were only observed investigating female odour samples and not empty camera trap stations, more echidnas (including known individual males) were attracted to female odours during the breeding season, than during the non-breeding season and we noted olfactory behaviours such as sniffing and tongue-flicking at the odour source. Male echidnas increase their mating opportunities by engaging in scramble competition and being sensitive to volatile and non-volatile female odour components, allowing them to locate potential mates and minimise search costs. Consequently, male ability to locate females may be sexually selected, but further bioassays, along with genetic information, is needed to investigate this further.

1 Introduction

In sexually reproducing species, individual fitness may be enhanced by the ability to assess the reproductive state of conspecifics (Johansson and Jones 2007; Thomas 2011; Shine and Mason 2012). Solitary and seasonally breeding animals also need to find dispersed conspecifics during a narrow window of opportunity, making chemical signals potentially more suitable for communication than other signal modes (Wyatt 2014). In mammals, chemical compounds may contain information on reproductive condition and receptivity to mating, for example, via changes in scent marking behaviour (e.g. *Felis nigripes*: Molteno et al. 1998; *Microtus pennsylvanicus*: Ferkin et al. 2004) or signal composition (e.g. *Elephas maximus*: Rasmussen et al. 1997; *Equus ferus caballus*: Mozūraitis et al. 2012). Males of several species use odour to locate

© Springer Nature Switzerland AG 2019
C. D. Buesching (Ed.): *Chemical Signals in Vertebrates 14*, pp. 71–80, 2019.
https://doi.org/10.1007/978-3-030-17616-7_6

females (e.g. *Rhinoceros unicornis*: Laurie 1982; Molteno et al. 1998) and assess female reproductive status (e.g. *Ailuropoda melanoleuca*: Swaisgood et al. 2002; Ferkin et al. 2004; reviewed in Thomas 2011). Therefore, the ability to detect and discriminate between females based on olfactory cues can allow males to minimise search costs associated with pursuing unreceptive females (Swaisgood et al. 2002; Johansson and Jones 2007; Thomas 2011).

Repeatable behavioural bioassays are useful for determining whether chemical variations in scent secretions (e.g. reflecting sex or seasonal differences) are salient to recipient individuals, while also controlling for visual, auditory or tactile cues (Thomas 2011; Drea et al. 2013; Wyatt 2014). For both mammalian and non-mammalian wildlife, such bioassays may be conducted in captivity, thus controlling for environmental conditions and allowing close monitoring of different behavioural responses, using either captive-bred animals (e.g. Ferkin et al. 2004; *Lemur catta*: Scordato and Drea 2007) or wild animals temporarily housed in captivity (e.g. *Hydrobates pelagicus*: Bonadonna and Sanz-Aguilar 2012; *Psammodromus algirus*: Martín et al. 2016). Alternatively, scent-based bioassays may be conducted in the field, particularly in cases where the animal of interest is threatened or cryptic, or not easily maintained in captivity (e.g. *Suricata suricatta*: Leclaire et al. 2013; *Ceratotherium simum*: Cinková and Policht 2015). Field-based bioassays, particularly those using camera traps and without requiring animal capture, may also have the advantage of being relatively non-invasive, allowing for natural behaviours to be observed within the context of the animal's home environment.

We used field trials to test whether short-beaked echidnas (*Tachyglossus aculeatus*; 'echidna') are attracted to conspecific odour cues, specifically those of females. The echidna is a usually solitary monotreme (Nicol et al. 2011) distributed throughout Australia and coastal New Guinea (Augee 2008). Olfactory signals are thought to be important for attracting and locating mates during the breeding season (June–September; Griffiths 1978). Males appear attracted to female scent (Boisvert and Grisham 1988; Rismiller 1992) and mating behaviour in captivity is characterised by males following females (Boisvert and Grisham 1988; Ferguson and Turner 2013). Female echidnas do not breed every year, resulting in a skewed operational sex ratio and intense male intra-sexual competition for access to mates (Nicol and Morrow 2012). In wild populations, such competition results in the formation of mating aggregations of one female with one or more males (Morrow et al. 2009), or even 'mating trains' of up to 10 males following a female in single file (Griffiths 1978; Rismiller 1992). In the Tasmanian subspecies, *T. a. setosus*, male echidnas may use female olfactory cues to locate hibernating females, in the absence of visual or auditory signals (Harris et al. 2012, 2014). Mating may even occur within a female's hibernaculum (Morrow and Nicol 2009). In non-breeding years, 'non-reproductive' females usually hibernate through the mating season (Nicol and Morrow 2012) undisturbed by males (Harris et al. 2016).

Although olfaction appears important for male echidnas to locate females, this hypothesis has not been tested in isolation of other potential cues. In field trials on Kangaroo Island, male *T. a. multiaculeatus* were attracted to hessian bags that had contained females recently found in a mating train (Rismiller 1992). However, because bags were placed near where the female had originally been found, the males that had been pursuing her could have used visual cues to locate the female over a short distance, rather than relying solely on odour. Unlike scent trials in captivity, field trials have the advantage of integrating an animal's ability to detect, localise and discriminate odour in the response. We used experimental field trials, while controlling for non-scent cues, to test the responses of wild echidnas, particularly males, to female odour during a single field season. We hypothesised that males would be attracted to odour from reproductive females, with the strongest response being observed during the breeding season.

2 Materials and Methods

We studied a wild population of echidnas on a grazing property in the Tasmanian southern midlands, approximately 55 km north of Hobart (42° 28'S, 142° 14'E; details see Nicol et al. 2011). In the 17 years leading up to this study, 262 echidnas (120 males, 99 females, 43 juveniles) were tagged in a 9 km^2 area of the field site. Population size and density are difficult to estimate, particularly as density seems to vary by habitat type, which is patchy at this field site (Nicol et al. 2011; Sprent and Nicol 2012). Up to approximately 15 individuals in any one year are fitted with radiofrequency transmitters for repeat sampling for hormones, odorants and behaviour observations (Nicol et al. 2011).

Although echidnas may communicate using multiple scent sources (Harris et al. 2012), in this study, we focussed on testing conspecific responses to odour cues present in cloacal secretions from reproductively active females. Cloacal secretions contain information on sex, reproductive status and individual identity (Harris et al. 2014), and scent-marking behaviour by females (Dobroruka 1960; Boisvert and Grisham 1988; Beard et al. 1992) suggests cloacal odour functions in communication, potentially to attract mates. Following our established protocols (Harris et al. 2012), we collected 56 cloacal secretions on cotton swabs from multiple captures of 15 adults, euthermic females (1–4 samples each) found in mating aggregations during the 2012 mating season (June–September). Oestrus is difficult to measure in this species, as female reproductive hormones including progesterone and estrogens are low or even below assay detection thresholds (Dean 2000; Morrow et al. 2017). For the purposes of this study, we did not time scent sample collection closely around ovulation. Rather, reproductively active females appear attractive to males throughout the breeding season, regardless of their oestrus or pregnancy state: female echidnas usually attract multiple males across several weeks, including prior to arousal from hibernation (Morrow and Nicol 2009) and during pregnancy (Harris et al. 2016; Morrow et al. 2017). Male harassment even continues after the female has entered a nursery burrow and begun egg incubation (Harris and Nicol 2014; Harris et al. 2016). Gas chromatography–mass spectrometry analyses show no detectable change in euthermic female chemical profiles during the mating period, including during pregnancy (Harris et al. 2016). Here, all 'scent donor' females showed mating activity within a week of sample collection. Swabs were stored in individual glass scintillation vials at –20 °C until used in scent trials.

We conducted scent trials at four camera trap sites in 2012, during the breeding season (*n* = 28 trials over 7 weeks between July and September 2012, until mating activity at our study site ceased; Morrow et al. 2009), and again during the non-breeding season (*n* = 28 over 11 weeks between October and December 2012). Camera traps were located at habitat features commonly associated with echidna mating aggregations (e.g. woodpiles, stumps, logs) on a 500 m transect within a central section of the study site. The section was chosen for its relatively high numbers of overlapping male and female home ranges, and uniform vegetation (open, dry eucalypt forest). Each camera trap consisted of two infrared motion-activated cameras (Reconyx PC800, Holmen, WI, USA; Scoutguard SC550, HuntingCamOnline Gadsden, SC, USA) mounted on wooden stakes or branches at a height of approximately 1 m, directed at a perforated 5 mL plastic container attached with plastic cable tie to a branch near ground level, containing a single female cloacal swab. We used two cameras to maximise the probability of detecting animals and to collect behaviour footage from different angles. To minimise novelty, we assembled camera traps sites without odours two weeks prior to conducting scent trials. During scent trials, we placed a cotton swab odour sample inside the plastic container and recorded visits by conspecifics for 3–7 nights. After each trial, the plastic container was replaced with a clean, empty container and the camera trap left unbaited for 3–4 nights. Each female odour sample was used once and assigned to a trial and location at random. We downloaded camera footage and changed camera batteries every 7–10 days. We calculated conspecific visit timing and duration based on camera timestamps, and identified known individual animals in the camera footage based on the location of transmitters, GPS loggers (CatTrack, Perthold Engineering, SC, USA) and colour-coded straws attached to the animal's spines. To control for scent trial duration, we calculated the number of echidnas detected per night, and tested for differences in the number of responders using Chi-squared tests with a Yates correction for small sample size. Statistical analyses were performed using Statistica software.

3 Results

We recorded seven visits by echidnas to camera traps baited with female cloacal odour and no visits to unbaited traps (total 592 camera nights, $\chi_1^2 = 3.92, P = 0.048$ after Yates correction; Fig. 1). Of the seven visits, three were by known individual males (one visit each) and none by females. Five visits (including all three visits by known males) occurred during the breeding season and two during the non-breeding season, although this seasonal difference was not significant, likely due to the small sample sizes ($\chi_1^2 = 2.89, P = 0.09$; Fig. 1). Mean visit duration tended to be longer in the breeding season (115 ± 122.5 s, *n* = 5, range: 2–348 s) than the non-breeding season

(24 ± 2.83 s, $n = 2$, range: 22–26 s), but also not significantly (two-tailed t-test, unequal variances, $t_3 = 1.70$, $P = 0.18$). Behaviours indicating olfactory interest (Fig. 2), such as tongue-flicking, scanning, probing and sniffing (defined as the echidna's beak being within 1–2 cm of the odour source) were only observed during the breeding season, and were exhibited by all animals except for one that was only filmed for two seconds. No animals filmed during the non-breeding season showed any interest in the female odour. Visit latency (time from a scent being placed to an echidna visiting a camera trap) varied from approximately 6 h to 6 days (mean latency during breeding season: 56.0 ± 50.7 h, range: 4–150 h; non-breeding season: 74.0 ± 18.0 h, range: 56–92 h; two-tailed t-test, unequal variances, $t_4 = 0.57$, $P > 0.5$). The variance in latency to approach during the breeding season was largely due to one male that approached approximately 150 h after the baited camera traps were set up, whereas all other visits occurred within 53 h. Olfactory behaviours and duration of visits did not seem to decline with increasing sample age.

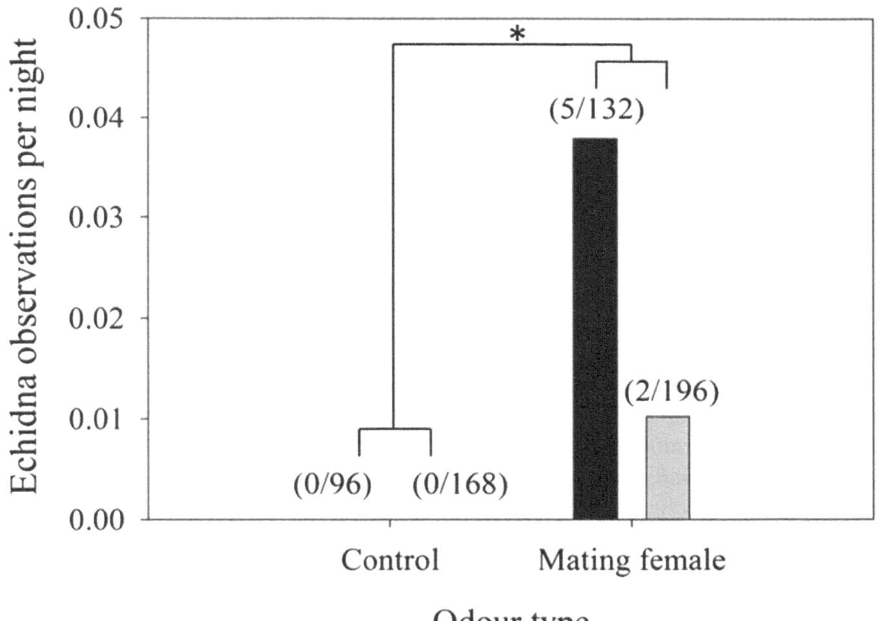

Fig. 1 Number of echidnas recorded per camera night during the 2012 breeding (black) and non-breeding (grey) seasons at camera traps baited with sample vials containing either nothing ('Control') or a female cloacal swab ('Mating female'). Parentheses: number of echidnas recorded per number of camera nights. Asterisk: $P < 0.05$

Fig. 2 Behaviour responses of an unknown echidna recorded at a camera trap baited with female odour, ~20 h after odour placement. Behaviours included scanning (top left, bottom left), probing and sniffing (top right), and tongue-flicking (bottom right). Stills taken from video footage recorded at 11.15 am on 14th August, 2012

4 Discussion

The results of this field-based pilot study suggest that male echidnas are attracted to conspecific female odour and could use odour cues to locate reproductive females (or other important resources) within their environment. Three different individual males were positively identified in camera footage during the breeding season, and we observed tongue-flicking, probing and sniffing behaviour directed only towards female odour sources, indicating animals were attracted to female odour and not the testing equipment. Broadly, the results of this study support the widely held view that echidnas use olfactory cues to locate conspecifics. However, given the low number of responses observed in this study, we recommend additional trials in both captive and wild populations to address more specific hypotheses about mate attraction in this species. For example, future trials comparing responses to odour from reproductive versus non-reproductive females (e.g. Swaisgood et al. 2002; Ferkin et al. 2004; *Loxodonta africana*: Bagley et al. 2006), or odours from males versus females (e.g. Scordato and Drea 2007; Cinková and Policht 2015), could show whether conspecifics are attracted to female odour alone, or to a broader species-specific signature. We unsuccessfully attempted controlled trials in captivity using wild individuals (Harris 2014), but bioassays using captive animals, such as zoo populations, might be more rewarding.

The results of this study are consistent with the broader hypothesis that male echidnas use odour cues to locate females during the breeding season. Our results support the previous qualitative study (Rismiller 1992) showing that males were attracted to hessian bags containing female scent. However, our trials decoupled the scent from the original female location so that we simultaneously tested both detection and localisation of conspecific odour, in the absence of visual cues. Male ability to locate females using odour alone could be particularly important in Tasmanian echidnas, as their mating behaviour differs to those of mainland subspecies. For example, females of mainland Australian subspecies scent mark in captivity (Dobroruka 1960; Boisvert and Grisham 1988; Beard et al. 1992), and long-lasting marks might function to increase the number of males that might encounter her odour. In contrast, male Tasmanian echidnas often seek out and mate with hibernating females that are concealed within shelter sites (Morrow and Nicol 2009), a behaviour seemingly driven by a unique combination of environmental and physiological constraints, a skewed operational sex ratio, intense male competition and the distribution of females (Nicol and Morrow 2012; Sprent and Nicol 2012; Morrow et al. 2016, 2017). Consistent with other solitary, widely dispersed animals, male ability to detect, locate and discriminate between females is likely to be under strong selection (e.g. Swaisgood et al. 2002; reviewed in Murphy 1998; Johansson and Jones 2007). In mammals with low sexual size dimorphism (including echidnas, see Nicol et al. 2011), males cannot monopolise females and instead engage in scramble competition, where males that are best able to locate females can increase their mating opportunities (Johansson and Jones 2007; Lane et al. 2009) and reduce time and effort wasted searching for non-receptive females or other males (Johansson and Jones 2007; Thomas 2011).

Although male echidnas might use sensitivity to female odour to increase their mating opportunities, we suggest that males are responding to sex-specific differences in conspecific odorants, rather than more subtle differences associated with ovulation. Such a tactic, while increasing a male's likelihood of encountering females, would also result in the patterns of mating behaviour observed in this study population: Females continue to attract males and show mating activity, including during pregnancy (Harris et al. 2016; Morrow et al. 2017). Similarly, gas chromatographic analyses indicated that female Tasmanian echidnas do not show the expected changes in odour cues with ovulation or pregnancy (Harris et al. 2016) seen in other species (Swaisgood et al. 2002; Ferkin et al. 2004; Bagley et al. 2006; Mozūraitis et al. 2012; *Lemur catta*: Crawford and Drea 2015). Controlled bioassays to examine whether male echidnas can discriminate between pregnant and non-pregnant females would be valuable, along with genetic information to determine paternity rates among males.

In this study, echidnas were recorded at camera traps baited with female odours exposed to the environment for up to 6 days. Scent longevity is likely to be important in solitary animals, where scents may be investigated irregularly by conspecifics and need to be effective over large distances and over time (Alberts 1992). Indeed, the scent signatures of other species have been shown to be stable over long periods of time (e.g. *Meles meles*, Buesching et al. 2002; *Lemur catta*: Greene et al. 2016). Echidna odorants may be infrequently encountered by conspecifics, as indicated by the low response rate observed in this study. Although the survey area was chosen for the relatively high numbers of overlapping home ranges, male home ranges may be over 100 ha, more than twice the size of females' and female home ranges are more exclusive (Nicol et al. 2011; Sprent and Nicol 2012), hence both sexes are thinly distributed in the environment. Echidna cloacal secretions contain a mixture of volatile and non-volatile components, which may help extend the life of their odorants (Harris et al. 2012) and allow communication between widely distributed conspecifics. Tongue-flicking and licking behaviour in other vertebrates suggest that non-volatiles are important components of female chemical cues, requiring direct contact for detection (e.g. Swaisgood et al. 2002; *Lemur catta*, Scordato et al. 2007; *Thamnophis sirtalis parietalis,* Shine and Mason 2012). Chemicals including non-volatiles may also contain more specific information on genetic identity (*Antechinus agilis,* Parott et al. 2007; *Lemur catta*, Charpentier et al. 2008). Given the value of controlled bioassays for informing and complementing analytical studies of mammalian semiochemistry (Drea et al. 2013; Wyatt 2014), bioassays should continue to be optimised and refined in this species to help reveal more about what information is salient to recipient individuals.

References

Alberts AC (1992) Constraints on the design of chemical communication systems in terrestrial vertebrates. Am Nat 139(Suppl.):S62–S89

Augee ML (2008) Family Tachyglossidae: echidnas. In: Van Dyck S, Strahan R (eds) The mammals of Australia. New Holland Publishers, Sydney, Australia, pp 36–39

Bagley KR, Goodwin TE, Rasmussen LEL et al (2006) Male African elephants, *Loxodonta africana,* can distinguish oestrous status via urinary signals. Anim Behav 71(6):1439–1445

Beard LA, Grigg GC, Augee ML (1992) Reproduction by echidnas in a cold climate. In: Augee ML (ed) Platypus and echidnas. The Royal Zoological Society of New South Wales, Mosman, Australia, pp 93–100

Boisvert M, Grisham J (1988) Reproduction of the short-nosed echidna, *Tachyglossus aculeatus*, at the Oklahoma City Zoo. Int Zoo Yearb 27(1):103–108

Bonadonna F, Sanz-Aguilar A (2012) Kin recognition and inbreeding avoidance in wild birds: the first evidence for individual kin-related odour recognition. Anim Behav 84(3):509–513

Buesching CD, Waterhouse JS, MacDonald DW (2002) Gas-chromatographic analyses of the subcaudal gland secretion of the European badger (*Meles meles*) part II: time-related variation in the individual-specific composition. J Chem Ecol 28(1):57–69

Charpentier MJE, Boulet M, Drea CM (2008) Smelling right: the scent of male lemurs advertises genetic quality and relatedness. Mol Ecol 17(14):3225–3233

Cinková I, Policht R (2015) Discrimination of familiarity and sex from chemical cues in the dung by wild southern white rhinoceros. Anim Cogn 18(1):385–392

Crawford JC, Drea CM (2015) Baby on board: olfactory cues indicate pregnancy and fetal sex in a non-human primate. Biol Lett 11:20140831

Dean KM (2000) Aspects of the reproductive biology of the echidna, *Tachyglossus aculeatus.* PhD thesis, University of Melbourne

Dobroruka VLJ (1960) Einige Beobachtungen an Ameisenigeln, *Echidna aculeata* Shaw (1792). Z Tierpsychol (trans: Polymeropoulos E) 17:178–181

Drea CM, Boulet M, Delbarco-Trillo J et al (2013) The "secret" in secretions: methodological considerations in deciphering primate olfactory communication. Am J Primatol 75:621–642

Ferguson A, Turner B (2013) Reproductive parameters and behaviour of captive short-beaked echidna (*Tachyglossus aculeatus acanthion*) at Perth Zoo. Aust Mammal 35(1):84–92

Ferkin MH, Lee DN, Leonard ST (2004) The reproductive state of female voles affects their scent marking behavior and the responses of male conspecifics to such marks. Ethology 110 (4):257–272

Greene LK, Grogan KE, Smyth KE et al (2016) Mix it and fix it: functions of composite olfactory signals in ring-tailed lemurs. R Soc Open Sci 3:160076

Griffiths M (1978) The biology of the monotremes. Academic Press, London, UK

Harris RL (2014) Chemical communication for reproduction in the Tasmanian short-beaked echidna, *Tachyglossus aculeatus setosus*. PhD thesis, University of Tasmania

Harris RL, Nicol SC (2014) Observations of breeding behaviour and possible infanticide in a wild population of Tasmanian echidnas (*Tachyglossus aculeatus setosus*). Aust Mammal 36 (1):108

Harris RL, Davies NW, Nicol SC (2012) Chemical composition of odorous secretions in the Tasmanian short-beaked echidna (*Tachyglossus aculeatus setosus*). Chem Senses 37(9): 819–836

Harris RL, Holland BR, Cameron EZ et al (2014) Chemical signals in the echidna: differences between seasons, sexes, individuals and gland types. J Zool 293(3):171–180

Harris RL, Cameron EZ, Davies NW et al (2016) Chemical cues, hibernation and reproduction in female short-beaked echidnas (*Tachyglossus aculeatus setosus*): implications for sexual conflict. In: Schulte B, Goodwin TE, Ferkin MH (eds) Chemical signals in vertebrates 13. Springer, Heidelberg, Germany, pp 145–166

Johansson BG, Jones TM (2007) The role of chemical communication in mate choice. Biol Rev 82(2):265–289

Lane JE, Boutin S, Gunn MR et al (2009) Sexually selected behaviour: red squirrel males search for reproductive success. J Anim Ecol 78(2):296–304

Laurie A (1982) Behavioural ecology of the greater one-horned rhinoceros (*Rhinoceros unicornis*). J Zool 196(3):307–341

Leclaire S, Nielsen JF, Thavarajah NK et al (2013) Odour-based kin discrimination in the cooperatively breeding meerkat. Biol Letters 9(1):20121054

Martín J, López P, Iraeta P et al (2016) Differences in males chemical signals between genetic lineages of the lizard *Psammodromus algirus* promote male intrasexual recognition and aggression but not female mate preferences. Behav Ecol Sociobiol 70(10):1657–1668

Molteno AJ, Sliwa A, Richardson PRK (1998) The role of scent marking in a free-ranging female black-footed cat (*Felis nigripes*). J Zool 245(1):35–41

Morrow G, Nicol SC (2009) Cool sex? Hibernation and reproduction overlap in the echidna. PLoS One 4(6):e6070

Morrow G, Andersen NA, Nicol SC (2009) Reproductive strategies of the short-beaked echidna—a review with new data from a long-term study on the Tasmanian subspecies (*Tachyglossus aculeatus setosus*). Aust J Zool 57(4):275–282

Morrow GE, Jones SM, Nicol SC (2016) Interaction of hibernation and male reproductive function in wild Tasmanian short-beaked echidnas *Tachyglossus aculeatus setosus*. J Mammal 97(3):852–860

Morrow GE, Jones SM, Nicol SC (2017) Frozen embryos? Torpor during pregnancy in the Tasmanian short-beaked echidna *Tachyglossus aculeatus setosus*. Gen Comp Endocr 244:139–145

Mozūraitis R, Būda V, Kutra J et al (2012) p- and m-Cresols emitted from estrous urine are reliable volatile chemical markers of ovulation in mares. Anim Reprod Sci 130(1–2):51–56

Murphy CG (1998) Interaction-independent sexual selection and the mechanisms of sexual selection. Evolution 52(1):8–18

Nicol SC, Morrow G (2012) Sex and seasonality: reproduction in the echidna (*Tachylossus aculeatus*). In: Ruf T, Bieber C, Arnold W et al (eds) Living in a seasonal world: thermoregulatory and metabolic adaptations. Springer, Heidelberg, Germany, pp 143–153

Nicol SC, Vanpé C, Sprent JA et al (2011) Spatial ecology of a ubiquitous Australian anteater, the short-beaked echidna (*Tachyglossus aculeatus*). J Mammal 92(1):101–110

Parott ML, Ward SJ, Temple-Smith PD (2007) Olfactory cues, genetic relatedness and female mate choice in the agile antechinus (*Antechinus agilis*). Behav Ecol Sociobiol 61(7):1075–1079

Rasmussen LEL, Lee TD, Zhang AJ et al (1997) Purification, identification, concentration and bioactivity of (*Z*)-7-dodecen-1-yl acetate: sex pheromone of the female Asian elephant, *Elephas maximus*. Chem Senses 22(4):417–437

Rismiller PD (1992) Field observations on Kangaroo Island echidnas (*Tachyglossus aculeatus multiaculeatus*) during the breeding season. In: Augee ML (ed) Platypus and echidnas. The Royal Zoological Society of New South Wales, Mosman, Australia, pp 101–105

Scordato ES, Drea CM (2007) Scents and sensibility: information content of olfactory signals in the ringtailed lemur, *Lemur catta*. Anim Behav 73(2):301–314

Scordato ES, Dubay G, Drea CM (2007) Chemical composition of scent marks in the ringtailed lemur (*Lemur catta*): glandular differences, seasonal variation, and individual signatures. Chem Senses 32(5):493–504

Shine R, Mason RT (2012) An airborne sex pheromone in snakes. Biol Lett 8(2):183–185

Sprent J, Nicol SC (2012) Influence of habitat on home-range size in the short-beaked echidna. Aust J Zool 60(1):46–53

Swaisgood RR, Lindburg DG, Zhang H (2002) Discrimination of oestrous status in giant pandas (*Ailuropoda melanoleuca*) via chemical cues in urine. J Zool 257(3):381–386

Thomas ML (2011) Detection of female mating status using chemical signals and cues. Biol Rev 86(1):1–14

Wyatt TD (2014) Pheromones and animal behaviour: chemical signals and signatures. Cambridge University Press, Cambridge, UK

How Diet Affects Vertebrate Semiochemistry

Jan Havlíček[1(✉)], Jitka Fialová[1], and S. Craig Roberts[2]

[1] Department of Zoology, Charles University, Viničná 7, 128 44 Prague, Czech Republic
jhavlicek@natur.cuni.cz
[2] Division of Psychology, University of Stirling, Stirling FK9 4LA, UK

Abstract. Diet appears to be a key factor influencing individual chemical profiles in various vertebrate species, and may thus affect perception by conspecifics. The specific chemical components in diet or their metabolites are either excreted in urine or faeces or from skin glands. If such chemicals are derived from food and cannot be synthesized by the organism, or if the corresponding foods are difficult to obtain, these dietary components may serve as cues to foraging ability and/or nutritional status of the individual. Several studies ranging from fish to mammals have found that food deprivation or quality of diet (e.g. protein content) affect olfactory attractiveness of the producer, as assessed by opposite-sex conspecifics. Finally, diet-associated compounds also influence perception of species recognition and individual odour signatures with potential to both strengthen and weaken recognition. These effects of diet are involved in various social contexts including mate choice, social affiliation, and kin and parent–offspring recognition. We discuss these phenomena from the perspective of signal evolution.

1 Introduction

A major source of variation in vertebrate body odour production is attributable to an individual's diet. The odour produced may include chemical compounds which are part of the diet or the metabolites of such compounds. One may, therefore, expect that involved chemical compounds will vary across individual species, depending on their feeding ecology. If such chemical compounds come from food resources which occur in low abundance, or from resources that are difficult to obtain, they may serve as cues to the foraging ability of the producer, shaping subsequent behaviour of conspecifics.

Diet can also shape body odour production indirectly by affecting nutritional status and related metabolic processes. For instance, in inadequately fed individuals, catabolic processes prevail, and this may be reflected in their odour profiles. Odorous metabolites may thus provide cues to the nutritional status of the producer. In contrast to the direct influence discussed above, such cues should be less dependent on specific components of the diet. As the odours result from metabolic processes which are shared across related species, one may expect relatively low variation across species.

In both scenarios described above, the diet-related chemical compounds can be perceived by conspecifics, and thus may affect their subsequent decision-making

© Springer Nature Switzerland AG 2019
C. D. Buesching (Ed.): *Chemical Signals in Vertebrates 14*, pp. 81–93, 2019.
https://doi.org/10.1007/978-3-030-17616-7_7

processes and behaviour. If such behaviour affects the fitness of the perceiver, then selection may be exerted on attentiveness to diet-related cues. However, it may also be the case that the production of certain diet-related compounds in odour do not provide useful information and are not attended to by conspecifics. Distinguishing between compounds which are attended to, or not, is a particular challenge.

For the sake of clarity, we employ the theoretical approach to communication proposed by Maynard Smith and Harper (2003). In their view, communication is sensu stricto carried out on the basis of signals. Signals are defined as a certain behaviour or structure that alters the behaviour of another organism, which evolved because of that effect, and which is effective because the receiver's response has also evolved. Signals are differentiated from cues, which are features of the world, animate or inanimate, that can be used by an animal as a guide to future action but did not evolve for that purpose. Moreover, Maynard Smith and Harper (2003) define a particular category of signal, which they term 'index signals', which are signals whose intensity is causally related to the quality being signalled and that cannot be faked. They also call these performance-based signals and point to their relation to fitness.

Below, we review the available empirical evidence on the effects of diet on odour production in various vertebrate taxa. We first focus on specific dietary components which may directly affect odour production. Subsequently, we review studies on indirect effects of diet, such as food deprivation and nutritional status. Then, we analyze evidence concerning the effects of diet on individual recognition. Finally, we discuss the reported effects in the light of theory concerning the evolution of communication.

2 Diet-Related Chemical Cues

The specific components of diet may affect how individuals are perceived by their conspecifics. Diet-related compounds might be directly perceived from excreta (faeces or urine) or secreted by skin glands. Whether the occurrence of such compounds in odour is simply a byproduct of digestive metabolism or was shaped by selection for communicative purposes is an important question. Therefore, the majority of studies have given particular focus on components of diet which are essential for the organism and/or are difficult to obtain, since it is more likely that these may serve as reliable indicators of some functionally relevant information, such as potential mate quality or the quality of its territory.

Various European lizard species possess femoral glands, products of which appear to be involved in mate selection and intrasexual competition. A series of experiments has explored how various food supplements affect the composition of femoral gland secretions in several species. Researchers primarily focused on the effects of provitamin D (cholestadienol) and vitamin E (α-tocopherol), which are essential nutrients involved in calcium metabolism (vitamin D), various antioxidative processes (vitamin E) and immunostimulation. Importantly, vitamin E cannot be synthesized by lizards and must be obtained from the diet. Vitamin D is synthesized from provitamin D in skin under solar radiation and is also partly obtained from diet. It was found that females prefer the odour of male European green lizards (*Lacerta viridis*) supplemented by vitamin E for 3 weeks over odour of non-supplemented controls. Furthermore, supplemented males excreted significantly higher abundances of vitamin E from their femoral glands, while no differences in other compounds were detected (Kopena et al. 2011). In another study, a positive association between secretion of vitamin E in femoral glands and cell-mediated immune response was reported in ocellated lizard males (*Lacerta lepida*); the secretion of vitamin E was also correlated with green coloration which is known to positively affect mate choice (Martín and López 2010). Supplementation of vitamin D similarly increased excreted provitamin D in the femoral glands of male Iberian rock lizards (*Lacerta monticola*), and females preferred odours from supplemented males (Martín and López 2006). Finally, Kopena et al. (2014) tested the relative contribution of carotenoids and vitamin E supplementation on femoral glands excretion of Iberian green lizards (*Lacerta schreiberii*). The results showed that secretion of males whose diet was supplemented by the carotenoids alone do not differ from the control males. However, supplementation by vitamin E alone, or combined with the carotenoids, significantly affected femoral secretions. Higher levels of vitamin E secretion were also correlated with brighter coloration.

In addition, there is some evidence suggesting that specific constituents of diet may affect human body odour. Havlicek and Lenochova (2006), for instance, focused on the consumption of red meat. They collected axillary odour samples from male donors after 14 days of a meat-rich diet and from the same donors when they refrained from consuming red meat. They found that odour of men while consuming meat was perceived by a panel of female raters as less attractive and more intense. However, interestingly, a different study recently reported the opposite, i.e. that the amount of meat consumed correlates positively with the attractiveness of body odour (Zuniga et al. 2017). The obvious discrepancy between these two studies might be a consequence of differences in the study design. For instance, the former study experimentally manipulated the level of meat consumed while the latter study is based on a between-subjects correlational design. However, it is also possible that meat consumption affects body odour attractiveness in a curvilinear rather than a linear fashion. More specifically, the amount of meat consumed in the study by Havlicek and Lenochova (2006) was comparatively high, affecting body odour negatively, while a positive association in the latter study may result from a much more moderate consumption. The reported frequency of consumed eggs, oils and fats was also positively related to body odour attractiveness, while consumption of seafood and carbohydrates appeared to have a negative effect. Zuniga et al. (2017) further found that attractiveness of body odour was positively correlated with skin yellowness, which is thought to be a useful cue of fruit and vegetable intake (although no significant association between reported consumption of fruit and vegetables was detected). This intriguing discrepancy thus awaits further studies, preferentially using experimental manipulation of the amount of fruits and vegetables consumed.

The most extensively studied aromatic food affecting human body odour is certainly garlic, which is well known for its specific smell due to its abundance of various sulphur-containing volatiles such as allicin (Block 1985). Previous studies found that ingestion of garlic influences odour of amniotic fluid and mother's milk, as assessed by adult panelists (Mennella and Beauchamp 1991; Mennella et al. 1995). It may also affect nursing behaviour of infants: ingestion of garlic capsules led to increased breastfeeding time in infants who had no previous experience with garlic in mothers' milk (Mennella and Beauchamp 1993).

Consumption of garlic is generally accompanied by its specific odour emanating from the mouth, which is usually perceived negatively by others. Therefore, it would be expected that consumption of garlic will similarly affect armpit odour. However, Fialová et al. (2016) found that garlic affects armpit odour positively. In a series of studies, the authors varied the amount and nature of garlic ingested (6 or 12 g, and garlic capsules containing an amount equivalent to 12 g) by male odour donors. The results revealed that consumption of the lower amount of garlic had no significant effect on perception of attractiveness, pleasantness and intensity of the armpit odours, but the higher amount and the capsules increased odour attractiveness and decreased its intensity. Currently, the exact mechanism for how odoriferous molecules from garlic are metabolized, and how they may affect body odour, is not fully understood. However, it should be noted that garlic has various positive effects on human health, including bactericidal properties and lowering oxidative stress (Ankri and Mirelman 1999; Banerjee et al. 2002). It is thus possible that odoriferous molecules from garlic do not affect armpit sweat directly, but rather affect its attractiveness indirectly, either by lowering oxidative stress or by affecting the armpit microflora that is primarily responsible for the armpit odour.

Finally, several case studies suggest that, like garlic, certain other aromatic foods may influence the odour of amniotic fluid and/or the newborn baby. In one case study, it was reported that a newborn baby had an odour resembling maple syrup, which is indicative of a rare metabolic disorder known as maple syrup syndrome, but it turned out that the smell was caused by fenugreek consumed by the mother shortly before delivery (Yalcin et al. 1999). Similarly, the body odour of a newborn baby smelling like cumin was attributed to the mother's diet (Hauser et al. 1985).

To summarize this section, available evidence indicates that various aromatic dietary constituents may affect excreted odours across various vertebrate taxa, although in most cases the mechanism of such odour changes is not well understood. Most simply, urine or faecal excretions may contain specific chemicals or metabolites directly derived from the diet, while more complex pathways might be expected in the case of odour produced from skin or specialized cutaneous glands. In such cases, the chemicals from diet are usually transported via the blood stream and might be metabolized further in the liver. An understanding of the underlying mechanisms is important for functional assessment of diet-related odours; the existence of specific transport mechanisms for the diet-related compounds, and their secretion from specialized glands, suggest that they evolved for communicative purposes and can thus be termed signals sensu Maynard Smith and Harper (1995). Furthermore, the secreted compounds must be derived from sources which either cannot be synthesized by the organism or are of rare origin. In both cases, they may serve as reliable indicators of the quality of the producer. These conditions appear to be met in several lizard species of the *Lacerta* genus, whose specialized femoral glands produce secretions affected by the quality of some dietary compounds such as vitamin E. In contrast to such signals, the influence of meat or garlic consumption observed in humans appears to be an indirect effect of their metabolism and there is currently no evidence for mechanisms that evolved specifically for communication purposes. These are therefore not strictly signals, but may nevertheless provide reliable cues to individual quality, such as current health in the case of garlic consumption.

3 Nutritional Status

Effects of nutritional status on the quality of excreted odour has been tested by manipulating caloric intake or even by complete food deprivation, and virtually all studies so far record conspecific responses in a mate choice context. In swordtail fish (*Xiphophorus birchmanni*), for instance, it was found that females prefer odour cues from well-fed males over odour of males that had been food-deprived for 5 consecutive days (Fisher and Rosenthal 2006); however, such preferences were not observed towards similarly treated females. The effect of food deprivation on body odour quality in mammals was tested in meadow voles (*Microtus pennsylvanicus*) by Pierce and Ferkin (2005), who deprived female voles for 24 h and tested whether it affects attractiveness of their odour to males. They found that males preferred odour of females fed ad libitum over those who were food-deprived. Reduced odour attractiveness of deprived females was still evident 24 h after refeeding, but not after 48 h. Interestingly, while severe deprivation has clear effects, partial food deprivation may be insufficient to affect female attractiveness, as Costello et al. (2009) withheld food from female house mice (*Mus musculus*) every other day for a week, finding no effect on male preferences.

Recently, Fialová et al. (in press) tested whether complete caloric restriction affects quality of body odour in a sample of women. They collected axillary odours before restriction, at the end of a 48 h fast, and then again 72 h after restoration of food intake. At the same three timepoints, they also collected blood samples to determine glucose and albumin levels. The odour samples were subsequently assessed for their attractiveness, pleasantness and intensity by a panel of male raters. As expected, the odour donors showed significantly lower levels of glucose and higher levels of albumin during food deprivation as compared to both the baseline and after food restoration, but no significant differences in odour attractiveness, pleasantness or intensity between baseline and deprivation. Interestingly, the odours collected 72 h after food restoration were rated as more attractive and less intense as compared to the baseline and food deprivation.

Moving on from effects of deprivation, another set of studies tested how the quality of diet affects odour production. In a carefully designed study, Ferkin et al. (1997) fed meadow voles (*Microtus pennsylvanicus*) a diet varying in protein content for 4 weeks, after which odours from the anogenital area, faeces and urine were used as stimuli. Both males and females preferred odours from those fed on high-protein diet, but only from opposite-sex individuals and not from individuals of the same sex. One may expect that urine would contain diet-related chemical compounds and would be therefore the most informative. Nevertheless, the urine samples appeared to be the least informative. A similar pattern was observed in another study which tested odour preferences in Nile tilapia (*Oreochromis niloticus*) (Giaquinto et al. 2010). Females preferred chemical cues of males fed for 20 days on a high-protein diet as compared to males on a low-protein diet, but showed no preference for the odour of females varying in their diet. In mosquito fish (*Gambusia holbrooki*), individuals spent more time in water emanating from a shoal fed on a high protein diet (Ward et al. 2011). However, in contrast to tilapia, preferences for odour cues of high quality diet might also be linked to non-mating cues since males did not show a strong preference for odour of females on high-protein diet, although they did tend to mate more frequently with females on the high quality diet.

Finally, the effect of high quality diet on mate preferences was also tested in woodland salamanders (*Plethodon cinereus*) (Chouinard 2012). Target males were fed either white worms (high quality) or fruit flies (low quality); these items vary primarily in the proportion of chitin, which reduces efficiency of digestion. It was observed that females prefer odour of males fed on high quality diet, who also produced higher quantities of protein in secretions from their mental and postcloacal glands. Chemicals produced in both glands have previously been shown to play an important role in courtship and territorial marking.

In sum, the majority of studies indicate that perceivers are sensitive to chemical cues of current nutritional status. This holds for both studies involving food deprivation and experimental manipulation of diet quality. Unfortunately, we know relatively little about the concrete mechanisms by which nutritional status affects the quality of excreted odours. Indeed, these mechanisms appear to vary from species to species, since they are directly assessed from urine or faeces in salamanders, while urine was only partially reliable in voles, compared to faeces or specialized anogenital glands.

4 Species Recognition

Many animals use odours for recognition between conspecific and heterospecific individuals and reliable species recognition is particularly important in sympatric species to prevent hybridization. Species recognition should be based on specific chemicals consistent among all individuals within the species. Nevertheless, variation in diet may disguise species-specific odour profiles which may lead to diet-dependent reproductive isolation. Such an effect was, for example, observed in fostering experiments with pups of spiny mice (*Acomys cahirinus*), who preferred odour cues from house mice (*Mus musculus*) fed on a familiar diet as compared to odours of their conspecifics fed on an unfamiliar diet (Doane et al. 1978; Porter and Doane 1977).

5 Individual and Kin Recognition

Odour-based individual recognition is employed in various contexts such as mating, cooperation or in recognition of parents and offspring, while olfaction also plays a role in differentiating kin and non-kin and even degree of consanguinity (Todrank et al. 1998). In this section, we describe briefly some main ways by which odours enable individual and kin recognition, and then proceed in the following section to describe how dietary cues may affect these processes.

A prerequisite for odour-based individual recognition is the uniqueness and stability of the individual odour, which is likely to be primarily genetically determined. There is robust evidence across vertebrates that individual odours are comprised of a unique (with the exception of monozygotic twins) chemical mixture (Wyatt 2014). That odours reflect the individual's genetic make-up is supported by studies on associations between odour similarity and level of biological relatedness (Roberts et al. 2005). Furthermore, individual recognition and mate preferences appear to be related to shared genes in the major histocompatibility complex (MHC), and, in some species, also to those coding for major urinary proteins (MUPs).

These effects are best known from studies of rodents. It was, for instance, found that rats and mice could discriminate urine odours of inbred individuals varying in a single MHC allele only (Brown and Schellinck 1992; Singh et al. 1987; Yamazaki et al. 1990). Furthermore, there is a growing body of evidence that MUPs, relatively small proteins of the lipocalin family are also secreted into urine and saliva and act as odour-binding molecules. MUPs are present in urine at concentrations approximately a million times higher than MHC fragments, exhibit a high degree of polymorphism (e.g. diversity of expression among wild mice might be as great as for MHC), and are able to associate with low-molecular-mass semiochemicals (Beynon and Hurst 2003). Polymorphism contributes to individuality of odour signatures and it was observed that differences in MUP profile allow mice to distinguish their own scent from that of other individuals (Hurst et al. 2001). Individual mice express a combination of several MUPs, and many different MUP patterns can be found even among mice from the same population (Payne et al. 2001). MUPs seem to be suited to communicate individual identity as they show stable patterns of expression, high concentrations and resistance to degradation.

Olfaction can also play a role in differentiating kin from non-kin (Todrank et al. 1998). In humans, Porter and Moore (1981) showed that mothers and siblings discriminate their relatives' scent at above chance levels. The effect does not seem to be due to the shared environment, as siblings were able to identify T-shirts worn by their sibling even though they had lived apart for a long time (Porter et al. 1986). In contrast, people are not successful in recognizing the body odour of spouses who (usually) are not related, but often share the same environment (Porter et al. 1985). It is assumed that kin recognition functions in inbreeding avoidance, directing altruism and/or facilitating parental care. The MHC, which plays a role in both immune processes and mate choice, could partly provide odour cues to relatedness. Most MHC genes express a high degree of polymorphism within a population, but similarity increases with relatedness (Havlíček and Roberts 2009). The degree of relatedness thus could be assessed through phenotypic matching, where organisms generalize from the stimuli of known kin or themselves to those of possible kin (Heth et al. 1998; Mateo and Johnston 2000). According to this mechanism, an individual learns its own phenotype and that of its kin, stores a representation of those traits to memory (creating a 'kin template'), and compares the phenotypes of animals subsequently encountered to this template (Sherman et al. 1997). Furthermore, young learn from an early age to recognize odours to which they are exposed in their nest and territory, probably using familiarity associations based on both genetic and environmental cues (Fletcher and Mitchener 1987). Mouse pups indeed prefer nest odours of maternal and sibling MHC type rather than unfamiliar MHC type (Yamazaki et al. 2000). Similarly, spiny mice (*Acomys cahirinus*) typically huddle together with their littermates; however, mice separated from birth did not huddle together whereas non-siblings reared together did (Porter et al. 1981).

5.1 Diet-Mediated Changes in Individual Recognition

There is growing evidence that diet may affect individual differences in odour profile and consequently might influence the level of correct identification. In humans, hand odours of unrelated women were significantly easier to discriminate than either monozygotic twins or female siblings. Moreover, monozygotic twins on a different diet were much easier to discriminate compared to twins on the same diet (Wallace 1977). Likewise, dogs discriminated monozygotic twins if they differed in their diet, but not if they were on the same diet, while they could discriminate dizygotic twins even on the same diet (Hepper 1988). More recent research has found that this depends on the level of training: specially trained German Shepherd dogs were able to distinguish individual scents of monozygotic twins despite the fact that they lived in the same environment and ate same food (Pinc et al. 2011).

The role of genetic and dietary factors was directly compared in the production of individual urinary odours in house mice (*Mus musculus*). It was found that rats discriminate more easily between the urinary odours of mice that differ in their diets than mice that differ at one locus of the MHC (Brown et al. 1996). In contrast, Phillips and Tang-Martinez (1998) found no significant effect of diet on recognition behaviours in prairie voles (*Microtus ochrogaster*), although they did report a non-significant trend for less aggressive behaviour to be observed between individuals on the same diet (Phillips and Tang-Martinez 1998).

5.2 Diet-Mediated Changes in Kin Recognition

Similarly to individual recognition, diet may also interfere with kin recognition. Decline in kin discrimination due to diet-related odour cues was observed in fish, namely, Atlantic salmon (*Salmo salar*) and brook trout (*Salvelinus fontinalis*). Juveniles discriminated kin and non-kin individuals when they shared a common diet with kin, but were not successful when kin were fed a different diet. Moreover, no clear preference was found for kin fed the same diet compared to kin fed a different diet. However, they preferred cues of familiar diet when kinship was absent (Rajakaruna and Brown 2006).

There is also some evidence suggesting that associations which are formed during early development might have evolved as a mechanism for kin recognition, as juveniles are usually surrounded by their parents or other relatives. This may explain why diet-related odours experienced early in life are subsequently preferred over the odours encountered later. It was, for instance, observed in the laboratory rat (*Rattus norvegious*) that pups appear to learn to associate the caecotrophe odour (caecotrophe is defecated semi-digested material which serves as a substrate for microbial growth) specific to the maternal diet with an approach response. This suggests an individually learned response of pups to the specific odour of the mother, and presumably also colony mates who likely consume similar diets. Moreover, the quantity and quality of diet is crucial for the emission and synthesis of the maternal attractant, as production of caecotrophe-carrying maternal odour is conditioned by consumption of large quantity of food (Leon 1975). Differences in diet may be useful in the identification of kin because these individuals should share similar diet and they would be more familiar with an odour that reflects a diet similar to their own.

5.3 The Role of Microflora

To finish, it is worth noting that odour properties are often influenced by interaction between diet and commensal microflora. Odour individuality is to some extent determined by the presence of commensal bacteria in many mammalian species (Schellinck et al. 1991; Singh et al. 1990; Taylor et al. 2003), and such bacteria might affect distinguishability (Schellinck et al. 1992, 1995; Yamazaki et al. 1990). It was also suggested that without the presence of bacteria, differences in MHC are not sufficiently expressed (Brown and Schellinck 1995; but see Schellinck et al. 1992; Yamazaki et al. 1990). It was hypothesized that gastrointestinal bacteria metabolize food items to produce various volatile molecules which could be delivered to urine and other odour sources (Singh et al. 1990). Individually specific odour could be produced by unique bacterial populations shaped either by genes or dietary and other environmental differences, or by a combination of both. Bacteria may influence the excreted products by metabolizing different food items, while at the same time diverse bacterial strains might metabolize the same diet differently, both mechanisms affecting the secondary metabolites produced. The gastrointestinal microflora is relatively stable, but major dietary changes can be responsible for its fluctuations. Thus, a change in diet may affect the gastrointestinal flora, both quantitatively and qualitatively, influencing, in turn, the type of volatiles produced (Turnbaugh et al. 2009; Dubos and Schaedler 1962). The microflora is established before weaning and depends on the infant's genotype, mother's antibody production and her commensal bacteria, the infant's immune system and its diet (Fanaro et al. 2003; van der Waaij 1986). The MHC type may influence the initial microflora which colonizes the gastrointestinal tract and becomes tolerated by the immune system. Both a genetic and environmental (acquirement of initial microflora from the mother) influence on microflora populations are further supported by the fact that flora is more similar in monozygotic than dizygotic twins, but significantly differ in unrelated adults who live together (Brown and Schellinck 1992; Holdeman et al. 1976; van de Merwe et al. 1983). At least in rats, individual odours could be the result of an interaction between MHC, gastrointestinal bacteria, and the diet. Then, odour of genetically identical individuals could be changed by diet (and by bacterial action) as shown in the study where rats successfully discriminated between odours of mice on different diets, but not on the same diet (Brown and Schellinck 1992).

6 Conclusions

Available evidence indicates that diet significantly affects olfactory perception of conspecifics across vertebrate taxa. If costly, the diet-related cues may provide information about individual quality. Therefore, most studies testing the effect of diet focused on mate preferences. Nevertheless, diet may also affect perception in other social contexts such as assessment of food availability in particular individuals or their groups. This might be especially important in species living in large groups such as some marine fish. As noted earlier by Henneken et al. (2017), the diet and its composition can be easily manipulated under controlled conditions and therefore its influences may provide an excellent model for research on mechanisms of production of chemical cues. For instance, several previous studies have investigated the effect of diet quality (e.g. protein content), but in most cases it is unclear whether the observed positive effect on receivers is due to the energetic value of the diet or specific metabolites from the diet which are secreted by the target individual. This question can be resolved by keeping energetic value of the diet constant, while varying its content. Further, future studies should not restrict their investigations to behavioural tests but should also investigate how chemical profiles are influenced by diet-related variation.

Studies on olfaction based kin or individual recognition suggest that shared diet may sometimes mask individual differences. Nevertheless, we currently do not know whether these results can be simply applied out of the lab. This is because under natural conditions, an individual's diet interacts with their genotype (and often also with the microflora) and creates a remarkably more complex system as compared to under laboratory conditions. In addition, there is also evidence suggesting that diet might be an integral part of olfactory individual recognition.

Acknowledgements. JH & JF are supported by the Czech Science Foundation grant [18-15168S].

References

Ankri S, Mirelman D (1999) Antimicrobial properties of allicin from garlic. Microbes Infect 1 (2):125–129. https://doi.org/10.1016/s1286-4579(99)80003-3

Banerjee SK, Dinda AK, Manchanda SC et al (2002) Chronic garlic administration protects rat heart against oxidative stress induced by ischemic reperfusion injury. BMC Pharmacol 2 (1):16. https://doi.org/10.1186/1471-2210-2-16

Beynon RJ, Hurst JL (2003) Multiple roles of major urinary proteins in the house mouse, *Mus domesticus*. Biochem Soc T 31:142–146. https://doi.org/10.1042/bst0310142

Block E (1985) The chemistry of garlic and onions. Sci Am 252(3):114–118. https://doi.org/10.1038/scientificamerican0385-114

Brown R, Schellinck HM (1992) Interactions among the MHC, diet, and bacteria in the production of social odors in rodents. In: Doty R, Müller-Schwarze D (eds) Chemical signals in vertebrates VI. Springer, Boston, MA, pp 175–181

Brown R, Schellinck H (1995) Effects of selective depletion of gut bacteria on the odours of individuality in rats. In: Apfelbach L, Reutter K, Weiler E et al (eds) Chemical signals in vertebrates VII. Oxford, Pergamon, pp 267–271

Brown RE, Schellinck HM, West AM (1996) The influence of dietary and genetic cues on the ability of rats to discriminate between the urinary odors of MHC-congenic mice. Physiol Behav 60(2):365–372. https://doi.org/10.1016/s0031-9384(96)80006-0

Chouinard AJ (2012) Rapid onset of mate-quality assessment via chemical signals in a woodland salamander (*Plethodon cinereus*). Behav Ecol Sociobiol 66(5):765–775. https://doi.org/10.1007/s00265-012-1324-5

Costello AK, Pultorak JD, Meikle DB (2009) Do male house mice (*Mus musculus*) discriminate between females that differ in nutritional status? Behav Process 82(2):119–125. https://doi.org/10.1016/j.beproc.2009.05.001

Doane HM, Porter RH, John T (1978) The role of diet in mother-infant reciprocity in the spiny mouse. Dev Psychobiol 11(3):271–277. https://doi.org/10.1002/dev.420110311
Dubos RJ, Schaedler RW (1962) The effect of diet on the fecal bacterial flora of mice and on their resistance to infection. J Exp Med 115(6):1162–1172. https://doi.org/10.1084/jem.115.6.1161
Fanaro S, Chierici R, Guerrini P et al (2003) Intestinal microflora in early infancy: composition and development. Acta Paediatr 92(1):48–55. https://doi.org/10.1111/j.1651-2227.2003.tb00646.x
Ferkin MH, Sorokin ES, Johnston RE et al (1997) Attractiveness of scents varies with protein content of the diet in meadow voles. Anim Behav 53(1):133–141. https://doi.org/10.1006/anbe.1996.0284
Fialová J, Hoffmann R, Roberts SC Havlíček J (in press) The effect of complete caloric intake restriction on human body odour quality. Physiol Behav
Fialová J, Roberts SC, Havlíček J (2016) Consumption of garlic positively affects hedonic perception of axillary body odour. Appetite 97:8–15. https://doi.org/10.1016/j.appet.2015.11.001
Fisher HS, Rosenthal GG (2006) Female swordtail fish use chemical cues to select well-fed mates. Anim Behav 72(3):721–725. https://doi.org/10.1016/j.anbehav.2006.02.009
Fletcher D, Mitchener C (eds) (1987) Kin recognition in animals. Wiley, New York
Giaquinto PC, da Silva Berbert CM, Delicio HC (2010) Female preferences based on male nutritional chemical traits. Behav Ecol Sociobiol 64(6):1029–1035. https://doi.org/10.1007/s00265-010-0918-z
Hauser GJ, Chitayat D, Berns L et al (1985) Peculiar odours in newborns and maternal prenatal ingestion of spicy food. Eur J Pediatr 144(4):403. https://doi.org/10.1007/bf00441788
Havlicek J, Lenochova P (2006) The effect of meat consumption on body odor attractiveness. Chem Senses 31(8):747–752. https://doi.org/10.1093/chemse/bjl017
Havlíček J, Roberts SC (2009) MHC-correlated mate choice in humans: a review. Psychoneuroendocrino 34(4):497–512. https://doi.org/10.1016/j.psyneuen.2008.10.007
Henneken J, Goodger JQD, Jones TM et al (2017) Diet-mediated pheromones and signature mixtures can enforce signal reliability. Front Ecol Evol 4:145. https://doi.org/10.3389/fevo.2016.00145
Hepper PG (1988) The discrimination of human odour by the dog. Perception 17(4):549–554. https://doi.org/10.1068/p170549
Heth G, Todrank J, Johnston RE (1998) Kin recognition in golden hamsters: evidence for phenotype matching. Anim Behav 56(2):409–417. https://doi.org/10.1006/anbe.1998.0747
Holdeman LV, Good IJ, Moore WE (1976) Human fecal flora: variation in bacterial composition within individuals and a possible effect of emotional stress. Appl Environ Microb 31(3):359–375
Hurst JL, Payne CE, Nevison CM et al (2001) Individual recognition in mice mediated by major urinary proteins. Nature 414(6864):631–634. https://doi.org/10.1038/414631a
Kopena R, Martín J, López P et al (2011) Vitamin E supplementation increases the attractiveness of males' scent for female European green lizards. PLoS ONE 6(4):e19410. https://doi.org/10.1371/journal.pone.0019410
Kopena R, López P, Martín J (2014) Relative contribution of dietary carotenoids and vitamin E to visual and chemical sexual signals of male Iberian green lizards: an experimental test. Behav Ecol Sociobiol 68(4):571–581. https://doi.org/10.1007/s00265-013-1672-9
Leon M (1975) Dietary control of maternal pheromone in lactating rat. Physiol Behav 14(3):311–319. https://doi.org/10.1016/0031-9384(75)90039-6
Martín J, López P (2006) Vitamin D supplementation increases the attractiveness of males' scent for female Iberian rock lizards. P Roy Soc B-Biol Sci 273(1601):2619–2624. https://doi.org/10.1098/rspb.2006.3619

Martín J, López P (2010) Multimodal sexual signals in male ocellated lizards *Lacerta lepida*: vitamin E in scent and green coloration may signal male quality in different sensory channels. Naturwissenschaften 97(6):545–553. https://doi.org/10.1007/s00114-010-0669-8

Mateo JM, Johnston RE (2000) Kin recognition and the "armpit effect": evidence of self-referent phenotype matching. P Roy Soc B-Biol Sci 267(1444):695–700. https://doi.org/10.1098/rspb.2000.1058

Maynard Smith J, Harper DGC (1995) Animal signals: models and terminology. J Theor Biol 177(3):305–311. https://doi.org/10.1006/jtbi.1995.0248

Maynard Smith J, Harper DGC (2003) Animal signals. Oxford University Press, Oxford, UK

Mennella JA, Beauchamp GK (1991) Maternal diet alters the sensory qualities of human milk and the nursling's behavior. Pediatrics 88(4):737–744

Mennella JA, Beauchamp GK (1993) The effects of repeated exposure to garlic-flavored milk on the nursling's behavior. Pediatr Res 34(6):805–808. https://doi.org/10.1203/00006450-199312000-00022

Mennella JA, Johnson A, Staley C et al (1995) Garlic ingestion by pregnant women alters the odor of amniotic fluid. Chem Senses 20(2):207–209. https://doi.org/10.1093/chemse/20.2.207

Payne C, Malone N, Humphries R et al (2001) Heterogeneity of major urinary proteins in house mice: population and sex differences. In: Marchlewska-Koj A, Lepri JJ, Müller-Schwarze D (eds) Chemical signals in vertebrates IX. Springer, Boston, MA, pp 233–240. https://doi.org/10.1007/978-1-4615-0671-3_31

Phillips ML, Tang-Martinez Z (1998) Parent-offspring discrimination in the prairie vole and the effects of odors and diet. Can J Zool 76(4):711–716. https://doi.org/10.1139/z97-242

Pierce AA, Ferkin MH (2005) Re-feeding and the restoration of odor attractivity, odor preference, and sexual receptivity in food-deprived female meadow voles. Physiol Behav 84 (4):553–561. https://doi.org/10.1016/j.physbeh.2005.02.003

Pinc L, Bartoš L, Reslová A et al (2011) Dogs discriminate identical twins. PloS ONE 6(6): e20704. https://doi.org/10.1371/journal.pone.0020704

Porter RH, Doane HM (1977) Dietary-dependent cross-species similarities in maternal chemical cues. Physiol Behav 19(1):129–131. https://doi.org/10.1016/0031-9384(77)90169-x

Porter RH, Moore J (1981) Human kin recognition by olfactory cues. Physiol Behav 27(3):493–495. https://doi.org/10.1016/0031-9384(81)90337-1

Porter RH, Cernoch JM, Balogh RD (1985) Odor signatures and kin recognition. Physiol Behav 34(3):445–448. https://doi.org/10.1016/0031-9384(85)90210-0

Porter RH, Tepper VJ, White DM (1981) Experiential influences on the development of huddling preferences and "sibling" recognition in spiny mice. Dev Psychobiol 14(4):375–382. https://doi.org/10.1002/dev.420140410

Porter RH, Balogh RD, Cernoch JM et al (1986) Recognition of kin through characteristic body odors. Chem Senses 11(3):389–395. https://doi.org/10.1093/chemse/11.3.389

Rajakaruna RS, Brown J (2006) Effect of dietary cues on kin discrimination of juvenile Atlantic salmon (*Salmo salar*) and brook trout (*Salvelinus fontinalis*). Can J Zool 84(6):839–845. https://doi.org/10.1139/z06-069

Roberts SC, Gosling LM, Spector TD et al (2005) Body odor similarity in noncohabiting twins. Chem Senses 30(8):651–656. https://doi.org/10.1093/chemse/bji058

Schellinck HM, Brown RE, Slotnick B (1991) Training rats to discriminate between the odors of individual conspecifics. Anim Learn Behav 19(3):223–233. https://doi.org/10.3758/bf03197880

Schellinck HM, Rooney E, Brown RE (1995) Odors of individuality of germfree mice are not discriminated by rats in a habituation-dishabituation procedure. Physiol Behav 57(5):1005–1008. https://doi.org/10.1016/0031-9384(94)00353-7

Schellinck HM, West AM, Brown RE (1992) Rats can discriminate between the urine odors of genetically identical mice maintained on different diets. Physiol Behav 51(5):1079–1082. https://doi.org/10.1016/0031-9384(92)90096-k
Sherman P, Reeve H, Pfenning D (1997) Recognition systems. In: Krebs J, Davies N (eds), Behavioural ecology, 4th edn. Blackwell Scientific, Oxford, pp 69–96
Singh PB, Brown RE, Roser B (1987) MHC antigens in urine as olfactory recognition cues. Nature 327:161–164. https://doi.org/10.1038/327161a0
Singh PB, Herbert J, Roser B et al (1990) Rearing rats in a germ-free environment eliminates their odors of individuality. J Chem Ecol 16(5):1667–1682. https://doi.org/10.1007/bf01014099
Taylor D, Daulby A, Grimshaw S et al (2003) Characterization of the microflora of the human axilla. Int J Cosmetic Sci 25(3):137–145. https://doi.org/10.1046/j.1467-2494.2003.00181.x
Todrank J, Heth G, Johnston RE (1998) Kin recognition in golden hamsters: evidence for kinship odours. Anim Behav 55(2):377–386. https://doi.org/10.1006/anbe.1997.0611
Turnbaugh PJ, Ridaura VK, Faith JJ et al (2009) The effect of diet on the human gut microbiom: a metagenomic analysis in humanized gnotobiotic mice. Sci Transl Med 1(6):6ra14. https://doi.org/10.1126/scitranslmed.3000322
van der Waaij D (1986) The apparent role of the mucous membrane and the gut-associated lymphoid tissue in the selection of the normal resident flora of the digestive tract. Clin Immunol 7(1):4–7. https://doi.org/10.1016/0197-1859(86)90029-4
van de Merwe JP, Stegeman JH., Hazenberg MP (1983) The resident faecal flora is determined by genetic characteristics of the host. Implications for Crohn's disease? A Leeuw J Microb 49 (2):119–124. https://doi.org/10.1007/bf00393669
Wallace P (1977) Individual discrimination of humans by odor. Physiol Behav 19(4):577–579. https://doi.org/10.1016/0031-9384(77)90238-4
Ward AJW, Herbert-Read JE, Simpson SJ (2011) Diets and decisions: the potential use of food protein cues in dietary, sexual and social decisions by mosquitofish. Anim Behav 82(4):783–790. https://doi.org/10.1016/j.anbehav.2011.07.010
Wyatt TD (2014) Pheromones and animal behavior: chemical signals and signatures. Cambridge University Press, Cambridge, UK
Yalcin SS, Tekinalp G, Ozalp I et al (1999) Peculiar odor of traditional food and maple syrup urine disease. Pediatr Int 41(1):108–109. https://doi.org/10.1046/j.1442-200x.1999.01005.x
Yamazaki K, Beauchamp GK, Curran M et al (2000) Parent-progeny recognition as a function of MHC odortype identity. P Natl Acad Sci USA 97(19):10500–10502. https://doi.org/10.1073/pnas.180320997
Yamazaki K, Beauchamp GK, Imai Y et al (1990) Odor types determined by the major histocompatibility complex in germfree mice. P Natl Acad Sci USA 87(21):8413–8416. https://doi.org/10.1073/pnas.87.21.8413
Zuniga A, Stevenson RJ, Mahmut MK et al (2017) Diet quality and the attractiveness of male body odor. Evol Hum Behav 38(1):136–143. https://doi.org/10.1016/j.evolhumbehav.2016.08.002

The Social Function of Latrines: A Hypothesis-Driven Research Approach

Christina D. Buesching[1(✉)] and Neil Jordan[2,3,4]

[1] Wildlife Conservation Research Unit, University of Oxford, Oxford, UK
christina.buesching@zoo.ox.ac.uk
[2] Centre for Ecosystem Science, University of New South Wales, Sydney, Australia
[3] Taronga Conservation Society Australia, Taronga Western Plains Zoo, Wildlife Reproduction Centre, Dubbo, Australia
[4] Botswana Predator Conservation Trust, Maun, Botswana

Abstract. The function of latrines is still debated in many mammals, and in most cases latrine use is likely multi-functional. Functional hypotheses can be broadly divided into five categories that are not all mutually exclusive: resource defence, information centre/advertisement, landmarks/orientation, parasite control, or predator–prey interactions. To standardise all investigations into latrine function across species, we present a hypothesis-driven research framework based on the following five key categories of data that can, and we argue should, be collected from latrines: (1) spatial distribution patterns; (2) temporal usage patterns; (3) individual visitation and contribution patterns (to individual latrines); (4) behaviour of visitors; and (5) scent signal content and longevity. We suggest that our proposed approach offers a much-needed standardised structure to guide investigation into the function of latrine use in mammals.

1 Introduction

Latrines are typically defined as accumulations of two to several hundred faeces resulting from the repeated use of the same defecation sites (Brown and Macdonald 1985). Faeces can be deposited in specially dug pits (e.g. European badgers *Meles meles*: Delahay et al. 2000) or—more commonly—directly on the surface. Latrine use has been reported in most mammal orders (for reviews see, e.g. Primates: Irwin et al. 2004; Lagomorphs: Sneddon 1991; ungulates: Mueller-Schwarze 1987; Musteloidea: Buesching and Stankowich 2017; Carnivora: Buesching and Jordan in press; marsupials: e.g. Rubial et al. 2011; Insectivora: Poduschka and Wemmer 1986) and has even been confirmed in dinosaurs (Fiorelli et al. 2013). Often these dedicated latrine sites are used not only for defecation and urination but also for the deposition of glandular secretions, resulting in composite latrines or middens. In many species, latrines are shared by several animals belonging to the same family/social group or to different (neighbouring) groups. Because conspecifics typically spend much time investigating excreta and secretions deposited by previous visitors (e.g. Stewart et al. 2002), the role of latrines in olfactory communication has long been accepted (Brown and Macdonald 1985). Nevertheless, despite the ever expanding field of semio-chemical analysis, few

© Springer Nature Switzerland AG 2019
C. D. Buesching (Ed.): *Chemical Signals in Vertebrates 14*, pp. 94–103, 2019.
https://doi.org/10.1007/978-3-030-17616-7_8

ecological studies discussing the function of latrine use take into consideration the olfactory information content known to be encoded in the faecal, urinary and secretory scent-marks.

A variety of hypotheses have been suggested to explain latrine use in different species, which can be broadly divided into five categories pertaining to (i) *resource defence* (encompassing territory, food, sleeping/breeding sites, mates), (ii) *information centre and reproductive advertisement*, (iii) *landmarks aiding in orientation*, (iv) *parasite avoidance*, or (v) *predator–prey interactions* (for a detailed review of hypotheses pertaining to latrine function see Buesching and Jordan in press). However, while some of the proposed functions, such as the roles as information centres and individual reproductive advertisement signals, would require rather differentiated olfactory messages to be conveyed at latrines, others (e.g. parasite avoidance) would require latrines only to be identifiable as suitable sites for defecation. In addition, some functions (such as the marking of territories and sleeping dens, landmarks) would benefit from long-lasting olfactory signals deposited in (quasi-)permanent sites, others (such as reproductive advertisement, marking of abundant but transient food resources) would best be fulfilled by short(er)-term signals left at currently active locations, while predator avoidance would best be achieved if latrine smell could be kept to a minimum altogether. While some functions are thus mutually exclusive, others could easily be served simultaneously.

In addition to the olfactory information content, five further categories of data can be collected when studying latrines: (1) their spatial distribution patterns; (2) their temporal usage patterns; (3) the visit and contribution patterns of individual animals; (4) the behaviour of visitors to and/or users of latrines; (5) the longevity, volatility and lifespan of the scent signals. The conclusions drawn about the function of latrines are thus very much influenced (and often restricted) by the data available.

Here, we aim to develop a research framework based on these six categories of data that might aid in testing specific hypotheses pertaining to the five broad theoretical paradigms that have been suggested to explain the function of latrines in the past.

2 Hypothesis-Based Predictions of Latrine Characteristics

2.1 Latrines as Resource Defence

Latrines are often explained as "advertisement of ownership" of one or several of four main resources (for a detailed discussion of the different hypotheses pertinent in this context, see Buesching and Jordan in press). Because, typically, only species that are either territorial or have stable home ranges use latrines, their function has been linked at least in part to territorial marking and the marking/defence of the resources contained within, namely food, sleeping/breeding sites, and sometimes mates (although mates are more typically defended by direct aggression/herding or allo-marking during mating/courtship) in almost all species. The traditional interpretation as a "scent-fence" (e.g. Hediger 1949), however, has been revised repeatedly since its conception to accommodate mounting observations in an ever increasing variety of species of territorial intrusions by non-residents. For example, it has been reformulated as the "scent matching hypothesis", which states that by comparing scent-marks encountered within the territory with the scent of an individual encountered within the territory, intruders can identify the resident through scent-matching (Gosling 1982), and so avoid costly confrontation (e.g. Gosling and McKay 1990). In this way, latrines are proposed to help establish a power asymmetry between territory-holder and intruder (Maynard-Smith and Parker 1976; Hammerstein 1981; Gosling 1982) according to the Payoff Asymmetry Hypothesis (Dawkins and Krebs 1978; Krebs 1982).

Either way, if latrine function is connected to territorial marking, we can predict the following (according to the five categories of data obtainable from latrines): Their *spatial distribution patterns* should optimise locations to intercept intruders and thus, dependent on costs of patrolling borders and faecal restraint (Buesching et al. 2016b) latrines should be located predominantly along (or near) the perimeter, and/or placed throughout the hinterland. Substrate and microclimate of latrine sites should be chosen to maximise detection (i.e. signal range increase through air flow/elevation) and signal longevity (e.g. by preventing scent-marks from drying out). Their *temporal usage patterns* should peak when intrusion threat is most intense and should be correlated with intruder encounters (especially interlopers threatening territory integrity). The *visit and contribution patterns of individual animals* should reflect primary marking activity by territory owners, be sex-dependent in species where intrusion or territoriality is sex-biased, and be higher in individuals with more to gain by territorial defence (e.g. breeding/dominant individuals). The *scent signals* should be of low volatility and potentially contain anti-microbial components to enhance signal longevity while simultaneously minimising marking costs. Scent-profiles should allow association between territory owner(s) and defended areas. Consequently, in gregarious species, scent-profiles should be group-specific (but could allow assessment of group-size through inclusion of individual-specific characteristics), whereas in solitary species signals should be individual-specific, allowing olfactory discrimination between group (own) odour from neighbours and strangers, and thus the identification of residents via scent matching. The *behaviour of visitors to and/or users of latrines* should indicate that intruders avoid scent-marked areas/retreat on encountering scents (scent-fence) or modify their interaction with owner(s) when they match scents in the environment to scents of encountered individual(s) (scent-matching hypothesis). Intruders should over-/countermark only when attempting to take over territories, although border latrines can be shared between neighbours and counter-marking can thus be common at borders.

Ultimately animals do not defend territories for space alone, but for the resources that they encompass. Although the general predictions outlined above are applicable to all 'resource marking' signals independent of the resource itself (i.e. territory or food/breeding/sleeping sites etc.), *latrine location* should be spatially tied to the respective resource (den, food patch etc.), and the *olfactory signal characteristics* may be adapted (i.e. shorter signal lifespan) to reflect transient food sources, short-term den use, etc. In the case of breeding dens (and, dependent on social structure, also sleeping sites), information about familiarity and kinship might be encoded to aid related individuals/offspring as well as sexual/social partners to recognise the 'owner' of the site. Particularly in the case of food marking, signal odour might encode information on the consumed food source, where marking can either function as resource defence (e.g.

Irwin et al. 2004), as a signal to self and others that the respective resource has been depleted (sensu Eisenberg and Kleimann, 1972), or to attract conspecifics to share valuable food (e.g. Zhou et al. 2015), where *behavioural responses* of conspecifics to the signals will vary accordingly.

2.2 Latrines as Information Centres and Reproductive Advertisement

Faeces and urine are naturally pungent metabolic waste products. As such, they serve to excrete indigestible dietary components that can then be used to identify what the animal has been eating (Ali et al. 2004), as well as endocrinological metabolites from corticosteroids that indicate stress levels or sex-steroid derivatives that can be used analytically (and thus likely in many cases also by the animals) to determine the animal's reproductive status (Palme et al. 2013). In addition, many species possess paired anal glands that open into the rectum and coat faecal deposits with glandular secretion that has been shown to encode a wealth of familial (e.g. Sun and Mueller-Schwarze 1998) and individual-specific (e.g. reviewed for Musteloidea in Buesching and Stankowich 2017) parameters that, together with other specialised scent glands (e.g. inter-digital glands: Wronski and Plath 2010; badger subcaudal gland: Buesching et al. 2002) provide a wealth of information about the latrine user.

Therefore, if latrines play a role in information exchange and individual advertisement (esp. in the context of reproduction), their *spatial distribution patterns* should maximise the likelihood of intercepting potential mates and/or reproductive rivals. Information centre latrines could be expected to be clustered around breeding grounds, near trails or throughout foraging areas. Their *temporal usage patterns* should peak during the breeding season/period of breeding pair formation. Marking activity should be increased when potential mates (or rivals) are present/in the vicinity. The *visit and contribution patterns* should differ significantly between individuals and vary according to dominance, reproductive status, somatic health, etc. The *chemical composition of the scent signals* should be highly volatile to ensure far-reaching signal output and encode sex difference, information on reproductive status and/or other fitness-related parameters, individual identity and possibly information on genetic and social affiliation (e.g. kinship/group-membership/pair bond, etc.). The relative placement of the scent-mark (top vs. lower) position can be important in signalling superiority (Jordan et al. 2011). Importantly, specific signal degradation patterns (e.g. Buesching et al. 2002) should encode information on signal age (i.e. time since deposition), to assess current signal validity (particularly in the case of female oestrus advertisement that might only be valid for a short time; i.e. hours more likely than days). The *behavioural responses* of conspecifics coming across these scent-marks should evidence heightened interest in all scent-marks. Responses should differ between individuals according to their respective physiological state (e.g. males in reproductive condition should react strongest/be most interested in marks left by oestrous females often employing their vomero-nasal organ in a characteristic flehmen response: Hart 1983). Thus, reactions should be sex-specific and depend on the physiological/reproductive characteristics of receiver, where reaction will vary with information content (i.e. individual characteristics of the marker. Individuals should prefer potential mates that scent-mark more. Subsequent social/reproductive interactions should be shaped by signal characteristics, where individuals that had overmarked the scent of others should be preferred as mates and be avoided more frequently as antagonists.

2.3 Latrines as Landmarks Aiding in Orientation

Latrines are often located at conspicuous landmarks, such as ditches, road or bridge crossings (e.g. grey wolf: Barja et al. 2004; North American river otters, *Lontra canadensis*: Roberts et al. 2008), or particularly big trees (e.g. common genet: Espírito-Santo et al. 2007). In many species, they are connected with well-travelled paths leading from sleeping sites to foraging grounds or watering holes (e.g. review in Gorman and Trowbridge 1989); an observation that has caused some authors to liken the function of latrines to sign posts along human roads.

If latrines predominantly serve as 'notes to self' to aid individuals in orientation, their *spatial distribution patterns* should be highly dependent on individual movement patterns, and not require specific detours to visit and maintain. They should be distributed throughout the territory/home range, placed at biologically important features (e.g. crossroads in paths, important (but potentially difficult to locate) resource patches, hazardous areas, etc.). Particularly in unfamiliar/difficult to navigate terrain, latrine position should be elevated to ensure far-reaching signals. Their *temporal usage patterns* should ensure continual scent, while some latrines in the same home range/ territory may become temporally or even permanently inactive/disused. As these landmarks may not be important to all individuals using the same area, *visit and contribution patterns* could be expected to differ considerably between individuals as well as between latrines encompassed in the same home range/territory. The *chemical composition of the scent signals* should encode individual identity to enable discrimination between own signal versus signal from another conspecific. Other encoded information could include signal age, travel direction, or even the stress level (e.g. adrenaline/cortico-steroid level) of the owner when the scent was deposited to warn/remind itself (and/or others) of an area-specific danger/hazard. In these 'notes to self', *behavioural responses* should differ between the marking individual and others, where only the depositing individual should be aided by this latrine in its spatial orientation, but other individuals should not be repulsed.

2.4 The Role of Latrines in Parasite Avoidance

Latrine use has long been suggested to reduce parasite load, particularly in primates including humans (reviewed in Curtis 2014; Weinstein et al. 2018). Nevertheless, there is currently no evidence that latrine use is actually associated with a measurable reduction in parasite load in wild mammals. In fact, in some species, such as latrines raccoons, *Procyon lotor*, likely have the opposite effect as evidenced by a series of studies on the spread of raccoon roundworm, *Baylisascaris procyonis* (e.g. Page et al. 1998; Logiudice 2001), where data from proximity-logging collars at 15 latrine sites implicated raccoon latrines as major foci for the infection and spread of *B. procyonis* (Hirsch et al. 2014).

If, however, latrine use has evolved as a way to avoid (re-) infection with gut parasites, their *spatial distribution* should reflect those areas in the home range/territory that are least valuable in terms of resources. Particularly sleeping and feeding sites should be avoided in latrine placement. Ideally, latrine sites would be chosen not to confer local (microclimatic) protection/, but to provide adverse conditions for excreted parasites. The *temporal usage patterns* should coincide with periods of high parasitic loads/infection risk and be related to the parasitic lifecycle. *Contribution patterns* should either not differ between individuals, or should be correlated positively with parasitic load of individuals (i.e. animals with lower than-average parasite loads should avoid using latrines frequented by high parasitic load individuals). The *signal scent profile* can be very basic, simply indicating that this is a latrine. Thus, the *behaviour* of latrine visitors should be such as to minimise contact with excreta (and thus (re-) infection with parasites) and should exclude sniffing/licking of marks/faecal deposits by previous visitors.

2.5 The Role of Latrines in Predator–Prey Interactions

A wide variety of prey species have been shown to avoid areas containing scent of their predators (e.g. Mella et al. 2014), and predators are typically attracted to the scent of their prey (Hughes et al. 2010). Thus, both might benefit from hiding/disguising their scent to avoid detection (e.g. reviewed in Banks et al. 2014), and even smaller predators might benefit from avoiding detection by larger predators in cases of intra-guild competition (e.g. Hughes et al. 2009).

If prey species defecate in latrines to avoid detection by predators or if predators use latrines to avoid detection by their prey ("eavesdropping"), *spatial distribution* should reflect the need to 'hide' the (smell of) excreta. Thus, latrines should be located in difficult to reach habitat, not conducive to the specific hunting strategies of the respective predator species/(or in the case of predator latrines the preferred habitats of their prey species) that minimises signal reach (i.e. visibility and scent diffusion). Overall number as well as size (in terms of area utilised) of latrines should be kept to a minimum, and latrines should not be located along obvious paths or landmarks. In prey species, *usage* should increase with greater predation risk and *temporal patterns* should reflect predator activity with prey species using latrines predominantly when predators are resting, whereas predators should avoid using latrines during peak activity of their prey. *Contribution patterns* should either not differ between individuals, or predators on the hunt/individual prey most vulnerable to predation and/or having experienced predation should exemplify the latrine use patterns described in this section. *Signal scent* should be minimal and be of low volatility and short lifespan, with prey species potentially mimicking scent characteristics of non-prey species and predators mimicking the scent of prey. Prey and predator *behaviour* at latrines should reflect the need to remain undetected through keeping visit duration and frequency to a minimum, without sniffing deposits by previous visitors and by not leaving glandular scent-marks in addition to the necessary ablutions. Particularly faecal deposits might be buried/covered with material. Nevertheless, prey species should investigate predator latrines and vice versa.

3 Conclusion

Currently, the vast majority of studies reporting latrine use in a specific focal species very often present only data on the spatial distribution patterns, e.g. in relation to territory boundaries, landmarks and resources and maybe digested food items (e.g. reviewed for carnivores in Buesching and Jordan in press). In contrast, only few studies investigate temporal variation in latrine usage of their study animals, and typically relate their findings to species-specific activity patterns and how these may be affected by, e.g. human disturbance (e.g. Tsunoda et al. 2018). Even fewer studies provide insights into inter-individual differences in visit and contribution patterns, which are then typically related to status (e.g. Jordan 2007) and differences in home range use (e.g. Kilshaw et al. 2009). It is therefore hardly surprising that the most widely suggested function of latrines pertains to resource defence in the context of territoriality and food sources. In fact, if only spatial data are collected, basically any latrine use pattern can be interpreted as territorial marking, which led Gosling and Roberts (2001) and many others to argue that the collection of spatial data alone is insufficient to infer latrine function conclusively (see also Buesching and Jordan in press).

As illustrated by the example of the European badger *Meles meles*, pronounced bias in data collection can thus result in having to re-evaluate established theory when more (and different) data become available (reviewed in Buesching and Stankowich 2017): While for many years, badgers were believed to be strictly territorial (reviewed in Buesching and Macdonald 2001; Roper 2010), maintaining a perimeter scent-fence of so-called border latrines frequented by members of adjoining neighbouring social groups and a grid of group-specific hinterland latrines, more recently, when data on the chemical composition of latrine-based anal(Tinnesand et al. 2015) and subcaudal scent-marks (Buesching et al. 2002, 2016a), but also on individual usage patterns (Kilshaw et al. 2009; Ellwood et al. 2017) became available, interpretation of border latrines changed from active (Delahay et al. 2000) to passive range exclusion (Stewart et al. 1997) and are now by many interpreted as information centres for self-advertisement (reviewed in Buesching and Macdonald 2001; Macdonald et al. 2015; Buesching et al. 2016b).

Nevertheless, most of the potential functions of latrine use discussed above are not mutually exclusive. And so, due to the complexity of information encoded in olfactory signals, it is likely that in most species latrines fulfil several of roles simultaneously. By designing studies to investigate the measurable parameters, we identified in this chapter, predictions based on their characteristics will allow functional interpretations of latrine use in the target species to be drawn more effectively than at present.

References

Ali HAM, Mayes RW, Lamb CS, Hector BL, Verma AK, Ørskov ER (2004) The potential of long-chain fatty alcohols and long-chain fatty acids as diet composition markers: development of methods for quantitative analysis and faecal recoveries of these compounds in sheep fed mixed diets. J Agricul Sci 142(1):71–78

Banks PB, Bytheway JP, Carthey AJ, Hughes NK, Price CJ (2014) Olfaction and predator-prey interactions amongst mammals in Australia. Carniv Aust: Past, Present Futur, 389

Barja I, de Miguel FJ, Bárcena F (2004) The importance of crossroads in faecal marking behaviour of the wolves (*Canis lupus*). Naturwissenschaften 91:489–492

Brown RE, Macdonald DW (eds) (1985) Social odours in mammals. Clarendon Press, Oxford

Buesching CD, Jordan NR (in press) The function of small carnivore latrines: case studies and a research framework for hypothesis-testing. In: San EDL, Sato JJ, Belant JL, Somers MJ (eds) Small carnivores: evolution, ecology, behaviour & conservation. Wiley Publishing

Buesching CD, Macdonald DW (2001) Scent-marking behaviour of the European badger (*Meles meles*): resource defence or individual advertisement? In: Chemical signals in vertebrates vol 9. Springer, Boston, MA, pp 321–327

Buesching CD, Stankowitch T (2017) Communication amongst the musteloids: signs, signals, and cues. In: Macdonald DW, Newman C, Harrington LA (eds) Biology and conservation of the musteloids (Badgers, Otters, Skunks, Raccoons and their kin). OUP

Buesching CD, Waterhouse JS, Macdonald DW (2002) Gas-chromatographic analyses of the subcaudal gland secretion of the European badger (*Meles meles*) part I: chemical differences related to individual parameters. J Chem Ecol 28(1):41–56

Buesching CD, Tinnesand HV, Sin Y, Rosell F, Burke T, Macdonald DW (2016a) Coding of group odor in the subcaudal gland secretion of the European badger *Meles meles*: chemical composition and pouch microbiota. In: Chemical signals in vertebrates vol 13. Springer, Cham, pp 45–62

Buesching CD, Newman C, Service K, Macdonald DW, Riordan P (2016b) Latrine marking patterns of badgers (*Meles meles*) with respect to population density and range size. Ecosph 7 (5):e01328

Curtis VA (2014) Infection-avoidance behaviour in humans and other animals. Tr Immunol 35 (10):457–464

Dawkins R, Krebs JR (1978) Animal signals: information or manipulation? In: Krebs JR, Davies NB (eds) Behavioural ecology: an evolutionary approach. Blackwell, Oxford, pp 282–309

Delahay RJ, Brown JA, Mallinson PJ, Spyvee PD, Handoll D, Rogers LM, Cheeseman CL (2000) The use of marked bait in studies of the territorial organization of the European badger (*Meles meles*). Mamm Rev 30(2):73–87

Eisenberg JF, Kleiman DG (1972) Olfactory communication in mammals. Ann Rev Ecol Syst 3:1–32

Ellwood SA, Newman C, Montgomery RA, Nicosia V, Buesching CD, Markham A, Mascolo C, Trigoni N, Pasztor B, Dyo V, Latora V (2017) An active-radio-frequency-identification system capable of identifying co-locations and social-structure: validation with a wild free-ranging animal. M Ecol Evol 8(12):1822–1831

Espírito-Santo C, Rosalino LM, Santos-Reis M (2007) Factors affecting the placement of common genet latrine sites in a Mediterranean landscape in Portugal. J Mammal 88:201–207

Fiorelli LE, Ezcurra MD, Hechenleitner EM, Arganaraz E, Taborda JR, Trotteyn MJ, ... , Desojo JB (2013) The oldest known communal latrines provide evidence of gregarism in Triassic megaherbivores. Sci Rep 3:3348

Gorman ML, Trowbridge BJ (1989) The role of odor in the social lives of carnivores. In: Gittleman JL (ed) Carnivore behavior, ecology, and evolution, vol 1. Cornell University Press, Ithaca, pp 57–88

Gosling LM (1982) A reassessment of the function of scent marking in territories. Z Tierpsychol 60:89–118

Gosling LM, McKay HV (1990) Competitor assessment by scent matching: an experimental test. Behav Ecol Sociobiol 26:415–420

Gosling LM, Roberts SC (2001) Testing ideas about the function of scent marks in territories from spatial patterns. Anim Behav 62(3):F7–F10

Hammerstein P (1981) The role of asymmetries in animal contests. Anim Behav 29:193–205

Hart BL (1983) Flehmen behavior and vomeronasal organ function. In: Chemical signals in vertebrates vol 3. Springer, Boston, MA, pp 87–103

Hediger H (1949) Säugetier-Territorien und ihre Markierung. Bijdr Tot Dierk 28:172–184

Hirsch BT, Prange S, Hauver SA, Gehrt SD (2014) Patterns of latrine use by raccoons (*Procyon lotor*) and implications for *Bayliscaris procyonis* transmission. J Wildl Dis 50:243–249

Hughes NK, Kelley JL, Banks PB (2009) Receiving behaviour is sensitive to risks from eavesdropping predators. Oecologia 160(3):609–617

Hughes NK, Price CJ, Banks PB (2010) Predators are attracted to the olfactory signals of prey. PLoS ONE 5(9):e13114

Irwin MT, Samonds KE, Raharison J-L, Wright PC (2004) Lemur latrines: observations of latrine behavior in wild primates and possible ecological significance. J Mammal 85(3):420–427

Jordan NR (2007) Scent-marking investment is determined by sex and breeding status in meerkats. Anim Behav 74(3):531–540

Jordan NR, Manser MB, Mwanguhya F, Kyabulima S, Rüedi P, Cant MA (2011) Scent marking in wild banded mongooses: 1. Sex-specific scents and overmarking. Anim Behav 81(1):31–42

Kilshaw K, Newman C, Buesching C, Bunyan J, Macdonald D (2009) Coordinated latrine use by European badgers, *Meles meles*: potential consequences for territory defense. J Mammal 90 (5):1188–1198

Krebs JR (1982) Territorial defence in the great tit (*Parus major*): do residents always win? Behav Ecol Sociobiol 11:185–194

Logiudice K (2001) Latrine foraging strategies of two small mammals: implications for the transmission of *Baylisascaris procyonis*. Amer Midl Nat 146:369–378

Macdonald DW, Newman C, Buesching CD (2015) Badgers in the rural landscape—conservation paragon or farmland pariah? Lessons from the Wytham Badger Project. In: Macdonald DW & Feber RE (eds) Wildlife conservation on farmland, vol 2: Conflict in the Countryside. Oxford University Press, Oxford, pp 65–95

Maynard-Smith J, Parker GA (1976) The logic of asymmetric contests. Anim Behav 24:159–175

Mella VS, Cooper CE, Davies SJ (2014) Behavioural responses of free-ranging western grey kangaroos (*Macropus fuliginosus*) to olfactory cues of historical and recently introduced predators. Austral Ecol 39(1):115–121

Muller-Schwarze D (1987) Evolution of cervid olfactory communication. In: Wemmer CM (ed) Biology and management of the Cervidae, vol 577. Smith Institute Press, Washington, DC, pp 223–234

Page LK, Swihart RK, Kazacos KR (1998) Raccoon latrine structure and its potential role in transmission of *Baylisascaris procyonis* to vertebrates. Amer Midl Nat 140:180–185

Palme R, Touma C, Arias N, Dominchin MF, Lepschy M (2013) Steroid extraction: get the best out of faecal samples. Wien Tierarztl Monatsschr 100(9–10):238–246

Poduschka W, Wemmer C (1986) Observations on Chemical Communication and its Glandular Sources in Selected Insectivora. In: Duvall D, Müller-Schwarze D, Silverstein RM (eds) Chemical signals in vertebrates vol 4. Springer, Boston, MA

Roberts NM, Crimmins SM, Hamilton DA, Gallagher E (2008) An evaluation of bridge-sign surveys to monitor river otter (*Lontra canadensis*) populations. Amer Midl Nat 160(2):358–363

Roper TJ (2010) Badger. The New Naturalist Library, Collins, London

Ruibal M, Peakall R, Claridge A (2011) Socio-seasonal changes in scent-marking habits in the carnivorous marsupial *Dasyurus maculatus* at communal latrines. Austral J Zool 58(5): 317–322

Sneddon IA (1991) Latrine use by the European rabbit (*Oryctolagus cuniculus*). J Mammal 72 (4):769–775

Stewart PD, Anderson C, Macdonald DW (1997) A mechanism for passive range exclusion: evidence from the European badger (*Meles meles*). J Theo Biol 184(3):279–289

Stewart PD, MacDonald DW, Newman C, Tattersall FH (2002) Behavioural mechanisms of information transmission and reception by badgers, *Meles meles*, at latrines. Anim Behav 63 (5):999–1007

Sun L, Müller-Schwarze D (1998) Anal gland secretion codes for relatedness in the beaver. Castor canadensis. Ethol 104(11):917–927

Tinnesand HV, Buesching CD, Noonan MJ, Newman C, Zedrosser A, Rosell F, Macdonald DW (2015) Will trespassers be prosecuted or assessed according to their merits? A consilient interpretation of territoriality in a group-living carnivore, the European Badger (*Meles meles*). PLoS ONE 10(7):e0132432

Tsunoda M, Kaneko Y, Sako T, Koizumi R, Iwasaki K, Mitsuhashi I, Saito MU, Hisano M, Newman C, Macdonald DW, Buesching CD (2018) Human disturbance affects latrine-use patterns of raccoon dogs. J Wildl Manag

Weinstein SB, Moura CW, Mendez JF, Lafferty KD (2018) Fear of feces? Tradeoffs between disease risk and foraging drive animal activity around raccoon latrines. Oikos

Wronski T, Plath M (2010) Characterization of the spatial distribution of latrines in reintroduced mountain gazelles: do latrines demarcate female group home ranges? J Zool 280(1):92–101

Zhou Y, Chen W, Buesching CD, Newman C, Kaneko Y, Xiang M, Nie C, Macdonald DW, Xie Z (2015) Hog badger (*Arctonyx collaris*) latrine use in relation to food abundance: evidence of the scarce factor paradox. Ecosphere 6(1):19

Part II

Olfaction in Humans

The Effects of Artificial Fragrances on Human Olfactory Communication

Caroline Allen[1](✉), Jan Havlíček[2], and S. Craig Roberts[3]

[1] School of Psychology, Newcastle University, Newcastle-upon-Tyne, NE1 7RU, UK
caroline.allen@newcastle.ac.uk
[2] Department of Zoology, Charles University, Prague, Czech Republic
[3] Division of Psychology, University of Stirling, Stirling, FK9 4LA, UK

Abstract. Humans were once considered to be microsmatic, but recent research suggests that we can use our sense of smell to detect important, socially relevant information about conspecifics. However, much of the research conducted to-date has investigated natural, fragrance free human body odours. While this is important in order to understand the evolution of olfactory communication in humans, it fails to account for the current (and historical) widespread use of artificial fragrances. In this chapter, we outline ways in which extraneous artificial fragrances may augment the underlying body odour, or 'odour space' of an individual, and how this might influence the perception of socially relevant information. In doing so, we describe some specific mechanisms for the interaction between body odours and artificial fragrances (blending and masking) and we discuss some recent work which has attempted to distinguish between these mechanisms.

1 Human Olfactory Communication and Fragrance Use

Olfactory communication is a key feature of social behaviour in most vertebrate species, and a wide range of information appears to be made available in this way. For example, social odours are known to contain cues of competitive ability (Huck et al. 1981; Rich and Hurst 1998), reproductive status (Clarke et al. 2009; Miranda et al. 2005), and genetic compatibility (Ilmonen et al. 2009; Ruther et al. 2009). While such examples of olfactory communication are well-documented in non-human species, there has been comparatively little attention, and some debate, about the nature and extent of olfactory communication in humans. Indeed, until relatively recently, humans have been considered by many to be functionally anosmic, or at least microsmatic (for a critique see McGann 2017; Schaal and Porter 1991).

Despite this, there is now accumulating evidence that human odours, like those of other non-human species, may also contain socially relevant information, and that this information can be perceived by the human nose. For example, humans are capable of discriminating between odours of males and females (Schleidt et al. 1981; Russell, 1976) and between odours of different individuals (Lord and Kasprzak 1989; Porter et al. 1983; Roberts et al. 2005; Weisfeld et al. 2003) and can perceive information about an individual's emotional state (de Groot and Smeets 2017; Fialová and Havlíček

© Springer Nature Switzerland AG 2019
C. D. Buesching (Ed.): *Chemical Signals in Vertebrates 14*, pp. 107–117, 2019.
https://doi.org/10.1007/978-3-030-17616-7_9

2012; Chen and Haviland-Jones 2000). There has also been a focus on the role of cues that, to varying extent, may be relevant to mate choice. People appear to attribute great importance to a partner's body odour (Havlíček et al. 2008), and body odours contain cues related to, for example, menstrual cycle stage (Havlíček et al. 2006; Singh and Bronstad 2001), genetic compatibility (Havlíček and Roberts 2009), health status (Moshkin et al. 2012; Olsson et al. 2014), diet quality (Fialová et al. 2013; Havlíček and Lenochova 2006), social status (Havlíček et al. 2005), and personality (Sorokowska 2013).

Despite this apparent wealth of information available in human body odour, we seem intent on doing all that we can to rid ourselves of it. In a number of cultures, the conscious detection of someone's body odour carries negative associations (e.g. Schleidt et al. 1981). As a result, the fragrance industry is worth billions of dollars. Personal use of fragranced products dates back to at least the ancient Greek and Egyptian civilisations (Stoddart 1990), and is widespread in contemporary culture. For example, one study found that 79% of women and 60% of men in the UK used a deodorant based product on a daily basis (Roberts et al. 2010).

Interestingly, researchers investigating the role of human body odours in social communication have usually attempted to avoid 'contamination' of body odour samples with extraneous fragrances, such as perfumes and scented deodorants. This is typically because they are interested in the extent to which naturally occurring human odours contain analogous socially relevant cues as found in other species. Consequently, much of what we know about olfactory communication in humans reflects body odour alone. It should be said that this is, we think, a reasonable, informative and important endeavour. However, by excluding such 'contamination', we create a situation that may not be representative of most people's experience, and inevitably miss an interesting opportunity to understand the effects of fragrances on everyday social communication. This is especially pertinent in terms of human mate choice, given that fragrances are often used to augment an individuals' attractiveness—as evidenced, for example, in the majority of perfume advertisements.

Here, we aim to specifically outline ways in which extraneous fragrances may affect an individual's underlying body odour and how they influence the perception by others of socially important information. We describe some possible mechanisms for the interaction between fragrance and body odour, and outline some recent work that has attempted to distinguish between them.

2 How do Fragrances Impact on Body Odours?

Even in the absence of extraneous fragrances, chemical analyses demonstrate that human body odours are complex mixtures of constituent compounds. Recent studies document nearly 2000 different volatile organic compounds (VOCs), particularly from breath and the skin (de Lacy Costello et al. 2014). Different people produce unique combinations of VOCs at variable concentrations, generating distinct odour profiles (Penn et al. 2007), coined 'odour signatures', which are shaped both by genes and environmental factors—for example, twins have similar but not identical VOC signatures (Sommerville et al. 1994). The most potent environmental factors are various aromatic compounds in diet and disease-related metabolites (Havlíček et al. 2017).

We are not yet able to describe which compounds within any given odour profile are responsible for communicating social cues used in human mate choice or other forms of interpersonal assessment. Indeed, in the relatively complex social odours of most vertebrates, as compared to some insects, any cues are likely to be represented by combinations of a multitude of compounds, or ratios between these, which renders the task still more challenging (Wyatt 2015). However, because of the fact that we *can* obtain socially relevant cues from an odour, we can reasonably surmise that these *must* be contained somewhere within the complex and multivariate characterisation of the odour in a GC trace—what we might call an individual's 'odour space'. We can further make the assertion that there are two distinctive segments of odour space (Fig. 1). First, there is a core segment which contains genetically mediated odour cues, for example, enabling discrimination of twins and relatives, and contributing to an individually distinctive odour. The core of odour space is also likely to contain cues of an individual's sex as well as other genetic markers that enable discrimination of dissimilarity between unrelated individuals. The second segment of odour space contains information that is more labile, cues that are environmentally determined such as diet, social or reproductive status, and current emotional state, some of which also influence mate preference judgments.

We can use this concept of odour space to understand how extraneous fragrances might influence the perception of an individual's body odour. The most common explanation for fragrance use is that it masks the underlying natural body odour, effectively overcoming and substituting it for an entirely different, socially acceptable and pleasurable smell. We can imagine this as an olfactory sheath, overlaid onto the naturally occurring body odour (see Fig. 2, left), through which only a small fraction of the body odour may escape, perhaps increasingly so over time. A more recent idea is that fragrances interact with the underlying body odour to produce a blend that carries a new and distinctive odour quality (Lenochová et al. 2012). It has been suggested that this process may explain at least some of the variation in fragrance preference between individuals: people may learn which fragrances suit their own body odour and which do not and select between fragrances to produce a pleasing mixture. The outcome of this selection may be that fragrances maintain or even enhance the underlying odour signature and the social information it contains. Using the model of odour space, this could happen in one of the two ways: fragrance selection may be driven by the interaction with the totality of the odour, or specifically with the genetically influenced core segment (see Fig. 2, right).

Of course, the two mechanisms are not mutually exclusive and some degree of masking (for example, in reducing intensity of the underlying odour) may occur even if blending drives specific fragrance choice. However, evidence is building for the blending hypothesis, and we describe some of this evidence in the next section (Roberts and Havlíček 2012; Havlíček and Roberts 2013).

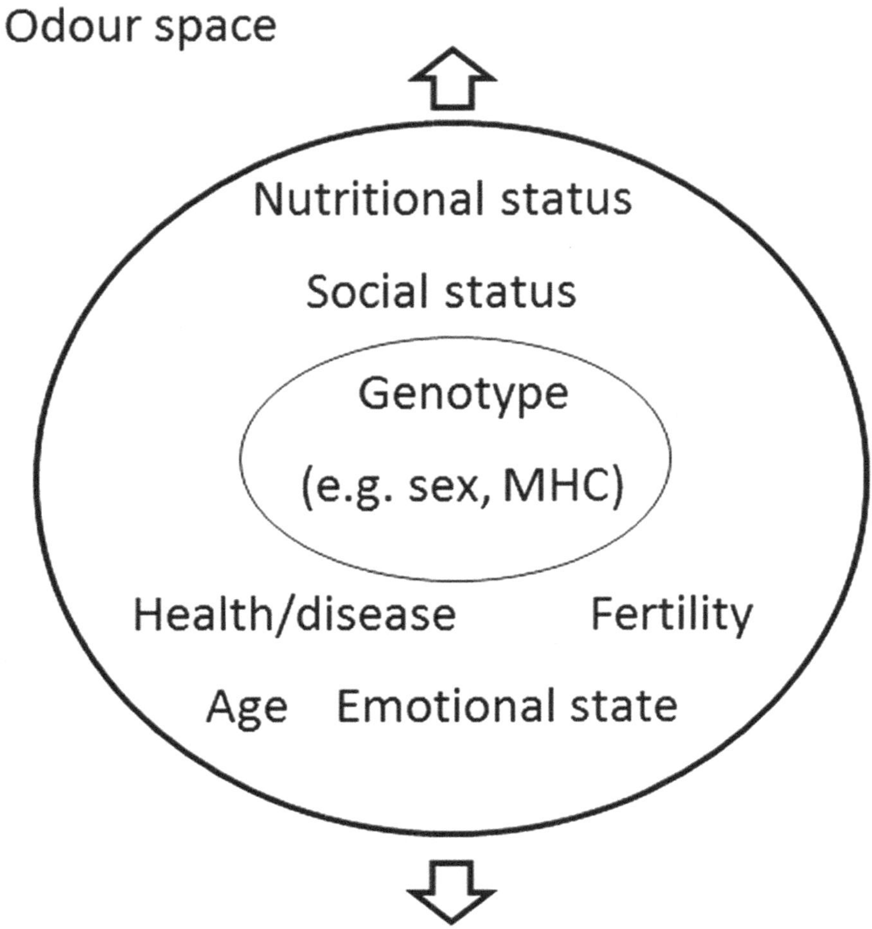

Fig. 1. A representation of body odour space. The larger circle notionally encompasses the entire naturally occurring odour profile, which is actually composed of hundreds of individual volatile compounds in varying abundance. Within this is contained a number of known environmentally associated social cues (e.g. nutritional status) as well as a (likely larger, though this is unknown) fraction of non-communicatory compounds. The smaller space represents a core of odour space that encodes genetically mediated odour compounds. Arrows denote the perceivable odour

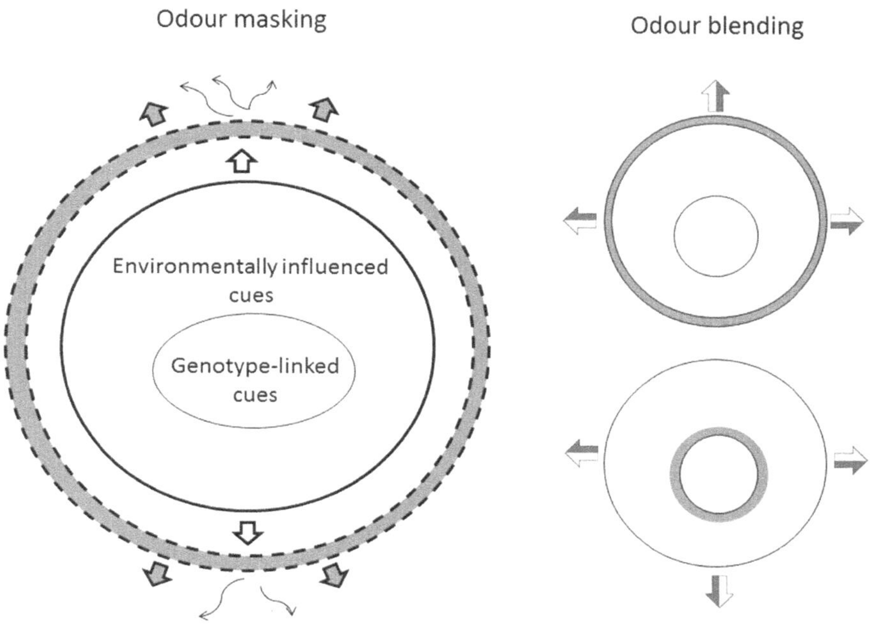

Fig. 2. Proposed mechanisms for the interaction between fragrance and body odour. In odour masking (left), an applied fragrance forms a perceptual sheath (grey shading) over the naturally occurring odour emanating from odour space (white arrows). Although some natural odour (small arrows) may still 'leak' through, the dominant smell perceived by others is that of the fragrance (grey arrows). In contrast, in odour blending (right), the odour perceived by others is a blend of the body odour (white) and fragrance (grey). Two possible blending mechanisms are depicted, that reflect how a suitable perfume might be chosen for a given odour. Above, the fragrance is chosen to create a blend with the entire odour profile; below, it is chosen to form a blend specifically with the genetically linked odour compounds (note these are extreme characterisations of blending mechanisms, in practice it is likely that only a proportion of genetically and environmentally influenced compounds are involved)

3 How do Fragrances Influence Odour Assessment?

The blending hypothesis owes its origin to research investigating fragrance use and biological markers for immune function. A number of experimental studies have been conducted to examine how human odour preferences may be influenced by gene sharing at genes within the major histocompatibility complex (MHC, which is formally referred to as HLA, human leukocyte antigen, in humans). The MHC is a gene family linked to immune functioning in vertebrates (Cooper and Alder 2006; Kelley et al. 2005). As heterozygosity at MHC genes is beneficial in terms of offspring viability and health, HLA-disassortative mating may also be beneficial as a factor in human mate choice (for an overview see Havlíček and Roberts 2009; Winternitz et al. 2017). A number of studies report that individuals tend to prefer the odour of an opposite-sex person who has a relatively dissimilar MHC profile to their own (Wedekind et al. 1995; Wedekind and Füri 1997; Roberts et al. 2008 but see Winternitz et al. 2017).

Building on this research, Milinski and Wedekind (2001) investigated the interaction between MHC genotype and fragrance use. They recruited 63 women and 74 men who had been typed for three HLA loci, and asked them to initially rate 36 perfume ingredients for self-use: 'would you like to smell like that yourself?'. In a second phase, participants were provided with a subset of 18 ingredients, and asked again about self but additionally about their partner's use: 'would you like your partner to smell like that?'. They found significant associations between the two most common HLA alleles and self-preferences in both rating tasks, but not for preferences for partner. On this basis, they suggested that this supports a hypothesis that fragrances are chosen to enhance the availability of HLA-related immunogenetic cues in mate choice. In further support of this hypothesis, Hämmerli et al. (2012) found that some of the more common MHC alleles could be used to subdivide a population into groups that expressed differential preferences for various perfume ingredients.

Using a different approach to further test this idea, Lenochová and colleagues (2012) found that mixtures of an individual's body odour with their fragrance of choice were rated as more pleasant than when mixed with a non-chosen fragrance, which was instead assigned to them by the experimenters, even when there was no difference in pleasantness of the fragrances alone. Furthermore, the degree of variance in ratings of the blends was maintained compared with variance in ratings of non-fragranced body odour samples, a finding that is predicted by the blending, but not the masking, hypothesis. Overall, these findings support the idea that idiosyncratic perfume preferences may be partly explained by producing a favourable blend of body odour and fragrance.

A more recent study offered partial support to this theory. Allen and colleagues (2015) conducted a study investigating the effect of fragrance on the ability to discriminate between individual body odours, a task that people often perform successfully when using unfragranced samples (Porter et al. 1983; Ferdenzi et al. 2010; Cernoch and Porter 1985). Men and women completed three triangle tasks, where they sniffed three odour samples, of which two were from the same person, and were asked to identify the odd one out. Each person did this once with unfragranced samples, once with samples that additionally contained the donors' own fragrance, and once with an experimenter assigned fragrance. It was found that discrimination performance was best when people were smelling unfragranced samples; however, in keeping with the findings of Lenochová et al. (2012), people performed significantly better when they were smelling samples blended with the self-selected fragrances as opposed to the assigned fragrance. This provides further evidence for the importance of fragrance choice and suggests that fragrances may be chosen to amplify a personal odour signature.

The evidence reviewed above indicates that fragrance and body odours create idiosyncratic blends, the quality of which might be difficult to predict from the individual components. One potentially practical consequence of these findings is that it might be more difficult to predict the suitability of a fragrance for someone else than it is to select one for yourself. However, one exception to this might be relatives, who to some extent share certain aspects of a body odour signature (e.g. Porter et al. 1985).

This may mean that we are better at selecting fragrances for a close family member than for a non-related individual. Based on this idea, Sobotková et al. (2017) recently tested whether perfume selected for a target male individual is perceived as more attractive when it has been chosen by themselves, a romantic partner, or a sister. The authors expected to find no difference in attractiveness of body odour and perfume blends selected by the target and the blends with perfumes selected by a sister. In contrast to this, it was expected that perfume odour blends which were selected by a romantic partner would be rated as less attractive when compared to the blend of body odour and perfume selected by the targets themselves. In fact, the results suggested that perfumes selected by sisters, and to some extent also by female romantic partners, were rated as more attractive when blended with the individual target's body odour than the blends which included the individual's chosen perfume. This surprising result might be a consequence of higher olfaction-related cognitive abilities in women as compared to men; women may be better than men at accurately assessing and choosing complimentary fragrances.

Studies have also been conducted which investigate the ways in which artificial fragrances may affect perceptions of masculinity and femininity. These traits are also potentially important for human mate choice (for an overview see Allen et al. 2016). Interestingly, they also reflect main fragrance classifications, suggesting that most fragrances are designed to mimic masculinity or femininity (Lindqvist 2012). It was for instance shown that the use of masculine perfume increased ratings of suitability for managerial positions (Sczesny and Stahlberg 2002), which are stereotypically considered to be masculine roles. However, this effect might be context dependent, as one study found that the addition of a feminine deodorant increased ratings of femininity given to female axillary samples, but that this was not the case for male samples incorporating masculine deodorants (Allen et al. 2016). When this was investigated further, it was found that men who had low ratings of masculinity without a fragrance showed increases of rated masculinity with the addition of a fragrance, but that men who were already highly masculine showed no change. This suggests that our evolved biological preferences for masculinity and femininity may well have shaped fragrance design (indeed, preference for masculine or feminine fragrances varies consistently with preferences for masculinity and femininity in mate judgment in other sensory modalities such as faces: Little et al. 2011). Evidence suggests that heterosexual women prefer moderate levels of masculinity (Rhodes et al. 2000), and so fragrances may be designed to only be moderately masculine; in contrast, there is no known evidence showing preferences for intermediate levels of femininity, and thus the femininity of fragrances is potentially unconstrained, or at least not constrained to the same level as masculinity in male fragrances.

As well as investigating the effect of fragrance in potential mating contexts, research has also been conducted which investigates the effect of fragrances on our attributions of personality characteristics. Previous research has found that judgements of neuroticism and dominance from body odour are correlated with these self-assessed personality characteristics (Sorokowska et al. 2012; Sorokowska 2013). In a follow-up study, researchers compared personality judgments of natural body odour samples and body odours collected from participants using cosmetics (Sorokowska et al. 2016). It was found that the congruent ratings of neuroticism and dominance were corroborated under the natural body odour condition. However, when participants were allowed to use cosmetics, the attributions of neuroticism were no longer correlated with the self-report measures. In contrast, the use of cosmetics did not affect the accuracy of attributions of dominance. The differential effect of fragrance in this case may stem from neuroticism being considered as a socially undesirable characteristic (Botwin et al. 1997), which may mean that people use cosmetics in order to positively affect the social impression that they make.

Recent work has further extended the ecological validity of human olfactory research by not only including fragrances but also carrying out 'live' assessments of an individuals' body odour, directly from the donor. Gaby and Zayas (2017) found that across multiple sessions individuals' ratings of odours without a fragrance were consistent, as were their ratings of odours when a fragrance was present, although despite this they found that social judgements (pleasantness, friendliness, and likeability) of unfragranced odours did not predict those given to fragranced odours, indicating that fragrance use had interfered with the communicative properties of the body odour.

Above, we have reviewed evidence suggesting that there is a direct influence of fragrances on perceivers. Nevertheless, fragrances may also affect others indirectly by changing mood and self-perception of the wearers themselves. It has previously been shown that raters tend to judge videos of individuals who wore fragrance as more attractive in comparison to those who did not, even though raters were not able to actually smell them (Roberts et al. 2009). This is perhaps due to lower ratings of self-confidence which was recorded in the individuals who used the placebo deodorant and were therefore not wearing a fragrance. Similar findings concerning the perception of self-confidence after fragrance application were reported by Higuchi and colleagues (2005) who also observed lower frequencies of some non-verbal behaviours such as self-touching.

The findings to-date suggest that there is no one rule to be applied when predicting the ways in which a fragrance may alter our perception of information present in body odours. Although the findings are intriguing, there is a need for replication (given the relative novelty of including fragrances in research) as well as expansion of testing with regards to the traits being investigated. For example, we know that perceptual qualities of body odour vary across the female menstrual cycle (Havlíček et al. 2006), but we have yet to investigate how fragrance use may impact upon our ability to detect these changes.

4 Conclusions and Recommendations

Cross-culturally, and in an analogous fashion to visual appearance, people tend to put a great deal of effort into augmenting their olfactory impression, especially in culturally sensitive contexts. In order to truly understand our current capacity for olfactory communication we must test under ecologically valid conditions—artificial fragrances should be included. Ecological conditions in the case of humans include cultural settings, and a quote from Plotkin (2010) further underlines the importance of this: 'Mankind's natural place is in culture, and culture is a part of human biology because it is our biology that gives us the ability to enter into culture'. Thus, we need to systematically investigate the effects of various artificial fragrances on detection of olfactory information which has been 'established' in previous studies: emotions, health status, genetic information, personality traits, fertility status, and biological sex. We also need to carefully consider the fragrances used. The research to-date which has incorporated fragrances uses a range of products, from perfumes to deodorants and antiperspirants. These products have different designs and purposes, using different varieties of (sometimes proprietary) active ingredients, and therefore it cannot be expected that they will interact with body odours in the same way. This chapter has focussed on deodorants and perfumes, but we live in a fragranced world and also use a multitude of other fragranced products on a daily basis (soap, shampoo, air fresheners, washing up liquid, etc.), which should also be considered in future research aiming to better understand the current role of olfactory communication in humans.

Acknowledgements. JH is supported by the Czech Science Foundation grant [18-15168S].

References

Allen C, Cobey KD, Havlíček J, Roberts SC (2016) The impact of artificial fragrances on the assessment of mate quality cues in body odor. Evol Hum Behav 37:481–489

Allen C, Havlíček J, Roberts SC (2015) Effect of fragrance use on discrimination of individual body odor. Front Psychol 6:1115

Botwin MD, Buss DM, Shackelford TK (1997) Personality and mate preferences: five factors in mate selection and marital satisfaction. J Pers 65(1):107–136

Cernoch JM, Porter RH (1985) Recognition of maternal axillary odors by infants. Child Dev 56:1593–1598

Chen D, Haviland-Jones J (2000) Human olfactory communication of emotion. Percept Mot Skills 91:771–781

Clarke PMR, Barrett L, Henzi SP (2009) What role do olfactory cues play in chacma baboon mating? Am J Primatol 71:493–502

Cooper MD, Alder MN (2006) The evolution of adaptive immune systems. Cell 124:815–822

de Groot JH, Smeets MA (2017) Human fear chemosignaling: evidence from a meta-analysis. Chem Senses 42(8):663–673

de Lacy Costello B, Amann A, Al-Kateb H, Flynn C, Filipiak W, Khalid T, … Ratcliffe NM (2014) A review of the volatiles from the healthy human body. J Breath Res 8:14001

Ferdenzi C, Schaal B, Roberts SC (2010) Family scents: developmental changes in the perception of kin body odor? J Chem Ecol 36:847–854

Fialová J, Havlíček J (2012) Perception of emotion-related odours in humans. Anthropologie 50:95–110

Fialová J, Roberts SC, Havlíček J (2013) Is the perception of dietary odour cues linked to sexual selection in humans? In: East ML, Denhard M (eds) Chemical signals in vertebrates XII. Springer, New York, pp 161–170

Gaby JM, Zayas V (2017) Smelling is telling: Human olfactory cues influence social judgments in semi-realistic interactions. Chem Senses 42:405–418

Hämmerli A, Schweisgut C, Kaegi M (2012) Population genetic segmentation of MHC-correlated perfume preferences. Int J Cosmet Sci 34:161–168

Havlíček J, Dvorakova R, Bartos L, Flegr J (2006) Non-advertized does not mean concealed: body odour changes across the human menstrual cycle. Ethology 112:81–90

Havlíček J, Fialová J, Roberts SC (2017) Individual variation in body odour. In: Buettner A (ed) Springer handbook of odor. Springer, New York, pp 943–955
Havlíček J, Lenochova P (2006) The effect of meat consumption on body odor attractiveness. Chem Senses 31:747–752
Havlíček J, Roberts SC (2009) MHC-correlated mate choice in humans: a review. Psychoneuroendocrinology 34:497–512
Havlíček J, Roberts SC (2013) The perfume-body odour complex: an insightful model for culture-gene coevolution? In: East ML, Dehnhard M (eds) Chemical signals in vertebrates XII. Springer, New York, pp 185–195
Havlíček J, Roberts SC, Flegr J (2005) Women's preference for dominant male odor: effects of menstrual cycle and relationship status. Biol Let 1:256–259
Havlíček J, Saxton TK, Roberts SC, Jozifkova E, Lhota S, Valentova J, Flegr J (2008) He sees, she smells? Male and female reports of sensory reliance in mate choice and non-mate-choice contexts. Pers Individ Differg 45:565–570
Higuchi T, Shoji K, Taguchi S, Hatayama T (2005) Improvement of nonverbal behaviour in Japanese female perfume-wearers. Int J Psychol 40:90–99
Huck WU, Banks EM, Wang S (1981) Olfactory discrimination of social status in the brown lemming. Behav Neural Biol 33:364–371
Ilmonen P, Stundner G, Thoss M, Penn DJ (2009) Females prefer the scent of outbred males: good-genes-as-heterozygosity? BMC Evol Biol 9:104
Kelley J, Walter L, Trowsdale J (2005) Comparative genomics of natural killer cell receptor gene clusters. PLoS Genet 1:129–139
Lenochová P, Vohnoutová P, Roberts SC, Oberzaucher E, Grammer K, Havlíček J (2012) Psychology of fragrance use: perception of individual odor and perfume blends reveals a mechanism for idiosyncratic effects on fragrance choice. PLoS ONE 7:e33810
Lindqvist A (2012) How is commercial gender categorization of perfumes related to consumers preference of fragrances? Proc Soc Behav Sci 65:370–374
Little AC, Connely J, Feinberg DR, Jones BC, Roberts SC (2011) Human preference for masculinity differs according to context in faces, bodies, voices, and smell. Behav Ecol 22:862–868
Lord T, Kasprzak M (1989) Identification of self through olfaction. Percept Mot Skills 69:219–224
McGann JP (2017) Poor human olfaction is a 19th-century myth. Science 356(6338)
Miranda A, Almeida OG, Hubbard PC, Barata EN, Canário AVM (2005) Olfactory discrimination of female reproductive status by male tilapia (*Oreochromis mossambicus*). J Exp Biol 208:2037–2043
Milinski M, Wedekind C (2001) Evidence for MHC-correlated perfume preferences in humans. Behav Ecol 12:140–149
Moshkin M, Litvinova N, Litvinova EA, Bedareva A, Lutsyuk A, Gerlinskaya L (2012) Scent recognition of infected status in humans. J Sex Med 9:3211–3218
Olsson MJ, Lundström JN, Kimball BA, Gordon AR, Karshikoff B, Hosseini N, Sorjonen K, Olgart Höglund C, Solares C, Soop A, Axelsson J (2014) The scent of disease: human body odor contains an early chemosensory cue of sickness. Psychol Sci 25:817–823
Penn DJ, Oberzaucher E, Grammer K, Fischer G, Soini HA, Wiesler D, …, Brereton RG (2007) Individual and gender fingerprints in human body odour. J R Soc Interface 4:331–340
Plotkin HC (2010) Evolutionary worlds without end. Oxford University Press, USA
Porter RH, Cernoch JM, Balogh RD (1985) Odour signatures and kin recognition. Physiol Behav 34:445–448
Porter RH, Cernoch JM, McLaughlin FJ (1983) Maternal recognition of neonates through olfactory cues. Physiol Behav 30:151–154

Rhodes G, Hickford C, Jeffery L (2000) Sex-typicality and attractiveness: are supermale and superfemale faces super-attractive? Br J Psychol 91:125–140
Rich TJ, Hurst JL (1998) Scent marks as reliable signals of the competitive ability of mates. Anim Behav 56:727–735
Roberts SC, Gosling LM, Carter V, Petrie M (2008) MHC-correlated odour preferences in humans and the use of oral contraceptives. Proceed Royal Soc B 275:2715–2722
Roberts SC, Gosling LM, Spector TD, Miller P, Penn DJ, Petrie M (2005) Body odor similarity in noncohabiting twins. Chem Senses 30:651–656
Roberts SC, Havlíček J (2012) Evolutionary psychology and perfume design. In: Roberts SC (ed) Applied evolutionary psychology. Oxford University Press, Oxford, pp 330–348
Roberts SC, Little AC, Lyndon A, Roberts J, Havlíček J, Wright RL (2009) Manipulation of body odour alters men's self-confidence and judgements of their visual attractiveness by women. Int J Cosmet Sci 31:47–54
Roberts SC, Miner EJ, Shackelford TK (2010) The future of an applied evolutionary psychology for human partnerships. Rev Gen Psychol 14:318–329
Russell MJ (1976) Human olfactory communication. Nature 260:520–522
Ruther J, Matschke M, Garbe L, Steiner S (2009) Quantity matters: Male sex pheromone signals mate quality in the parasitic wasp Nasonia vitripennis. Proc R Soc B: Biol Sci 276:3303–3310
Schaal B, Porter R (1991) "Microsmatic humans" revisited: the generation and perception of chemical signals. Adv Study Behav 20:135–199
Schleidt M, Hold B, Attili G (1981) A cross-cultural study on the attitude towards personal odors. J Chem Ecol 7:19–31
Sczesny S, Stahlberg D (2002) The influence of gender-stereotyped perfumes on leadership attribution. Eur J Soc Psychol 32:815–828
Singh D, Bronstad PM (2001) Female body odour is a potential cue to ovulation. Proc R Soc B: Biol Sci 268:797–801
Sobotková M, Fialová J, Roberts SC, Havlíček J (2017) Effect of biological relatedness on perfume selection for others: preliminary evidence. Perception 46:498–515
Sommerville BA, Wobst B, McCormick JP, Eggert F, Zavazava N, Broom DM (1994) Volatile identity signals in human axillary sweat: the possible influence of MHC class I genes. Adv Biosci 93:535–538
Sorokowska A (2013) Seeing or smelling? Assessing personality on the basis of different stimuli. Pers Individ Differ 55:175–179
Sorokowska A, Sorokowski P, Havlíček J (2016) Body odor based personality judgments: the effect of fragranced cosmetics. Front Psychol 7:530
Sorokowska A, Sorokowski P, Szmajke A (2012) Does personality smell? Accuracy of personality assessments based on body odour. Eur J Pers 26:496–503
Stoddart M (1990) The scented ape. Cambridge University Press, Cambridge
Wedekind C, Füri S (1997) Body odour preferences in men and women: do they aim for specific MHC combinations or simply heterozygosity? Proc R Soc B: Biol Sci 264:1471–1479
Wedekind C, Seebeck T, Bettens F, Paepke AJ (1995) MHC-dependent mate preferences in humans. Proc R Soc B: Biol Sci 260:245–249
Weisfeld GE, Czilli T, Phillips KA, Gall JA, Lichtman CM (2003) Possible olfaction-based mechanisms in human kin recognition and inbreeding avoidance. J Exp Child Psychol 85:279–295
Winternitz J, Abbate JL, Huchard E, Havlícek J, Garamszegi LZ (2017) Patterns of MHC-dependent mate selection in humans and nonhuman primates: a meta-analysis. Mole Ecol 26 (2):668–688
Wyatt TD (2015) The search for human pheromones: the lost decades and the necessity of returning to first principles. Proc R Soc B: Biol Sci 282:20142994

The Human Mammary Odour Factor: Variability and Regularities in Sources and Functions

Benoist Schaal(✉), Sébastien Doucet, Robert Soussignan, Magali Klaey-Tassone, Bruno Patris, and Karine Durand

Developmental Ethology and Cognitive Psychology Group, Centre for Smell, Taste and Food Science, CNRS (UMR 6265), Université de Bourgogne-Franche Comté, Dijon, France
benoist.schaal@u-bourgogne.fr

Abstract. In the course of evolution, human mothers have been, and still are, under strong selective pressure to induce their newborns' colostrum ingestion promptly after birth. As a concentrate of nutrients, passive immunity, antioxidants, growth factors and symbiotic microbiota, colostrum functions as the evolved antidote to ubiquitous pathogens and threats of neonatal exhaustion. Under such constraints, any means to speed up colostrum/milk intake can only have been beneficial to neonatal viability and adaptive life onset along evolutionary time. The areolar-nipple areas of human lactating females emit lacteal substrates conveying chemostimuli that are attractive and release mouthing and sucking in infants. Current information about areolar glands of Montgomery is summarized here, in terms of variability/regularity in number and distribution, of behavioural activity of their secretion, and of the adaptive value of their occurrence. It is concluded that a majority of lactating women investigated so far bear from 1 to >40 glands/areola, among which some emit a visible secretion. Reciprocally, a majority of neonate infants react to the odour of these secretions. The number of areolar glands correlates with infants' behaviour at the breast, indicating that areolar phenotypic variability has potential fitness consequences on the infant and the mother. Ideas for future studies are outlined to advance our understanding of how human areolar phenotypic variability relates to life history characteristics.

1 Constraints on Mammary Odour Signalling

Various adverse situations can affect breastfeeding initiation in human newborns, highlighting the delicacy of the mechanisms underlying its achievement (e.g. Winberg 2005). For instance, the initiation of nutritive sucking may be delayed from hours to days after birth by local representations and related prescriptions on how neonate infants should be raised. Such delays that impede the transfer of colostrum have been, and still are, far from uncommon in both traditional and highly medicalized societies. For example, in rural India a 1996 study indicates that not only did mothers postpone initiation of breastfeeding (only 0.3% of them allowed sucking within the first post-partum hour), but additionally they replaced the colostrum with artificial prelacteal

© Springer Nature Switzerland AG 2019
C. D. Buesching (Ed.): *Chemical Signals in Vertebrates 14*, pp. 118–138, 2019.
https://doi.org/10.1007/978-3-030-17616-7_10

formulas (e.g. Banapurmath et al. 1996; see also Abuidhail 2014). Such delays in putting infants at the breast and avoidance of feeding them colostrum have been recorded on all continents and across time up to now, even in medically advanced societies (e.g. Reissland and Burghart 1988; Barkow and Hallett 1989; Odent 2010; Candau and Schaal 2015).

Besides traditional and medical beliefs, other causes can interfere with optimal breastfeeding initiation, such as maternal inexperience. For example, 49, 22 and 17% of normal term-born newborns in a 2003 Californian sample behaved suboptimally toward the breast on the day of birth and on days 3 and 7, respectively, despite adept advice and assistance being conveyed. These disadvantageous outcomes were strongly associated with primiparity (Dewey et al. 2003). A related French study reached similar conclusions (Michel et al. 2007). Follow-up studies of home births in rural Ghana link such early delays in breastfeeding initiation to subsequent infant morbidity and mortality. In such poorly medicalized conditions, a 1-day postponement in the initiation of breastfeeding explained 16% of neonatal death before day 28; moreover, a delay of only 1 h post-birth to engage breastfeeding explained 22% of neonatal mortality (Edmond et al. 2006). Causal links between early (<24 h) versus late timing of breastfeeding initiation and infant health outcome are multiple and non-exclusive, including (among other nutritive and sensory effects): increased protective exposure to colostral bioactive compounds (immunoglobulins, antioxidants); mother-to-infant transfer of non-pathogenic microbiota; colostrum-induced decrease of intestinal permeability to pathogens; stimulation of mucosal lymphoid tissue through maternal-infant contact. In addition, colostral ingestion lowers exposure risk to pathogens conveyed in prelacteal feeds (e.g. Uruakpa et al. 2002; Debes et al. 2013).

In addition to the denial of early access to the breast imposed by many cultural contexts, inexperienced newborns face intrinsic difficulty to first orient and orally seize a nipple, thus increasing potential threats to viability. Nonhuman and human (historical societies and modern hunter-gatherers) primates have been shown to globally lose 27% of their offspring during the first year of life, a remainder of 47% failing to survive to puberty (Volk and Atkinson 2013). Such early mortality is causally complex and multifaceted (e.g. Lawn et al. 2005; Kinney et al. 2010), with prevailing infectious causes (e.g. 36% in Lawn et al. 2005; see also Edmond et al. 2007). Some authors (Volk and Atkinson 2013) have linked part of this high mortality rate of human infants to the neonates' problematic engagement of suckling. Covariations in the regional scheduling of breastfeeding initiation and records of neonatal mortality corroborate this view (e.g. Debes et al. 2013). Further, the daily risk of infantile mortality is very high in the first postnatal week, with an even higher peak within the 48 h following birth (Lawn et al. 2005). Thus, human mothers are under an extremely strong selective pressure to induce their newborns to promptly suck their breasts and to transfer colostrum to them as soon as possible after birth. Although modest in quantitative terms, colostrum is indeed qualitatively peculiar as a concentrate of nutrients, passive immunity, antioxidants, growth factors and symbiotic microbiota functioning together as the evolved antidote to omnipresent pathogens and threats of neonatal exhaustion (e.g. Hale and Hartmann 2007).

2 Mammary Signalling and the Neonate's Receptive Range

In such highly constrained conditions over evolutionary time, any means that speed up localisation and oral acceptance of the nipple (or *papilla mammae* or *mamilla*) by the newly born infant can only have been beneficial to viability and adaptive life onset. In that line, the most obvious adaptive way certainly is the direct guidance and assistance mothers or other helpers offer the infant when presented to the breast for the first few times. However, aforementioned evidence (Dewey et al. 2003; Michel et al. 2007) indicates that maternal handling of the infant is often suboptimal in initial breastfeeds, especially in inexperienced first-time mothers. The ultimate decision to positively orient, orally grasp the nipple, and express the full latching response is therefore inherent to infants themselves. Multiple observations of the 'breast-crawl' document indeed that human newborns can autonomously reach the breast when they are given time and optimal environmental conditions (e.g. Widström et al. 1987, 1990, 2011; Righard and Alade 1990; Christensson et al. 1992; Varendi et al. 1994; Klaus 1998; Cantril et al. 2014; Girish et al. 2013). To accomplish this early performance, infants rely on a range of motor abilities directed by their senses including somesthesis, with tactile and thermal components, chemosensation, with olfaction and taste, and perhaps even vision (as the human areola-papilla is pigmented contrastedly by dermal melanocytes). These multiple guidance cues are natural features of the breast that have been shown to separately affect neonatal head orientation and oro-motor responses.

Briefly, newborn infants display bilateral rooting movements on their track to the breast and, when reaching it with their chin, they express oral gapes that generally end in seizing a nipple (e.g. Prechtl 1958; Widström et al. 1990). These head sweeping movements expose their perioral tactile receptors to the textural contrasts between the skin of the breast and the softer tissue of the areola, and then between the areolar epidermis and the firmer erectile tissue of the nipple. Such tactile discrepancies may facilitate the infant's directional progression toward the nipple and the labial grasping of it. Otherwise, the metabolic activity of the lactating breast induces a thermal contrast between its epidermal surface and the rest of the chest (Burd et al. 1977). Due to the vasoactivity of Haller's subareolar vascular plexus before nursing episodes, additional thermal contrasts exist between the areola and the rest of the breast, and between the areola and the nipple. These local thermal contrasts were established long ago by thermographic studies (Vuorenkoski et al. 1969; Kimura and Matsuoka 2007) and recently confirmed by surface thermometry (breast versus areola temperature difference: 0.56 °C; Zanardo and Straface 2015; Zanardo et al. 2017). If they are detectable to neonates (which has never been empirically attested) these tactile and thermal differences could operate as proximal cues to finalize their localisation of the areola. Another, more distally conspicuous cue of the areolae for awake infants is their visual contrast with surrounding skin due to their strong pigmentation (e.g. Stolnicu 2018).

Such areolar physical cues are complemented or synergized by chemical cues. Post-parturient lactating women may release from there a specific body odour, areolar skin glands being activated by their physiological stage-specific endocrine balance. For example, at that time sebaceous glands excrete then higher amounts of sebum (Burton et al. 1973). Apocrine glands may also produce a lactation stage-specific odour mixture that appears to be detected and differentiated by neonates (Cernoch and Porter 1985; Porter et al. 1992). But an especially attractive odour is given off by the lactating breast. For example, breastfed infants (aged 15 days) prefer the breast odour of an unfamiliar lactating mother over her axillary odour (Porter et al. 1992) and bottle-fed infants tend to prefer the breast odour of an unfamiliar nursing mother to the odour of the formula to which they were conditioned since birth (Porter et al. 1991). Further, when the natural odour of the breast is masked by coating nipples with extraneous odorants (Schaal 1988) or eliminated by washing the nipple-areolar regions (Varendi et al. 1994), the neonates' orientation and searching responses are inhibited or disorganized. Otherwise, when the natural odour is merely blocked by applying a scentless transparent plastic film on the breast, nursing infants display significantly reduced rooting, sucking and mouthing responses, and begin crying sooner (Doucet et al. 2007; see below).

In sum, the areolar-nipple zones of human lactating females convey a range of natural stimuli of diverse nature that may play, among other functions, the roles of guidance beacons and motivational cues which end in efficient oral seizing and initiation of nutritive sucking. These breast cues conveyed by different sensory channels probably operate conjointly. For example, the heightened surface temperature of the areolar discs may regulate the evaporation of odorous substrates from areolar skin glands or remnants of milk, and may therefore maximize and synchronize mammary odour signals in time with the infants' demand for suckling. The mother's reception of distal signals emitted by the crying infant do indeed result in increased temperature of the breast (Vuorenkoski et al. 1969), bringing time-adjusted thermal amplification of areolar odour impressions.

3 Sources of Mammary Odours: The Areola as a Scent Organ

Briefly summarized (for reviews, cf. Schaal and Porter 1991; Schaal et al. 2008; Schaal 2010; Koyama et al. 2013; Schaal and Al Aïn 2014), the human areolar region harbours multiple glandular sources giving off secretions that potentially contribute to the local odour mixture. These are composed of elementary skin glands of all types, which are present in the areolar and papillary dermis. Minute eccrine sweat glands and sebaceous glands are distributed in the areolae (Montagna and MacPherson 1974), while nipples are endowed with simple, hairless sebaceous and apocrine glands, which ducts open distally and in the ostia of the lactiferous ducts (Perkins and Miller 1926; Montagna and Yun 1972; Montagna and MacPherson 1974; Vorherr 1974; Koyama et al. 2013). Increased excretion of sebum has been reported during lactation (Burton et al. 1973), which may contribute to the lactation-specific breast odour to which infants are responsive.

But, more remarkably, the areolar surface is speckled with small bumps, in the form of 1–2 mm-wise structures contrasting with the surrounding areolar skin in elevation and/or pigmentation. Although first described by the Paduan anatomist Morgagni (1719), these structures were later recognized by (and named after) the obstetrician Montgomery (1837) as typical of the gestational and lactational areola. These Montgomery's glands become then notably enlarged as small areolar protuberances and sometimes emit a visible milky secretion. The functional entity of the gland of Montgomery is formed by sebaceous glands associated with isolated mammary acini (Montagna and Yun 1972; Smith et al. 1982). The partial lactational nature of the gland was confirmed by the fact that milk-specific lactose is a constituent of Montgomerian secretion (Naeslund 1957). Further, Montgomerian secretion conveys some odorous volatile compounds also evidenced in colostrum and milk (Schaal et al. 2008).

Finally, the most profuse sources of potentially odorous constituents are obviously colostrum and milk, which are ejected from the milk ducts opening at the top of the nipple. This mainstream lacteal secretion carries multiple volatile compounds in free and conjugated form (e.g. Loos et al. 2017). Collectively, these compounds are efficient in eliciting positive head orientation, mouthing and sucking motions in newborn infants (Soussignan et al. 1997; Mizuno and Ueda 2004; Marlier and Schaal 2005). The nature of these responses is detailed below.

To sum up, the human areolar-nipple structure combines multiple sources of free and conjugated odorous constituents carried in milk, sebum, and apocrine and eccrine sweat. Sebum, as well as the fat fraction of milk, may play the role of fatty fixatives that slow down the evaporative timing of volatiles (e.g. Regnier and Goodwin, 1977). Further, enzymes or bacteria conveyed in milk (e.g. Cabrera-Rubio et al. 2012), as well as salivary enzymes left on the areola by the suckling infant, may promptly alter non-volatile or conjugated substrates into odour-active volatiles, giving off additional odour cues in a way that synchronizes with nursing. Finally, the vasoactive susceptibility of the subareolar vascular plexus may work as a thermal diffusion device which operation is mediated by the mother's emotional responses to seeing, hearing or smelling the infant (e.g. Vuorenkoski et al. 1969). Such co-localized elementary sources of odorous substrates and emission mechanisms are typical of the mammalian design of a chemo-communicative structure (e.g. Adams 1980; Albone 1984; Wyatt 2014), a feat that was well recognized by early German anatomists who ascribed the human breast the function of a *Duftorgan*, a scent gland (Schiefferdecker 1922; Schaffer 1937).

4 Inter-individual Variability of Areolar Odour Sources

Despite its recognized potential for semiochemical mediation, the areolar region of lactating females has attracted little systematic investigation along these lines. The first to describe it in more detail reported the morphological changes of the areola during pregnancy and came out with an endowment range of 12–20 glands (although it is not clear whether these counts are per areola or per individual; Montgomery 1837). So far, three modern studies have investigated the number and distribution of areolar glands in periparturient lactating human females.

In the first of these studies, 64 post-parturient women were intensively trained to identify and locate their areolar glands on both breasts, which they did with high reliability (Schaal et al. 2006). Every pigmented skin elevation (excluding moles) was termed 'areolar gland' (AG) without further histological enquiry (previous histological studies had indeed attested that such visible areolar protuberances, despite variability in size and pigmentation, did combine sebaceous units and lactiferous acini). All but one women in that sample bore AG (average AG number/areola: 8.9, standard deviation

(SD): 5.6; range/areola: 0–38 AG) with 83% of them having 1–15 AG per areola. In a second report on 121 new mothers (Doucet et al. 2012), digital pictures were taken from both breasts for subsequent numeration and distributional analyses of AG. In this case, the mean number of AG per breast was 10.39 (S.D.: 9.04; range: 0–48.5). The majority of these mothers (96.7%) were endowed with AG, 80.2% bearing 1–20 AG/areola (4 of them were AG-less). Finally, a third study (Seidel et al. in preparation) using the same photographic method reported the quasi-general prevalence of AG (98.5%) in a sample of 67 mothers (average number: 8.6 AG/areola, SD: 5.8; range: 0–25 AG/areola; 89.6% bearing 1–15 AG/areola). In all these studies, the AG number was unrelated with breast side, areolar surface, mother's parity and newborns sex. Further, the AG count appeared intra-individually stable between postpartum days 1 and 3.

In some mothers, a fraction of these AG was seen to emit very small amounts of a latescent secretion. In study 1 (Schaal et al. 2006), 12 out of 64 mothers had at least one secretory AG. In study 2 (Doucet et al. 2012), 34 out of 121 women had one or both breasts endowed with at least one AG emitting a noticeable secretion. This latter subgroup of 34 mothers did not differ from the rest of the sample in terms of parity and infant sex, and their secretory AG count was unaffected by breast side, postpartum timing (within 84 h post-birth) or the 24 h cycle. The rate of women bearing secretory AG might be considered low (24.8%) in view of the massive evidence of AG in the samples of studies 1 and 2. However, although visible secretions confirm the functional activity of the AG, unnoticeable oozing from more or all AG cannot be excluded. Thus, an obvious areolar fluid emission may not be necessary for the AG to influence infant responsiveness. In fact, mothers bearing secretory AG did not differ from mothers who do not bear such secretory AG on several maternal (i.e. breastfeeding frequency, latency to lactation onset) and infant variables (i.e. infant breastfeeding score) (Doucet et al. 2012).

The numeric polymorphism of AG number may be balanced by their distribution on the areolae. While AG appear to be randomly spread at first sight (Fig. 1), a closer distributional analysis on the women's samples of study 1 and 2 (Schaal et al. 2006; Doucet et al. 2012) highlights average higher density of AG on the lateral areolar quadrants, and the AG that are obviously secretory are more numerous on the upper lateral quarter (Fig. 2). The uneven distribution of AG on the areolar disc, whether secretory or not, corresponds to the typical positioning of the nursing infant's nose on the areolae. Thus, AG are in the best spatial arrangement to direct related secretions under the nursling's nose.

However, the other areolar quadrants also harbour AG, suggesting that the putative communicative role is inclusive of other functions of AG related to the optimization of sucking and nipple integrity. The hypothesis of the communicative prevalence of AG would be strengthened if their secretory output would be higher at the time of initial breastfeeds or later, right before each suckling episode, their release being potentially stimulated by antecedent infant signals. So far, only suggestive results were obtained on the latter issue, as more women tended to exhibit secretory AG on postpartum days 1–3 than on days 15 or 30 (Schaal et al. 2006).

1

2

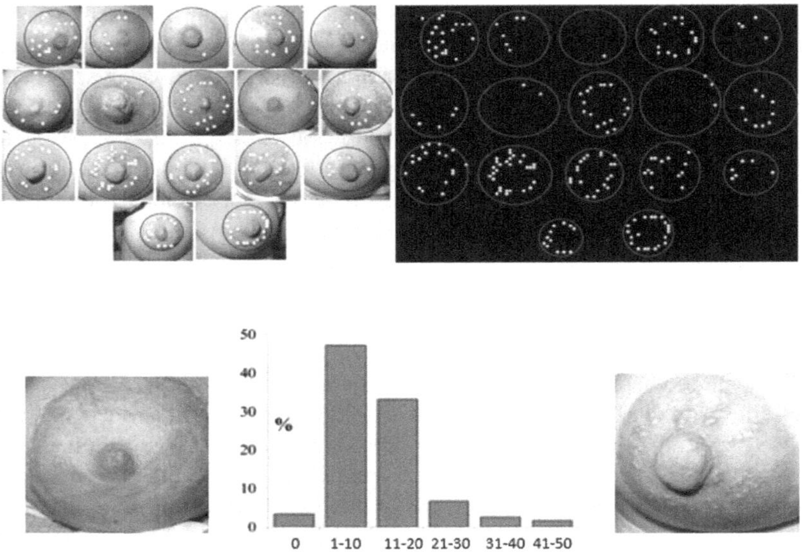

Fig. 1. **1** The variability of the areolar phenotype in a sample of human lactating breasts. Note the inter-individual variation in areolar diameter, number of areolar glands (AG) and their spatial distribution, highlighted in the black panel. Due to various scales, this illustration should be taken only as visually indicative. **2** Frequency (%) distribution of mean counts of AG per areola between birth and postpartum day 3 (sample size: 121 women). Inserts: pictures of extreme AG endowment in the present sample (left: 0 AG; right: >31 AG). After data from Doucet et al. (2007)

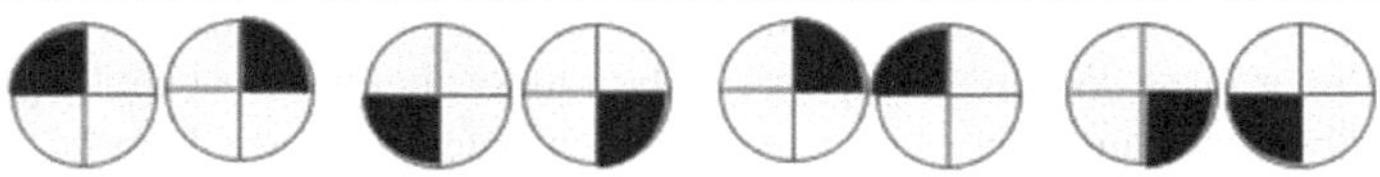

Areolar Quadrant				
AG	2.89* (2.62)	2.96* (3.22)	2.23 (2.34)	2.29 (2.75)
Secretory AG	3.74* (4.96)	1.83 (2.96)	2.33 (4.04)	1.48 (3.38)

Fig. 2. Distribution of areolar glands (AG) and of secretory AG on the areolar surface divided into 4 quadrants (upper and lower external, upper and lower internal). Mean numbers and standard deviations (SD); *$p < 0.05$. After data from Doucet et al. (2012)

5 Areolar Secretions and Their Effects on Newborns

The possible effects of mammary, including areolar, secretions have been investigated in a variety of different ways. First, a series of pioneering studies, starting with Macfarlane (1975), have drawn the big picture: providing evidence that infants detect and react to their mother's breast odour but without assessing how various sources of the lactating breast form the whole odour mixture given off. Collectively, these studies showed that a gauze pad carried in contact with the breast for several hours conveys enough 'signal value' to alter the attentional and directional responsiveness of newborn infants (Russell 1976; Schaal et al. 1980; Schaal 1986; Makin and Porter 1989; Varendi and Porter 2001).

A second series of studies attempted to separate the perceptual effects of mainstream milk within the whole breast odour mixture. Brain responses recorded through EEG or NIRS confirmed that neonates, born either term or preterm, can detect the odour of human milk or colostrum and discriminate it from orange juice (Yasumatsu et al. 1994), vanilla (Bartocci et al. 2000), or formula milk odours (Aoyama et al. 2010). But behavioural investigations further demonstrated that infants tested on postpartum days 0–14 clearly react to the odour of human milk alone by either arousal modulation, attraction, orientation or appetence displayed orally (Mizuno and Ueda 2004; Marlier and Schaal 2005; Nishitani et al. 2009). Human milk odour is behaviourally effective in neonates never exposed to it before the test, either because they were exclusively fed with a bovine milk-based formula (Marlier and Schaal 2005) or because they were born preterm (Bingham et al. 2003, 2007; Raimbault et al. 2007). Thus, mainstream milk as a separate source of odorous substrates might explain the global attraction-appetence responses of neonates to the lactating breast odour.

In an attempt to further dissociate the components of this intricate mammary odour cocktail, one study assessed the responses of neonates to the mother's breast after the different putative odour sources were selectively blocked (Doucet et al. 2007). To do so, morphologically, and hence semiochemically differentiable breast areas were overlaid with a scentless plastic film (Fig. 3). Diverse apertures made in this plastic cover were designed to separate the areolar contribution to the total breast odour from the contributions of either the nipple or fresh mainstream milk. Unrestrained 3-day-old newborns were then held close to the fully covered, fully uncovered or selectively uncovered, breast (Fig. 3). As opposed to the fully covered breast which was assumed to be odourless, the whole odour of the uncovered breast affected significantly the infants' arousal state (increased eye opening) and feeding-preparatory oral responses (mouthing, licking and sucking) (Fig. 3). However, when comparing the odour of the olfactorily intact (uncovered) breast with the "fractionated" breast odours exposing the infants only to the areola versus only to the nipple versus only to milk odours, the infants' differential responses did not reach significance (although their response to the areola-only odour appeared similar to that elicited by the whole breast odour; Fig. 3).

Thus, the present experiment failed to reveal whether a mammary subregion contributed more than another to the whole odour cocktail of the lactating breast. Either areola, nipple and milk odours convey quasi-equivalent perceptual qualities due to shared constituents [even though distinct milks (areolar versus mainstream) are involved, similar odorous constituents may be present in all considered breast subdivisions; cf. Schaal et al. 2008] or from cross-contamination (the sucking infant spreading and mixing the different milks on the entire areolar-nipple area), or did these odour "fractions" bear equivalent reward value despite being perceptually dissimilar (as shown in Delaunay-El Allam et al. 2006). Anyway, more "radically" isolated exocrine components of breast secretions and finer grained analyses are needed to further assess the source of the attractive-appetitive odour factor in the mammary odour melange.

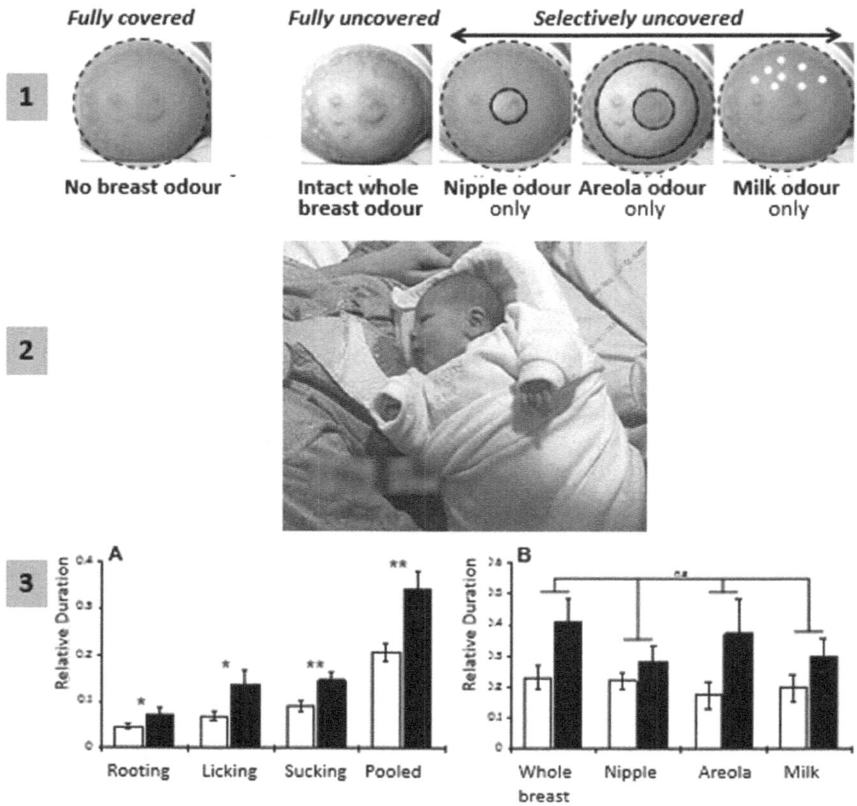

Fig. 3. Design and results of the experiment aiming to behaviourally "fractionate" the potentially discernible odour sources of the breast. **1** The distinct conditions of breast covering with a scentless plastic film and the corresponding odour assumed to be delivered to the neonate. **2** The testing situation which consisted of holding awake newborns in their habitual nursing posture in close proximity, but without contact with their mother's breast. **3A** Relative duration of infants' cephalic (i.e. rooting) and oral (i.e. licking, vacuum sucking) responses, and pooled responses, when facing the non-odorous (fully covered) breast (white bars) versus the olfactorily intact (fully uncovered) breast (black bars). B. Duration of infants' pooled head and oral responses when facing the non-odorous (fully covered) breast (all white bars) versus the whole breast odour (fully uncovered), the nipple odour only, the areolar odour only and milk odour only (all black bars). Significance: $*p < 0.05$; $**p < 0.01$; ns: non-significant. After data from Doucet et al. (2007)

Such exocrine compounds fully isolated from the breast were later assayed in exposing 3-day-old neonates to the odour of the latescent secretion taken directly from AG (Doucet et al. 2009). The native AG secretion was presented on a glass stick. AG secretion components, viz. mainstream milk and sebum from forehead skin, were also presented on a glass stick, assuming *provisionally* that areolar milk and sebum are compositionally similar to mainstream milk and frontal sebum. In addition, the above three stimuli (donated by unfamiliar lactating females) were tested for specificity against control stimuli, including water, vanilla, cow milk and cow milk-based formula. Infant reactive measures consisted in mouth–face movements, as well as variations of heart and respiratory parameters.

As a result, pure AG secretion elicited much higher rates of oral–facial responses than all other stimuli, including mainstream milk and forehead sebum (Fig. 4). This oro-facial response release was both immediate during the 10 s AG secretion presentation and lasted for the next 10 s period. In terms of responding infants among the 19 tested, 12 reacted to AG secretion, while only 5 reacted to human milk, and 4 to sebum [4 reacted to water as the blank].

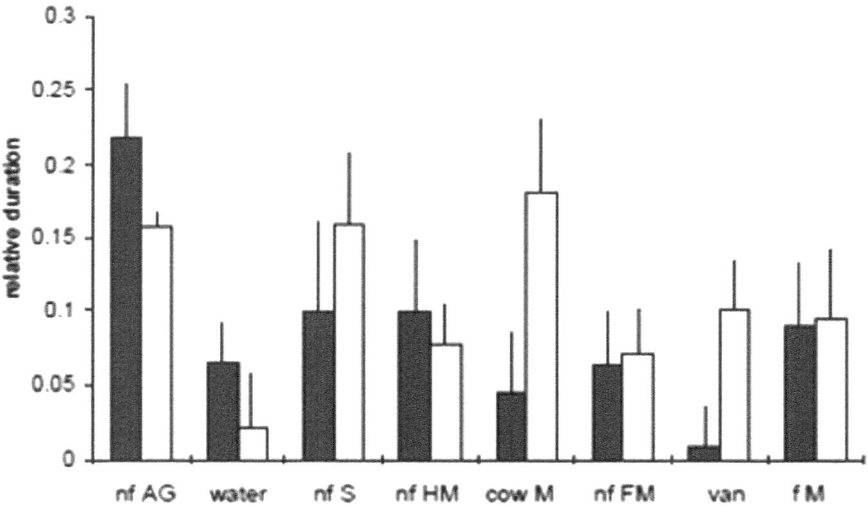

Fig. 4. Mean (± standard error) relative durations of newborns' oro-facial responses during (10 s stimulus period; black bars) and after (10 s post-stimulus period; white bars) presentation of various odour stimuli (AG: secretions of areolar glands; S: sebum; HM: human milk; cow M: cow's milk; FM: formula milk; van: vanillin; M: milk; f: familiar; nf: non-familiar; $n = 19$). Note that the areolar gland secretion (nf AG) elicits significantly more oro-facial responses than all other stimuli during the stimulus presentation period (black bars). After data from Doucet et al. (2009)

The AG secretion was also clearly more behaviourally active than the heterospecific substrates, cow's milk and unfamiliar cow's milk-based formula, and the arbitrary odorant vanilla [5, 3, and 1 responding infants among 19, respectively]. Thus, the behavioural response caused by the AG secretion cannot be interpreted as resulting from the effects of any odorant or of stimulus novelty. Finally, the AG secretion from an unrelated mother released longer oro-facial responses than the familiar milk, either maternal or formula, to which they were exposed at each feed since birth (only 3 responsive infants among 19), suggesting that the intrinsic reinforcing power of unfamiliar AG odour is much stronger than the reinforcing power attached to the familiar milk through repeated alimentary conditioning.

The isolated AG secretion also influenced the responsiveness of the autonomic nervous system. It increased the infants' respiratory rate and inspiratory amplitude more than all other stimuli, including mainstream milk and forehead sebum, the theoretically most resembling components of AG secretion. Otherwise, this effect of AG odour on the respiratory variables was not dependent on the mode of infant feeding (breast or bottle), supporting further the notion that AG odour reactogenic potency does not depend on previous exposure to a lactating breast. Thus, the isolated secretion from the AG conveys volatile compounds the effects of which on neonatal behavioural and autonomous responsiveness are stronger than mainstream milk or frontal sebum.

In sum, this study supports the hypothesis that the human AG secretion carries an odour factor that strongly impinges on human neonates. Specifically, AG odour appears more behaviourally effective than the odours of isolated human milk or sebum. Our premise that the lacteal and sebaceous fractions of AG secretions could be reflected in mainstream milk and frontal sebum may not be accurate, or, to be behaviourally effective, both fractions need to be mingled within the same mixture as is often the case in mammalian chemical signals (Albone 1984; Wyatt 2014). Importantly, in Doucet et al.'s (2009) study, the rate of infants' previous exposure to breast-related stimuli did not affect their response to the odour of AG secretion, suggesting that extensive postnatal exposure is not required for AG odour to become reactogenic. Further investigation is obviously needed to settle these questions, especially in verifying whether AG secretions and mainstream milk convey similar or distinct volatile signals to infants.

6 Adaptive Effects of Areolar Glands on the Infant and the Infant-Mother Dyad

The above ethology-inspired assays highlight that infants detect and react to the effluvium released from the breast by their own mother or by any lactating woman. They bring little insight into how these early emissive and reactive abilities are involved in adaptive processes in the context of real breastfeeds. Other ways of investigating infant responses to mammary odours may complement the aforementioned experimental approaches. They consist in recording reports from particular observers, such as mothers or midwives, who are indeed privileged witnesses of infant behaviour in the hours/days following birth. Not only do they scrutinize infants' condition quasi-continuously, but also, with adequate tutoring, they can evaluate the quality, quantity (intensity), patterning, and timing of infants' actions and interactions in the breastfeeding context. Thus, informed maternal or clinical observations may be a worthy approach to gauge the functional value of olfactory stimuli in infants' life, which is unmanageable in experimental settings. Such "intimate" reports of infant responsiveness to mammary odours have been implemented in two instances.

In the first of these reports, infants' behaviour toward the breast was related to their mothers' endowment in AG (Schaal et al. 2006). Mothers, blind to the hypotheses of the study, rated their nursing infant's behaviour through an *ad hoc* 4-item rating scale applied on days 1–3 after birth. They evaluated the infant's behaviour in terms of (i) duration of bilateral rooting movements, (ii) latching speed, (iii) avidity/intensity of sucking, and (iv) general arousal state while nursing. The outcome variable was the average of a given item's scores for each breast, and then the average over postpartum days 1–3. A global score of maternal perception of infant behaviour aggregated the separate scores of the 4 items. As a result, the mothers' AG count (irrespective of their apparent secretory status) correlated negatively with infants' rooting duration ($r = -0.61$, $n = 17$, $p < 0.05$) and positively with nipple grasping speed ($r = 0.55$, $p < 0.05$). The AG count also correlated positively with the infants' intensity of sucking ($r = 0.40$, $p < 0.05$). Finally, the global score of mother's perception of infant activity at the breast correlated positively with AG number ($r = 0.69$, $p < 0.05$).

The correlations between AG number and the above 4 items of infant behaviour were further examined daily over the first 3 postnatal days. The mean AG number correlated (i) negatively with the mother-reported frequency of rooting activity on days 1 and 2, but no more significantly on day 3 ($r = -0.56$, -0.50 and -0.36; $p < 0.01$, <0.05, and >0.05, respectively); (ii) positively with the speed of nipple grasping on days 2 and 3 ($r = 0.49$ and 0.56, $p < 0.05$); and (iii) positively with sucking intensity on day 1 ($r = 0.64$, $p < 0.01$). Thus, mothers whose areolae are richer in AG appear to report shorter latencies to their infants' nipple seizing and higher sucking activity during the daily breastfeeds, but these correlations appear to be stronger on the two first postnatal days.

A second study sought to check further the above hypothesis in assessing the relationship of AG counts with the neonates' behaviour during their earliest breastfeeds (Seidel et al. in preparation), i.e. when they initiate coping with the novel task of localizing their mother's areola-nipple and acting on it. The newborns' behaviour was assessed by a midwife using the validated Infant Breastfeeding Assessment Tool (IBFAT; Matthews, 1988), itemized along (i) alertness and readiness to feed, (ii) rooting, (iii) fixing (latching) and (iv) sucking pattern (all coded 0–3 on 4-point Likert scales). The aggregated items provided discrete IBFAT scores (range: 0–12) for each of the first 4 breastfeeds that occurred within 24 h from birth, the first one happening in the delivery room (mean time since birth: 53 ± 36 min). Then, a global IBFAT score was defined as the mean of these discrete feed-specific scores. Thus, the present study assessed whether the maternal endowment in AG was influential on the infants' performance during the first four breastfeeding episodes.

In these conditions, mothers having higher (than the median of the group = 8) AG counts reported a higher global IBFAT score as compared to mothers bearing lower AG counts [mean global IBFAT score: 9.9 ± 1.2 versus 8.8 ± 2.3, respectively; $F(1, 63) = 3.96$; $p = 0.05$]. Interestingly, the mothers' AG endowment interacted with their experience of mothering a previous child: a 2-way ANOVA considering parity and AG count yielded a marginally significant effect [$F(1, 63) = 3.36$; $p = 0.07$]. In other words, newborns from primiparous-and-lowly-AG-endowed mothers tended to evince a lower mean global IBFAT score than those of primiparous-and-highly-AG-endowed mothers and of multiparous-and-highly-AG-endowed mothers ($p = 0.03$ and 0.01, respectively). Finally, the mean global IBFAT score of newborns of multiparous-and-highly-AG-endowed mothers was not significantly different from the ones of the newborns from the other above groups. When considering how the AG count relates with infant behaviour at the separate first 4 breastfeeds, it appeared that mothers bearing higher AG number reported higher IBFAT score in their infant than mothers bearing lower AG counts at all 4 breastfeeds, but the difference was significant only on feeds 1 and 4.

These results indicate that a mother's endowment in AG relates somehow with the adaptive sucking responses of infants at the breast. However, multiple other factors qualify this relation. Among these, mothers' parity (i.e. the number of previous births) is of prominent influence (it is noteworthy that the above groups of mothers were balanced in size and infant sex ratio, as well as in newborns' birth weight and gestational age, and in mothers' age, labour duration and time to first breastfeeding). The impact of parity is indeed omnipresent in the maternal physiology and behaviour of mammals, including humans (e.g. Thoman et al. 1970; Fleming 1990; Hrdy 1999). In particular, first-time mothers appear less optimal in their assistance to the newborn, especially in the first episodes of nursing (Dewey et al. 2003; Michel et al. 2007), leading their infants to be more contingent upon their own sensitivity and behavioural efficiency to achieve successful nipple localization and latching. In this context, primiparous mothers endowed with more AG (compared with primiparae endowed with less AG) may be more efficient in their (chemo)sensory guidance to the nipple.

A logical consequence of the above results would be that, if AG activate nursing infants, they should also thrive more efficiently. Indeed, infants of mothers having higher AG counts exhibit lesser average weight loss between birth and day 3 as compared to infants of mothers with lower AG number. Thus, the amount of AG is tied to fitness consequences in terms of infants' weight regain after birth, an effect that is expectedly again qualified by parity (Schaal et al. 2006; Doucet et al. 2012): Passive (areolar odour) and active (experience-dependent care) affordances of the maternal organism interact in the infants' breastfeeding competence, modulating early postnatal body weight fluctuations. This link between maternal AG and neonatal short-term weight gain can be understood as resulting from the reciprocal action of infant's and mother's physiology/behaviour: more AG leads to more efficient sucking, which leads to extra nipple stimulation, commanding earlier onset of lactation, increased release of milk, and finally enhanced infant growth, which in turn feeds back onto previous processes. Data from two studies confirm indeed that the AG count correlates negatively with the timing of lactation stage 2 onset: the more mothers bear AG, the earlier they do evince 'coming in' of milk (Schaal et al. 2006; Doucet et al. 2012). Finally, through their reactogenic power on newborn infants and the mediation of induced sucking, chemical cues emanating from the areolar-nipple regions may orchestrate in parallel lactational physiology and adaptive effects that favour empathetic feelings and bonding on the part of mothers (Bourvis et al. 2018).

7 Conclusion: Variability and Regularity in the Human Mammary Chemosignalling System

The above account reveals the polymorphism of the human mammary phenotype regarding traits such as total number of AG, number of secretory AG, distribution of AG on the areolae, which all may participate in areolar chemosignalling. On the other side, there is also variability/regularity in the behavioural activity that these volatile compounds elicit in newborn infants.

7.1 Variability/Regularity in Areolar Chemo-Emission

Three replication studies showed that the amount and spatial patterning of AG varies between women. Despite this phenotypic variability, the majority of women in the studied samples (85–90%) bore 1 to more than 40 AG on each areola. In about 1 in 4 or 5 mothers (depending on the study), some of these AG were seen to emit a secretion in minute amounts (several tenths of μL between two breastfeeds). These secretory AG appear located 'strategically' under the nose of infants in typical nursing position (i.e. over the external/upper quadrant of the areolar disc), a coincidence suggesting their potential involvement in communication.

However, the chemosignalling service of AG may be ancillary to other basic functions encompassing integrity of areola-nipple against sucking-induced pain, soreness and corrosion, as well as creation of the seal permitting the negative pressure of sucking. Thus, if any, the communicative function of areolar chemistry must be understood in the wider context of these mechanical or protective roles. One may expect that a glandular system that assigns conjointly protective, sealant and signalling functions to the interface structure that delivers milk to offspring be of highest survival value. For the sake of species perenniality, such an interface structure should be positively selected and, hence, an optimally designed AG pattern should be well conserved across individuals.

The current state of research makes it hard to draw an optimum AG endowment in either amount, arrangement, activity or timing. Limited evidence indicates that a certain amount of AG has fitness consequences in terms of initial infant growth and the onset of stage-2 lactation. But, although AG secretions are especially reactogenic for infants, we have no insight on the number of secretory AG necessary to elicit adaptive responses *in the nursing newborn*. Probably only one, well-positioned AG would be enough to boost orientation, mouth gaping, latching and sucking. In an optimal AG design, one may also expect intra-individual phenotypic plasticity of emission of AG secretions, in the sense that their activity should be maximal right after partum in order to boost infants' responses at the initial breastfeed attempts. Such an early adaptive consequence of AG could also work on the mother's side, improving preservation of the papillary epidermis when it is maximally exposed to sucking-related stress. All these points need further observational and experimental scrutiny.

The reasons and mechanisms underlying the variability in areolar skin structure and functioning are unclear. They could rest first on methodological issues, such as our sampling of secretory AG made outside episodes of breastfeeding, their activity (and the odour-amplifying thermal properties of the areola) being partly conditional upon infant's physical presence. Another limitation of current investigation may reside in the restriction of extant studies to Caucasian participants. One cannot exclude that certain ethnic groups may express higher variability in areolar features than others (e.g. in a similar way that axillary glandular endowment differs between Asiatic, African and Caucasian populations; e.g. Rawlings 2006; Schaal and Porter 2001; Wang 2018; cf. also Prokop-Prigge et al. 2015, 2016). More distal explanations of human AG variability may be sought in an evolutionary relaxation of selective pressure on areolar structure and functions. The variable glandular endowment of the mammary outlet may indeed reflect biocultural compensation of suboptimal areolae by the enrolment of allomothering or wet nurses along hundreds of generations (e.g. Badinter 1982; Hrdy 2009; Roulin and Heeb 1999) and, since the domestication of dairy species 8,000–10,000 years ago (e.g. Vigne and Helmer 2007; Evershed et al. 2008), the use of heterospecific milks and substitutive milk-delivery devices (e.g. Delahaye 1990; Dasen 2003). Thus, we hypothesize that biocultural evolution may have buffered the adverse effects of lower extreme cases of variability of the areolar-papillary structure, especially in populations consuming dairy products collected from the cow or other mammals.

Future studies on AG and related secretions should consider comparing populations of different genetic and cultural backgrounds. Based on the above argument, one may hypothesize that non-medicalized hunting-gathering populations could have a different AG design than populations relying on dairy farming. Also, whether there is a particular AG design that is more effective for mother-to-infant chemosignalling may be addressed in studying extreme cases of AG numbers. Mothers (especially primiparae) bearing AG-defective areolae could thus be investigated in terms of infant's breastfeeding and growth performances or, on the maternal side, nipple integrity, course of lactation, delay to supplement the infant needs with nonhuman milk-feeds. Another point for future studies would be whether AG function at a higher rate (i.e. with more secretory AG) during the colostral period, when breastfeeding is initiated and nipples/areolae need both greater protection and olfactory advertisement to infants; alternatively, does AG function extend much longer than the initial days/weeks/months of lactation? Such a case of intra-individual phenotypic plasticity in mothers is of great theoretical interest as it relates to events that are both causal and consequential on the infant's activity. Finally, it would be of comparative interest to analyse whether other primate lactating females, especially anthropoids, have also evolved infant-directed chemosignalling mechanisms around, on or within their nipples.

7.2 Variability/Regularity in Neonatal Response to AG Secretion

In the so far single study to assess neonatal response to the odour of pure AG secretion (Doucet et al. 2009), 12 on 19 (63%) were reactive according to the response criterion. Although this is a notable proportion, it also points that between- and within-individual variability occurs in neonatal responsiveness to mammary odour cues. This highlights the difficulty in designing an optimal functional assay for olfaction in human newborns.

The behavioural responses elicited in neonates by the volatiles emitted by the lactating breast pertain to several categories, including attention (temporary inhibition of crying, eye opening), head orienting, whole-body approach and activation of the face, mouth, and tongue. These basic responses are not all expressed in experimental conditions that do not mimic the context of nursing. However, several studies, among which the aforementioned study by Doucet et al. (2007), show that all these items are sequentially displayed when the behavioural testing occurs in ecological conditions, i.e. when motivated infants are positioned in proximity to the breast or a representative odour stimulus (see also Varendi et al. 1994; Mizuno and Ueda 2004). However, feeding-ready infants are tricky to assess for choice or preferences, as their response repertoire promptly ends in disorganized fussing and crying (and then the ability to put an end to crying may be another criterion of efficiency for a given odour stimulus). Psychophysiological measurements (e.g. heart and respiratory parameters, metabolic or electrical brain activity) are less compatible with such ecological testing, as movements generally induce recording 'artefacts' that mitigate interpretation of autonomic or central variables. These latter variables can however be assessed in sleeping infants as indices of detection and differential processing of odour stimuli (e.g. Soussignan et al. 1997). Thus, infant odour bioassays encompass a range of behavioural measurements that are variably expressed depending on the ecological validity of the test setting. In sum, convergent ecological and non-ecological approaches are needed to establish a complete or optimal multi-bioassay for human newborns.

Another point is that inter-individual variability in responsiveness to mammary odour cues is not exceptional in mammalian newborns, even in case of species which neonates do exclusively depend on olfaction to locate a nipple (e.g. Hudson and Distel 1995; Schaal et al. 2003; Schaal 2010). For example, 8.7 and 7.2% of rabbit newborns do not respond when experimentally exposed to the mammary pheromone on post-partum days 1 and 3, respectively (Coureaud et al. 2007), although they are critically dependent on it to survive. In contrast to rabbit neonates that are exclusive olfacters (their auditory and visual tracts being not functional), sensorily precocial human infants perceive the environment through olfaction in the same time as through the other sensory systems including audition and vision (Schaal and Durand 2012; Durand et al. 2013). Thus, the early multisensory ability that is typical of human infants may be another inherent source of response variability to areolar odour cues.

In sum, future studies are needed to replicate human neonatal responses to areolar or lacteal chemosignals, using multiple indices of attention, motivation and selective appetence (see also Loos et al. 2019). One may expect that species with precocial newborns, like those of humans who have a complete, although immature sensorium, are more variable in their responsiveness to such maternal cues. *In fine*, the plasticity range of sensory drivers of neonatal behaviour may be a determining factor of the structure of the species-specific maternal milk-transfer interface.

Another point of interest for future studies would be the impact of AG on infants beyond the colostral period: do they influence infant behaviour only during a limited postnatal window ensuring the transition until infants engage other sensory or cognitive adaptive means, or do they provide a prolonged stimulating effect during weeks or months of breastfeeding? In addition, do AG secretions have a role in the maintenance of sucking activity during the whole breastfeeding episode rather than only guiding the infant to the nipple and stimulating onset of sucking?

In any case, the areolar polymorphism of human lactating females evidenced here reveals the considerable plasticity of human infants to adapt to their nursing mother's breast. One point will be to establish how an earlier maturing sensory modality such as olfaction (relative to later-maturing senses like vision) contributes to this initial plasticity in the ways of acquiring the life-saving colostrum and milk. Another point relates to current theoretical debates relating inter-individual variability of a bodily feature and the function of it, with the frequent postulate that high morphological variability of a feature implies a lack of function or of selection thereof (e.g. Kelly et al. 2018). In contrast, Houle (1992) suggests that morphological traits supposed to be under strong selection tend to exhibit more variability than those assumed to be under weak selection. If this principle is relevant for the phenotypic variability of physiological or behavioural traits, the interindividually variable endowment in skin glands' of the human areola during lactation may relate to functionality and strong selection for functionality.

Acknowledgements. The Centre National de la Recherche Scientifique (CNRS), the Regional Council of Bourgogne-Franche-Comté, and ANR grant "Milkodor" 15-CE21-0009-01 (to BS) supported the writing of this chapter. The authors gratefully thank Tristram D. Wyatt and Elisabeth III Hertling for useful comments on an earlier draft.

References

Abuidhail J (2014) Rural Jordanian mothers' beliefs, knowledge and practices of postnatal care. Qual Primary Care 22:285–293

Adams MG (1980) Odour-producing organs of mammals. Symp Zool Soc London 45:57–86

Albone E (1984) Mammalian semiochemistry: the investigation of chemical signals between mammals. Wiley, Chichester

Aoyama S, Toshima T, Saito Y, Konishi N, Motoshige K, Ishikawa, Nakamura K, Kobayashi M (2010) Maternal breast milk odour induces frontal lobe activation in neonates: A NIRS study. Early Hum Dev 86:541–545

Badinter E (1982) Myth of motherhood: an historical view of the maternal instinct. Souvenir Press Ltd., London

Banapurmath CR, Nagaraj MC, Banapurmath S, Kesaree N (1996) Breastfeeding practices in villages of central Karnataka. Indian Pediatr 33:477–479

Barkow JH, Hallett AL (1989) The denial of colostrum. In: Barkow JH (ed) Darwin, sex and status: biological approaches to mind and culture. Toronto University Press, Toronto, pp 301–309

Bartocci M, Winberg J, Ruggiero C, Bergqvist LL, Serra G, Lagercrantz H (2000) Activation of olfactory cortex in newborn infants after odor stimulation: a functional near-infrared spectroscopy study. Pediatr Res 48:18

Bingham PM, Abassi S, Sivieri E (2003) A pilot study of milk odor effect on nonnutritive sucking by premature newborns. Arch Pediatr Adol Med 157:72–75

Bingham PM, Churchill D, Ashikaga T (2007) Breast milk odor via olfactometer for tube-fed, premature infants. Behav Res Meth 39:630–634

Bourvis N, Singer M, Saint Georges C, Bodeau N, Chetouani M, Cohen D, Feldman R (2018) Pre-linguistic infants employ complex communicative loops to engage mothers in social exchanges and repair interaction ruptures. Royal Soc Open Sci 5:170274

Burd LI, Dorin M, Philipose V, Lemons JA (1977) The relationship of mammary temperature to parturition in human subjects. Am J Obstet Gynecol 128:272275–273278

Burton JL, Shuster S, Cartlidge M, Libman LJ, Martell U (1973) Lactation, sebum excretion and melanocyte-stimulating hormone. Nature 243:349–350

Cabrera-Rubio R, Collado MC, Laitinen K, Salminen S, Isolauri E, Mira A (2012) The human milk microbiome changes over lactation and is shaped by maternal weight and mode of delivery. Am J Clin Nutr 96:544–551

Candau J, Schaal B (2015) Human neonatal exposure to colostrum as a biocultural experience. In: Talk at the 48th annual meeting of the international society for developmental psychobiology, San Sebastian, Spain, 20–23 July 2015

Cantrill RM, Creedy DK, Cooke M, Dykes F (2014) Effective suckling in relation to naked maternal-infant body contact in the first hour of life: an observation study. BMC Pregn Childbirth 14:20

Cernoch JM, Porter RH (1985) Recognition of maternal axillary odors by infants. Child Dev 56:1593–1598

Christensson K, Sileses C, Moreno L, Belaustequi A, De La Fuente P, Lagercrantz H (1992) Temperature, metabolic adaptation and crying in healthy newborns cared for skin-to-skin or in cot. Acta Paediatr Scand 8:488–503

Coureaud G, Fortun-Lamothe L, Langlois D, Schaal B (2007) The reactivity of neonatal rabbits to the mammary pheromone as a probe for viability. Animal 1:1026–1032

Dasen V (2003) Naissance et petite enfance dans l'Antiquité. Editions Universitaires & Göttingen, Vandenhoek-Ruprecht Verlag, Fribourg

Debes AK, Kohli A, Walker N, Edmond K, Mullany LC (2013) Time to initiation of breastfeeding and neonatal mortality and morbidity: a systematic review. BMC PublHealth 13 (Suppl 3):S19

Delahaye MC (1990) Tétons et tétines, Histoire de l'allaitement. Editions Trame-Way, Paris

Delaunay-El Allam M, Marlier L, Schaal B (2006) Learning at the breast: preference formation for an artificial scent and its attraction against the odor of maternal milk. Infant Behav Dev 29:308–321

Dewey KG, Nommsen-Rivers LA, Heinig MJ, Cohen RJ (2003) Risk factors for suboptimal infant breastdeefing behavior, delayed onset of lactation, and excess neonatal weight loss. Pediatrics 112:607–619

Doucet S, Soussignan R, Sagot P, Schaal B (2007) The 'smellscape' of mother's breast: effects of odour masking and selective unmasking on neonatal arousal, oral and visual responses. Dev Psychobiol 49:129–138

Doucet S, Soussignan R, Sagot P, Schaal B (2009) The secretion of areolar (Montgomery's) glands from lactating women elicits selective, unconditional responses in neonates. PLoS ONE 4:e7579

Doucet S, Soussignan R, Sagot P, Schaal B (2012) An overlooked aspect of the human breast: Areolar glands in relation with breastfeeding pattern, neonatal weight gain, and dynamics of lactation. Early Hum Dev 88:119–128

Durand K, Baudouin JY, Lewkowicz D, Goubet N, Schaal B (2013) Eye-catching odors: familiar odors promote attention and sustained gazing to faces and eyes in 4 month-old infants. PLoS ONE 8:e7067

Edmond KM, Kirkwood BR, Amenga-Etego S, Owusu-Agyei S, Hurt LS (2007) Effect of early infant feeding practices on infection-specific neonatal mortality: an investigation of the causal links with observational data from rural Ghana. Am J Clin Nutr 86:1126–1131

Edmond KM, Zandoh C, Quigley MA, Amenga-Etego S, Owusu-Agyei S, Kirkwood BR (2006) Delayed breastfeeding initiation increases risk of neonatal mortality. Pediatrics 117:e380–e386

Evershed RP, Payne S, Sherratt AG, Copley MS, Coolidge J, Urem-Kotsu D, … Akkermans PM (2008) Earliest date for milk use in the near East and Southeastern Europe linked to cattle herding. Nature 455(7212):528

Fleming AS (1990) Hormonal and experiential correlates of maternal responsiveness in human mothers. In: Krasnegor NA, Bridges RS (eds) Mammalian parenting: biochemical, neurobiological and behavioral determinants. Oxford University Press, New York, pp 184–208

Girish M, Mujawar N, Gotmare P, Paul N, Punia S, Pandey P (2013) Impact and feasibility of breast crawl in a tertiary care hospital. J Perinatol 33:288–291

Hale TW, Hartmann PE (eds) (2007) Textbook of human lactation. Halle Publishing, Amarillo, TX

Houle D (1992) Comparing evolvability and variability of quantitative traits. Genetics 130: 195–204

Hrdy SB (1999) Mother nature. Pantheon Press, New York

Hrdy SB (2009) Mothers and others. The evolutionary origins of mutual understanding. Belknap-Harvard University Press, Cambridge, MA

Hudson R, Distel H (1995) On the nature and action of the rabbit nipple-search pheromone: a review. In: Apfelbach R, Müller-Schwarze D, Reuter K, Weiler E (eds) Chemical signals in vertebrates, vol 7. Elsevier Science, London, pp 223–232

Kelly AJ, Dubbs SL, Barlow FK, Zietsch BP (2018) Male and female nipples as a test case for the assumption that functional features vary less than nonfunctional by-products. Adaptive Hum Behav Physiol 4:344–353

Kimura C, Matsuoka M (2007) Changes in breast skin temperature during the course of breastfeeding. J Hum Lact 23:60–69

Kinney MV, Kerber KJ, Black RE, Cohen B, Nkrumah F et al (2010) Sub-Saharan Africa's mothers, newborns, and children: where and why do they die? PLoS Med 7(6):e1000294

Klaus M (1998) Mother and infant: early emotional ties. Pediatrics 102(Suppl E1):1244–1246

Koyama S, Wu HJ, Easwaran T, Thopady S, Foley J (2013) The nipple: a simple intersection of mammary gland and integument, but focal point of organ function. J Mammary Gland Biol Neopl 18:121–131

Lawn JE, Cousens S, Zupan J, Team NSS (2005) 4 million neonatal deaths: when? where? why? Lancet 365:891–900

Loos HM, Reger D, Schaal B (2019) The odour of human milk: its chemical variability and detection by newborns. Physiol Behav 199:88–99

Loos HM, Doucet S, Védrines F, Sharapa C, Soussignan R, Durand K, Büttner A, Schaal B (2017) Responses of human neonates to highly diluted odorants from sweat. J Chem Ecol 43:106–117

Macfarlane AJ (1975) Olfaction in the development of social preferences in the human neonate. Ciba Found Symp 33:103–117

Makin JW, Porter RH (1989) Attractiveness of lactating female's breast odors to neonates. Child Dev 60:803–810

Marlier L, Schaal B (2005) Human newborns prefer human milk: conspecific milk odor is attractive without postnatal exposure. Child Dev 76:155–168

Matthews MK (1988) Developing an instrument to assess infant breastfeeding behaviour in the early neonatal period. Midwifery 4:154–165

Michel MP, Gremmo-Féger G, Oger E, Sizun J (2007) Étude pilote des difficultés de mise en place de l'allaitement maternel des nouveau-nés à terme, en maternité: incidence et facteurs de risque. Arch Pédiatr 14:454–460

Mizuno K, Ueda A (2004) Antenatal olfactory learning influences infant feeding. Early Hum Dev 76:83–90

Montagna W, MacPherson EE (1974) Some neglected aspects of the anatomy of human breasts. J Invest Dermatol 63:10–16

Montagna W, Yun JS (1972) The glands of montgomery. Br J Dermatol 86:126–133

Montgomery WF (1837) An exposition of the signs and symptoms of pregnancy, the period of human gestation, and signs of delivery. Sherwood, Gilber and Piper, London

Morgagni GB (1719) Adversaria Anatomica Omnia. A.J.A. Langerak, *Lugduni Batavorum*. https://archive.org/details/joannisbaptistae1723morg/page/n7

Naeslund J (1957) The function of montgomery's tubercles. Acta Obst Gynecol Scand 36: 460–467

Nishitani S, Miyamura T, Tagawa M, Sumi M, Takase R, Doi H, Moriushi H, Shinohara K (2009) The calming effect of a maternal breast milk odor on the human newborn infant. Neurosci Res 63:66–71
Odent PLG (2010) Early infant feeding and neonatal survival in Nepal: breastfeeding, colostrum and discarding of the first milk. PhD thesis, University College, London
Perkins OC, Miller AM (1926) Sebaceous glands in the human nipple. Am J Obstet Gynecol 11:789–794
Porter RH, Makin JW, Davis LB, Christensen KM (1991) An assessment of the salient olfactory environment of formula-fed infants. Physiol Behav 50:907–911
Porter RH, Makin JW, Davis LB, Christensen KM (1992) Breast-fed infants respond to olfactory cues from their own mother and unfamiliar lactating females. Infant Behav Dev 15:85–93
Prechtl HFR (1958) The directed head turning response and allied movements of the human baby. Behaviour 13:212–242
Prokop-Prigge KA, Greene K, Varallo L, Wysocki CJ, Preti G (2016) The effect of ethnicity on human axillary odorant production. J Chem Ecol 42:33–39
Prokop-Prigge KA, Mansfield CJ, Parker MR, Thaler E, Grice EA, Wysocki CJ, Preti G (2015) Ethnic/racial and genetic influences on cerumen odorant profiles. J Chem Ecol 41:67–74
Raimbault C, Saliba E, Porter RH (2007) The effect of the odour of mother's milk on breastfeeding behaviour of premature neonates. Acta Paediatr 96:368–371
Rawlings AV (2006) Ethnic skin types: are there differences in skin structure and function? Int J Cosmet Sci 28:79–93
Regnier FE, Goodwin M (1977) On the chemical and environmental modulation of the pheromone release from vertebrate scent marks. In Müller-Schwarze DM, Mozzell MM (eds) Chemical signals in vertebrates, vol 1. Plenum Press, pp 115–133
Reissland N, Burghart R (1988) The quality of a mother's milk and the health of her child: Beliefs and practices of the women of Mithila. Soc Sci Med 27:461–469
Righard L, Alade MO (1990) Effect of delivery room routines on success of first breast-feed. Lancet 336:1105–1107
Roulin A, Heeb P (1999) The immunological function of allosuckling. Ecol Lett 2:319–324
Russell MJ (1976) Human olfactory communication. Nature 260:520–522
Schaal B (1986) Presumed olfactory exchanges between mother and neonate in humans. In Le Camus J, Cosnier J (eds) Ethology and psychology. Privat, I.E.C, Toulouse, pp 101–110
Schaal B (1988) Olfaction in infants and children: developmental and functional perspectives. Chem Senses 13:145–190
Schaal B (2010) Mammary odor cues and pheromones: mammalian infant-directed communication about maternal state, mammae, and milk. Vitam Horm 83:81–134
Schaal B, Al Aïn S (2014) Chemical signals 'selected for' newborns in mammals. Anim Behav 97:289–299
Schaal B, Durand K (2012) The role of olfaction in human multisensory development. In: Bremner A, Lewkowicz D, Spence C (eds) Multisensory development. Oxford University Press, Oxford, UK, pp 29–62
Schaal B, Porter RH (1991) Microsmatic humans revisited: the generation and perception of chemical signals. Adv Study of Behav 20:135–200
Schaal B, Doucet S, Sagot P, Hertling E, Soussignan R (2006) Human breast areolae as scent organs: morphological data and possible involvement in maternal-neonatal co-adaptation. Dev Psychobiol 48:100–110
Schaal B, Montagner H, Hertling E, Bolzoni D, Moyse R, Quichon R (1980) Les stimulations olfactives dans les relations entre l'enfant et la mère. Reprod Nutr Dev 20:843–858
Schaal B, Coureaud G, Langlois D, Ginies C, Sémon E, Perrier G (2003) Chemical and behavioural characterization of the mammary pheromone of the rabbit. Nature 424:68–72

Schaal B, Doucet S, Soussignan R, Francke W, Weibchen G, Rietdorff M (2008) The human breast as a scent organ: structures, secretions, chemistry and possible functions in mother-infant interactions. In: Hurst JL, Beynon RJ, Roberts SC, Wyatt TD (eds) Chemical signals in vertebrates, vol 11. Berlin, Springer, pp 325–335

Schaffer J (1937) Die Duftorgane des Menschen. Wiener Klin. Wochenschr. 20:790–796

Schiefferdecker P (1922) Die Hautdrüsen des Menschen und der Säugetiere. Zoologica 27:1–154

Smith JD, Peters TG, Donegan WL (1982) Montgomery's areolar tubercle. A light microscopic study. Arch Pathol Lab Med 106:60–63

Soussignan R, Schaal B, Marlier L, Jiang T (1997) Facial and autonomic responses to biological and artificial olfactory stimuli in human neonates: re-examining early hedonic discrimination of odors. Physiol Behav 62:745–758

Stolnicu S (2018) Histology of the normal breast, normal changes, and abnormalities of breast development. In: Practical atlas of breast pathology. Springer, Cham, pp 1–25

Thoman EB, Turner AM, Leiderman PH, Barnett CR (1970) Neonate-mother interactions: effects of parity on feeding behavior. Child Dev 41:1103–1111

Uruakpa FO, Ismond MA, Akobundu EN (2002) Colostrum and its benefits: a review. Nutr Res 22:755–767

Varendi H, Porter RH (2001) Breast odour as the only maternal stimulus elicits crawling towards the odour source. Acta Paediatr 90:372–375

Varendi H, Porter RH, Winberg J (1994) Does the newborn baby find the nipple by smell? Lancet 344(8928):989–990

Vigne JD, Helmer D (2007) Was milk a "secondary product" in the old world Neolithisation process? Its role in the domestication of cattle, sheep and goats. Anthropozoologica 42:9–40

Volk AA, Atkinson JA (2013) Infant and child death in the human environment of evolutionary adaptation. Evol Hum Behav 34:182–192

Vorherr H (1974) The breast. Morphology, physiology, and lactation. Academic Press, New York

Vuorenkoski V, Wasz-Hockert O, Koivisto E, Lind J (1969) The effect of cry stimulus on the temperature of the lactating breast of primipara. Experientia 25:1286–1287

Wang S (2018) Understudied skin characteristics awaiting genetic breakthroughs. J Invest Dermatol 19:S101–S102

Widström AM, Wahlberg V, Matthiesen AS (1990) Short-term effects of early suckling and touch of the nipple on maternal behaviour. Early Hum Dev 21:153–163

Widström AM, Ransjö-Arvidson AB, Christensson K, Matthiesen AS, Winberg J, Uvnäs-Moberg K (1987) Gastric suction in healthy newborn infants effects on circulation and developing feeding behaviour. Acta Paediatr 76:566–572

Widström AM, Lilja G, Aaltomaa-Michalias P, Dahllöf A, Lintula M, Nissen E (2011) Newborn behaviour to locate the breast when skin-to-skin: a possible method for enabling early self-regulation. Acta Paediatr 100:79–85

Winberg J (2005) Mother and newborn baby: mutual regulation of physiology and behaviour—a selective review. Dev Psychobiol 47:217–229

Wyatt TD (2014) Pheromones and animal behaviour: chemical signals and signature mixtures, 2nd edn. Cambridge University Press, Cambridge, UK

Yasumatsu K, Uchida S, Sugano H, Suzuki T (1994) The effect of the odour of mother's milk and orange on the spectral power of EEG in infants. J UOEH 16:71–83

Zanardo V, Straface G (2015) The higher temperature in the areola supports the natural progression of the birth to breastfeeding continuum. PLoS ONE 10:e0118774

Zanardo V, Volpe F, Luca F, Straface G (2017) A temperature gradient may support mother-infant thermal identification and communication in the breast crawl from birth to breastfeeding. Acta Paediatr 106:1596–1599

Cross-Cultural Approaches to Better Understand Chemical Communication in Humans

Camille Ferdenzi(✉)

Centre de Recherche En Neurosciences de Lyon, CNRS UMR 5292, INSERM U1028, Université Claude Bernard Lyon 1, Centre Hospitalier Le Vinatier, 95 Bd Pinel, 69675 Bron Cedex, France
camille.ferdenzi@cnrs.fr

Abstract. Social communication is one of the major functions of olfaction. Body odors appear to be extremely important in human interpersonal relationships, as shown by the amount of energy we spend every day to control our own "olfactory image", but also by the serious consequences of olfactory loss on the quality of social interactions. However, many aspects of the social function of olfaction remain as a mystery in humans. Among the many different directions that should be explored to increase our knowledge in this domain, cross-cultural comparisons through the study of differences (cultural specificities) and similarities (universal features) seem to be an interesting option. Universal abilities to perceive and emit one or several particular compounds believed to be involved in a given aspect of chemical communication may indicate a particular relevance at the level of the human species. Of course, there might also be some variations according to culturally shaped differences in rules of social interactions in space (theory of proxemics), and, by extension, in attitudes toward body odors.

1 The Importance of Social Odors in Human Relationships

The human sense of smell is one of the least explored senses for which there is still much to discover. This intuitive and emotional sense has often been considered as hardly "intellectual" and has therefore received little esteem, because it brings us back to our animality (Le Guérer 2002). In addition, work in neuroanatomy and genetics has suggested that progressive degradation of olfactory function has occurred during evolution and that, compared to other species known for their developed sense of smell (e.g., dogs, rats: Quignon et al. 2005; Wackermannová et al. 2016), humans would have limited olfactory abilities (Schaal and Porter 1991). Today, this belief is largely questioned, thanks to the relatively recent multiplication of works showing the fine performance of the human sense of smell. For example, we are perfectly able to follow an olfactory track, similar to the ability of dogs (Porter et al. 2007), and we are potentially able to discriminate more than one trillion different odors (Bushdid et al. 2014). At present, poor human olfaction is just an old myth debunked (McGann 2017).

© Springer Nature Switzerland AG 2019
C. D. Buesching (Ed.): *Chemical Signals in Vertebrates 14*, pp. 139–152, 2019.
https://doi.org/10.1007/978-3-030-17616-7_11

Considering all the everyday contexts where our olfactory sense steps in that have been targeted by experimental studies, at least three major functions of olfaction in humans can be identified (Stevenson 2010): regulation of food intake, alarm or detection of danger (fire, gas leak: Santos et al. 2004), and social communication. The latter function, which includes emotional contagion, parent–child relationships, and reproduction/partner choice, remains the least well known (see next section). There is empirical evidence though that odors biologically produced by the body are not just some unwelcome emissions that we try to remove or mask with cosmetic products: they are important to us and meaningful in our daily interactions with others, especially the ones close to us. This can be appreciated at the societal level, for example, through the growing number of mainstream initiatives around human smells: in artistic performances (olfactory dating https://smell.dating, performance experiments https://etaturgent.wordpress.com/love-sweat-love), commercial applications (like the recreation of the scent of a loved one http://kalain.fr), as well as in the literary http://www.nez-larevue.fr and cinematographic http://www.jupiter-films.com/film-odorat-l-57.php entertainment industry.

Beyond these noteworthy social trends, scientific studies provide solid reasons to believe that there is chemical communication through biological odors in humans. For instance, biological odors may be involved in mother–infant communication: mothers' areolar glands produce secretions that are attractive to newborns (Doucet et al. 2009) and that are positively associated with breastfeeding success (Doucet et al. 2012). In adults, it has been shown that body odors are involved in the communication of emotions. Namely, the responses of receivers to the odors collected during the experience of fear (skydiving, watching scary videos or before an academic exam for example) differ from those to body odors collected in a control state (for a meta-analysis, see de Groot and Smeets 2017): amydgala activation in an fMRI study and sharper emotion detection in ambiguous faces (Mujica-Parodi et al. 2009), recruitment of more neuronal resources in an EEG study (Pause et al. 2010) and higher facial muscle activity indicating negative affect (de Groot et al. 2014). The exploration of the positive emotion communication is, however, in its early stages (de Groot et al. 2015). In the context of the choice of a partner, people declare that body odor is a key factor in a person's attractiveness when they are interviewed about the importance of several sensory and behavioral characteristics of a potential partner (Herz and Inzlicht 2002). Being unable to perceive these odors (patients with olfactory disorders) has serious adverse effects on the quality of social interactions (i.e., insecurity about personal hygiene, social isolation and social anxiety, lower number of sexual partners and lower partnership security; see review in Croy et al. 2014). Additionally, preferences have been shown for biological odors that signal the quality of a partner. For instance, people tend to prefer MHC-dissimilar mates based on their body odor (Wedekind et al. 1995; Chaix et al. 2008), which is interpreted as a mechanism favoring the production of MHC-heterozygous offspring more resistant to pathogens. Information about health status is also communicated through human body odor: compared to healthy donors, the odor of sick ones is more aversive to conspecifics (Olsson et al. 2014) and activates more intensely brain networks involved in odor processing (and in multisensory integration; Regenbogen et al. 2017). Fertility status is another relevant information in partner choice. It seems to be advertised by body odor, as shown by men's preference for odor samples collected in women during the fertile phase of their menstrual cycle

(Singh and Bronstad 2001; Kuukasjarvi et al. 2004). Finally, recent works suggest that chemical signals in tears and in the secretions of the sweat glands on our hand influence our social behavior. More specifically, sexual arousal of men decreases when they are exposed to the odor women's tears (Gelstein et al. 2011). In addition, men and women sniff their own hands more after hand-shaking (Frumin et al. 2015), but in the current state of knowledge, this behavior remains difficult to link with partner choice mechanisms. Nevertheless, the latter studies open stimulating perspectives with these almost unexplored odor sources of the human body.

2 Limits in the Study of Human Chemical Communication

Anyone interested in the implication of odors in human interactions and especially human reproduction may soon realize that we have only very partial knowledge of it, for three main reasons.

First, the fascination with pheromones has given rise to significant drifts. After the concept of pheromones[1] appeared (Karlson and Lüscher 1959), such substances have been identified in several species including mammals (Schaal et al. 2003; Roberts et al. 2010), and at the same time researchers have speculated about the existence of sexual pheromones in humans without rigorously demonstrating it. Many studies in humans have focused on androgen steroids, such as androstadienone, but this focus—to the detriment of hundreds of other components of human biological odor—obviously lacks scientific legitimacy (Wyatt 2015). Indeed, studies examining the effects of these androgenic steroids on human behavior have been widely criticized for their methodological weaknesses such as lack of satisfactory control condition or replication of results (Doty 2010); the positive effect of these male compounds on the attractiveness of men evaluated by women is not very repeatable (Lundström and Olsson 2005; Saxton et al. 2008). The effects are also non-sex-specific (Ferdenzi et al. 2016b) where sex pheromones would be expected to exhibit clear sex specificity. Neurobiological approaches also challenge this sex specificity, for example for the activation of the hypothalamus (Savic et al. 2001; Burke et al. 2012) or the psychophysiological responses to these molecules (Bensafi et al. 2004). In sum, whether the existence of pheromones (sexual or other) in humans is presumed or not, everything remains to be discovered in this field.

Second, knowledge of the nature of human biological odors remains incomplete. Human body odor is a complex mixture of hundreds of volatile compounds, produced by the skin microbiome (particularly *Corynebacteria*, *Proprionibacteria*, *Staphylococcus*) through the transformation of precursors secreted by excretory skin glands (Leyden et al. 1981; Wilke et al. 2007). These volatile compounds are predominantly alcohols, carboxylic acids, ketones, aldehydes, esters, and hydrocarbons. Human body odor both has a fixed and a variable fraction. The fixed fraction is genetically determined as indicated by twin studies (Roberts et al. 2005; Kuhn and Natsch 2009) and is

[1] Substances which are secreted to the outside by an individual and received by a second individual of the same species, in which they release a specific reaction, for example, a definite behavior or a developmental process (Karlson and Lüscher 1959).

used as a marker of individuality (Penn et al. 2007) with, for example, applications in forensic science (Marchal et al. 2016). The variable fraction in contrast depends on the individual's diet (Havlicek and Lenochova 2006; Fialova et al. 2016), health status (Olsson et al. 2014), age (Mitro et al. 2012) and, in women, also on the phase of their menstrual cycle (Singh and Bronstad 2001). Most of the studies cited above use perceptual approaches, and overall, there is very little evidence on the chemical bases of these variations. Explorations of the human scent using analytical chemistry are scarce and they often respond to very specific issues. These are mainly commercial, in the field of cosmetics (identifying the compounds responsible for body malodor with the aim to neutralize them: Haze et al. 2001; Starkenmann et al. 2005; Natsch et al. 2005), or public health issues (limiting the spread of diseases such as malaria, carried by insects that are attracted to the human smell: Dormont et al. 2013). It must be acknowledged that the complexity and variability of human body odor makes it difficult to study, and that there are additional debates about sampling and/or chemical analysis methods (Curran et al. 2005, 2006; Preti et al. 2006). One last point must be emphasized here: the knowledge of the human odor is mainly confined to one particular production area: the armpit. This limitation is explained in part by cosmetic motives, as mentioned above, but also by the fact that this region of the body is one of the richest in apocrine glands (sources of odor), is little contaminated by exogenous odors in comparison with other more exposed parts of the body, and is easy (and discreet) to sample. However, other sources of odors than the armpit are relevant to study, including areas of the body that are rich in sebaceous glands (Stoddart 1990; Wyatt 2015) that are so far almost completely ignored. The involvement of these glands in chemical communication in humans is, however, likely and has been suggested in mother–child interactions (Doucet et al. 2012).

Third, although an interdisciplinary approach is necessary to study the role of odors in human interactions (Wyatt 2015), it is most often lacking. Indeed, to increase our knowledge of human body smells, and to understand how they are integrated into social interactions, research needs to combine several areas of expertise, like: (1) the psychology of perception, to characterize how human odors—or some of their molecular components—are perceived quantitatively (detection threshold) and qualitatively (description); (2) evolutionary biology and psychology, in order to make assumptions about the adaptive functions of odors in the human species; (3) analytical and synthetic chemistry, to characterize the molecules that constitute the human odor and possibly synthesize some of them to test their effect on human behavior; (4) affective neuroscience, to identify the neural circuits recruited in the processing of such sensory stimuli that have a strong biological significance; (5) social cognition, to determine how olfaction contributes—on a perceptual level—to more general processes that allow us making sense of people; finally, (6) the methodology of collection/storage/use of body odors, since they form a category of odors quite particular given their source (human) and their instability (related to bacterial activity). Combining these different areas of expertise is a challenge but is necessary to develop our knowledge of the human odor and to better understand chemical communication in humans.

3 How Could Cross-Cultural Approaches Increase Our Understanding of Chemical Communication?

Although there is many different directions to explore to clarify the role of human biological odor in social interactions, here I choose to develop one specific perspective: the cross-cultural approach. Comparisons between populations of different countries and different cultural backgrounds allow identifying differences (or cultural specificities) and similarities (or universal features), each of these being useful to test hypotheses regarding the conservation of perceptual/emission characteristics in the human species and their variations, as detailed hereafter.

3.1 Cross-Cultural Study of Odor Perception

Cultural variations are, with gender, one of the major factors of interindividual variability in olfactory perception (Herz 2009). Historically, two reference studies (Pangborn et al. 1988; Wysocki et al. 1991) investigated on a very large-scale hedonic perception of smells, and found that higher pleasantness ratings were given in countries where the odors were encountered more frequently or were associated with more positively connoted contexts (products or uses); a certain degree of consensus between cultures can be found, on the contrary, for the negative evaluation of the smells recalling decaying organic matter, feces and body odors (European, Asian, and American populations: Schleidt et al. 1988; Schaal et al. 1997). More recently, using a verbal measure of odor-related affective feelings—which is much more comprehensive than just hedonicity—the Emotion and Odor Scales (Chrea et al. 2009; Ferdenzi et al. 2013), we confirmed some of the previous results. For example, more positive feelings were associated with odors (e.g., wintergreen) in cultures where they are more positively connoted (present in candies and sodas in North America versus muscle balms in Europe). Also, the similarity in affective responses generally reflected geographic proximity. But most importantly, there were cultural variants and invariants in the very nature of the words chosen as relevant to describe odor-elicited feelings by participants of the six studied geographic areas (Switzerland, United Kingdom, United States, Brazil, China, and Singapore). Indeed, several affective categories were recurrent (Disgust/Irritation, Happiness/Well-being, Sensuality/Desire, Energy, Soothing/Peacefulness and Hunger/Thirst), probably because they are linked with adaptive universal functions of olfaction such as danger avoidance, ingestion, and social communication. Culture-specific dimensions also turned up, like Nostalgia in Brazil and Spirituality in Singapore, which were directly related with rituals and habits in these cultures (for more details, see Ferdenzi et al. 2011, 2013).

In this context, cultural comparisons of verbal descriptions of body odors or of feelings elicited by body odors remain rare (Schleidt et al. 1981, 1988). Yet, tools are now available to measure in a more sophisticated way (i) body odor perception, with a specific vocabulary suited to this category of smells (fruity, acid, spicy, musky, etc., see Wedekind et al. 2007; Allen et al. 2018) and (ii) affective responses to body odors, with the above described emotional scales (Ferdenzi et al. 2013). The latter scales have already shown that Sensuality/Desire is a recurrent feeling that people from several countries over the world consider relevant to describe their affective feelings in response to smell. More detailed affective profiles could be analyzed in the future to describe how cultures differ or are similar in their affective responses to body odors—with the help of the culture-specific or the "universal" UniGEOS scales (Ferdenzi et al. 2013), or even the shortened version ScentMove® (Porcherot et al. 2010). Anthropological studies suggest a strong modulation of body odor perception by social cognition (Ferdenzi et al. 2016a), especially in the relationships between groups with different cultural, ethnic, or religious backgrounds: expectations, beliefs, and stereotypes are leading to negative attitudes toward the odor individuals from what is called the outgroup ("foul smelling"): the hypothesized function of such negative attitudes being to maintain interpersonal distance with a potential source of threat and to maintain boundaries between groups (Classen 1992; Largey and Watson 2006). In the context of mate choice and partner attractiveness, much is to be discovered in terms of body odor chemical composition and perception, as said above, but there is just as much (if not more) to explore in terms of how body odor perception is modulated by social cognition. For example, what are our expectations in terms of how a partner should smell? What happens if our expectations are not fulfilled? How is our perception of body odors modulated by a context (social, cultural), by our stereotypes about categories of people (sex, age, etc.)? This is an important piece of the puzzle that can be valuably investigated with the multidisciplinary approach proposed in Sect. 2.

Finally, there are other important considerations to take into account at a lower level of processing of the biological odors, namely the individual ability to detect body odor compounds. Interindividual variability is a strong characteristic of odor perception in general (e.g., in terms of sensitivity and pleasantness), and it has been demonstrated recently that this phenotypic variability has biological underpinnings. For example, genetic studies have shown that variants on the olfactory receptor (OR) genes OR7D4, OR11H7P, OR2J3, or OR5A1 influence how the odor of androstenone, isovaleric acid, cis-3-hexen-1-ol and β-ionone, are perceived, respectively (Keller et al. 2007; Menashe et al. 2007; McRae et al. 2012; Jaeger et al. 2013). Furthermore, from one individual to another many OR genes can be in an intact (functional) or pseudogene (nonfunctional) form, leading every individual to possess its own personal repertoire of OR genes (Mainland et al. 2014). However, it seems reasonable to think that if human body odor contains one or several compounds that are relevant for mate choice or even pheromone candidates, then OR genes responsible for their detection would be highly conserved in the human species, and would not be part of the highly variable OR genes mentioned above (Wyatt 2015). In light of this, the only compounds of human body odor that have been extensively studied and have been considered at some point as potential human sex pheromones, androstenone and androstadienone, appear as relatively improper candidates. Indeed, at least one in five people cannot detect them (specific anosmia: Amoore 1991; Ferdenzi et al. 2016b), and there is a significant variability in their perception according to the variant of OR7D4 possessed by the individuals (Keller et al. 2007). This puts in doubt their real relevance at the scale of the human species. These considerations would be worth extending with cross-cultural approaches to determine whether some body odor compounds of interest is perceived in a stable manner across cultures, and if not, whether deviations in their perception (nondetection, altered perceptual or behavioral response to it) are associated with alterations in parameters such as partner choice, satisfaction with the relationship, reproductive success, or fertility, at a population level.

3.2 Cross-Cultural Study of Odor Emission

On the side of odor emission, cross-cultural studies are scarce. Anthropological research suggests that body odor differs according to ethnicity based on group's verbal description of outgroups' odor (see reports about white and black people in Classen 1992). Whether this anecdotal evidence has biological and perceptual grounds is not very well known: this question remains poorly studied undoubtedly because it can be considered as "politically incorrect". In this context, our knowledge is limited to the following: A significant part of the human population (80–95% of Asian people) possesses a nonfunctional variant of the ABCC11 gene, which has a key function in the formation of body odor (among others): compared to European and African people who almost all (97–100% of them) possess the functional variant of the gene, they thus smell less strongly (Martin et al. 2010). They produce less androgen odorous compounds, thereby contributing to the idea that these molecules may not be as relevant in human mate choice as we once thought. Differences in the number, size, and secretory activity of axillary apocrine sweat glands were compiled in a review paper by Hart (1980), comparing between Asian, African and Caucasian people: he concludes that African people have the largest apocrine glands and highest levels of sweat secretion, and that Asian people have less apocrine glands, and thus a lower level of axillary odor than the other ethnic groups. Differences in perceived quality (intensity, pleasantness, and descriptions) or in chemical composition between the body odors of different ethnic groups have rarely been experimentally tested. Recent studies using cultural comparisons of verbal descriptions of body odors or of feelings elicited by body odors are only at their preliminary stage (Parma and Soares 2017); one study shows that the emission of some carboxylic acids differs between East Asian, Caucasian, and African-American ethnic groups but without direct link with ABCC11 genotype (Prokop-Prigge et al. 2016). There is no doubt that differences do exist: for example, elephants can discriminate, based on smell, between two Kenyan ethnic groups (Bates et al. 2007). Although it is an ethically and politically delicate subject to study, increasing the knowledge of this aspect of social olfaction would contribute to the understanding of chemical communication in humans: especially, in spite of the differences mentioned above, examination of the cultural invariants in the odor produced could help to identify paradigms that may have a function at the level of the human species.

3.3 The Influence of Culturally Shaped Rules of Social Interactions: Proxemics

In some cases, cultural differences in odor perception—and especially social odor perception—have been explained by possible differences in the culturally shaped rules of interpersonal distances. Namely, based on the proxemics theory (Hall 1966), it can be hypothesized that individuals belonging to a so-called "non-contact culture" such as northern Europe, North America, and Asia, would respond to body odors in a more negative (less tolerant) way than people living in southern Europe, Latin America,

Africa, and the Middle East, which are defined as "contact cultures". The former maintain greater interpersonal distance when interacting with people, have lower sensory involvement (including olfaction), minimal eye contact and little touching. This is possibly associated with a less positive perception of sensory cues in public and private interactions. In a cross-cultural study investigating how people's odor collected on T-shirts would be perceived by Japanese (noncontact) and Italian adults (contact), the Japanese gave lower pleasantness ratings than the Italians: this confirms the above hypothesis about such a relation between individual rules of social interactions in space and attitudes toward body smells (Schleidt et al. 1981). Another research involving children from Finland and France used the Children's Olfactory Behaviors in Everyday Life questionnaire (COBEL) to compare both groups' attention and reactivity to odors (Ferdenzi et al. 2008). Finnish children, whose social interactions are supposed to involve less contact than those of French children according to Hall's proxemics theory, tended to be more oriented toward odors in general than French children; more specifically, they were more aware of people's natural odor as well as body-related odors (personal hygiene products). Although other northern–southern Europe comparisons provided no differences in attitudes toward odors (Germans versus Italians: Schleidt et al. 1981), here we found that, in accordance with the above hypothesis, the more noncontact northern culture was more reactive to social odors.

The proxemics theory (Hall 1966) also proposes that individuals possess different levels of optimal interpersonal distances that are associated with different types of social interactions. These can be described as concentric circles around the individual: the closest delimits the intimate space (reserved for lovers, children), the next one with a larger radius delimits the personal space (for friends, family), the next one with an even larger radius delimits the social space (for acquaintances), and the largest one delimits the public space. The size of these spaces (radius values) is believed to be shaped by culture and to determine the way people interact with each other and with their environment (and thus how they build and organize their environment). This theory could constitute a very useful framework for the study of chemical communication.

Indeed, using this framework leads to think human chemical communication as a function of the type of odor (defined by its quality but also its intensity and thus range of dispersion), its source (body areas, more or less involved in different types of social interactions), and the type of social interaction. The range of action of the odors, depending on their intensity or on their saliency in a background mixture of other odorant compounds, probably is an important parameter to consider. For example, attractiveness through body odor, or through specific body odor compounds, has almost essentially been investigated with axillary samples or synthesized molecules from the axilla: although this choice is defendable (as explained in Sect. 2, knowledge about human body odor is mainly based on axilla/malodors), the focus appears mistakenly oriented. These odors may actually have a range of action larger than the intimate and personal spaces, and may serve functions as warning signals against pathogens (sickness, lack of hygiene) or other types of threat (through emotional contagion). It is reasonable to put forward that odors involved in partner choice and mate attractiveness would be less intense (not the highest peaks on the chromatograms stemming from chemical analysis of the odor), that they would rather be found within a more proximal interpersonal space, and that they would be produced by body areas that are in close contacts during intimate interactions. For example, the odors from the facial area, associated with apocrine but also sebaceous glands, deserve attention just like other areas such as the breast areola involved in a very proximal type of interaction (mother–infant interaction during breast feeding; Doucet et al. 2012). One can also find it useful to define which body parts are involved in which types of interaction (hand shaking, kissing), whether they convey odors and possibly whether/how these are sampled by the interactant (consciously or not) like in Frumin et al.'s (2015) study about hand shaking. Finally, the proxemics framework is not only valuable to rethink chemical communication in humans in terms of nature of odors, of odor sources and of behavioral involvement of body parts: it is also interesting because it allows, in a second phase, to apprehend the variability related to the variation in proxemics, either between cultural groups (with cross-cultural approaches) or more subtly from one individual to another (according to personality style, and differences in attitudes toward the body, for example).

4 Conclusion

Although there is increasing evidence that the odors conveyed by the human body are a matter of importance to many of us, and undoubtedly have an impact on our behavior in social interactions, the scientific understanding of human chemical communication is still in its early stages. To develop our knowledge about how human odors are involved in human attractiveness in particular and in other social functions more generally, it is crucial to dedicate a significant research effort to body odor chemical composition, but also to the exploration of various significant odor sources other than the axilla. This is essential to extend the researchers' view beyond malodors, which are the most studied. A multidisciplinary approach of chemical communication combining psychology and chemistry but also genetics, social sciences, and evolutionary sciences, is vital to understand the nature and functions of biological body odors and their variations. This chapter supports the idea that an interesting option to have a real overview of the question is to conduct cross-cultural investigations of human chemical communication. This could allow identifying universal abilities to perceive and emit odorant compounds that are possibly relevant (e.g., for partner choice), namely properties that are common to humans in general independently of their culture and environment, because they are involved in the social function of olfaction at the level of the human species. Further, analyzing whether the alteration of these properties (e.g., no perception, no emission) in some individuals has deleterious consequences on social behavior would help understanding the social function of olfaction. Cross-cultural considerations would also allow having a finer understanding of chemical communication in the way it could point, besides its invariant parts, some variations according to cultural influences. Especially, variants in attitudes toward body odors according to rules of social interactions in space (proxemics) seem to exist and deserve attention in future studies. The theory of proxemics is also stimulating for within-culture use, as a framework for identifying specific odors and odor sources that are likely to be involved in specific types of social interactions.

References

Allen C, Havlíček J, Williams K, Roberts SC (2018) Perfume experts' perceptions of body odors: toward a new lexicon for body odor description. J Sens Stud 33:e12314. https://doi.org/10.1111/joss.12314

Amoore JE (1991) Specific anosmias. In: Getchell TV, Doty RL, Bartoshuk LM, Snow JB (eds) Smell and taste in health and disease. Raven Press, New York, pp 655–664

Bates LA, Sayialel KN, Njiraini NW, Moss CJ, Poole JH, Byrne RW (2007) Elephants classify human ethnic groups by odor and garment color. Curr Biol 17:1938–1942. https://doi.org/10.1016/j.cub.2007.09.060

Bensafi M, Tsutsui T, Khan R, Levenson RW, Sobel N (2004) Sniffing a human sex-steroid derived compound affects mood and autonomic arousal in a dose-dependent manner. Psychoneuroendocrinology 29:1290–1299. https://doi.org/10.1016/s0306-4530(04)00046-0

Burke SM, Veltman DJ, Gerber J, Hummel T, Bakker J (2012) Heterosexual men and women both show a hypothalamic response to the chemo-signal androstadienone. PLoS ONE 7: e40993. https://doi.org/10.1371/journal.pone.0040993

Bushdid C, Magnasco MO, Vosshall LB, Keller A (2014) Humans can discriminate more than 1 trillion olfactory stimuli. Science 343:1370–1372. https://doi.org/10.1126/science.1249168

Chaix R, Cao C, Donnelly P (2008) Is mate choice in humans MHC-dependent? PLoS Genet 4: e1000184

Chrea C, Grandjean D, Delplanque S, Cayeux I, Le Calvé B, Aymard L, Velazco MI, Sander D, Scherer KR (2009) Mapping the semantic space for the subjective experience of emotional responses to odors. Chem Senses 34:49–62. https://doi.org/10.1093/chemse/bjn052

Classen C (1992) The odor of the other: olfactory symbolism and cultural categories. Ethos 20:133–166

Croy I, Nordin S, Hummel T (2014) Olfactory disorders and quality of life—an updated review. Chem Senses 39:185–194. https://doi.org/10.1093/chemse/bjt072

Curran AM, Rabin SI, Prada PA, Furton KG (2005) Comparison of the volatile organic compounds present in human odor using SPME-GC/MS. J Chem Ecol 31:1607–1619

Curran AM, Rabin SI, Prada PA, Furton KG (2006) On the definition and measurement of human scent: response by Curran et al. J Chem Ecol 32:1617–1623. https://doi.org/10.1007/s10886-006-9096-x

de Groot JHB, Smeets MAM (2017) Human fear chemosignaling: evidence from a meta-analysis. Chem Senses 42:663–673. https://doi.org/10.1093/chemse/bjx049

de Groot JHB, Semin GR, Smeets MAM (2014) Chemical communication of fear: a case of male-female asymmetry. J Exp Psychol Gen 143:1515–1525. https://doi.org/10.1037/a0035950

de Groot JHB, Smeets MAM, Rowson MJ, Bulsing PJ, Blonk CG, Wilkinson JE, Semin GR (2015) A sniff of happiness. Psychol Sci 26:684–700. https://doi.org/10.1177/0956797614566318

Dormont L, Bessière J-M, Cohuet A (2013) Human skin volatiles: a review. J Chem Ecol 39:569–578. https://doi.org/10.1007/s10886-013-0286-z

Doty RL (2010) The great pheromone myth. John Hopkins University Press, Baltimore

Doucet S, Soussignan R, Sagot P, Schaal B (2009) The secretion of areolar (Montgomery's) glands from lactating women elicits selective, unconditional responses in neonates. PLoS ONE 4:e7579. https://doi.org/10.1371/journal.pone.0007579

Doucet S, Soussignan R, Sagot P, Schaal B (2012) An overlooked aspect of the human breast: Areolar glands in relation with breastfeeding pattern, neonatal weight gain, and the dynamics of lactation. Early Hum Dev 88:119–128. https://doi.org/10.1016/j.earlhumdev.2011.07.020

Ferdenzi C, Rouby C, Bensafi M (2016a) The social nose: importance of olfactory perception in group dynamics and relationships. Psychol Inq 27:299–305. https://doi.org/10.1080/1047840x.2016.1215207

Ferdenzi C, Delplanque S, Atanassova R, Sander D (2016b) Androstadienone's influence on the perception of facial and vocal attractiveness is not sex specific. Psychoneuroendocrinology 66:166–175. https://doi.org/10.1016/j.psyneuen.2016.01.016

Ferdenzi C, Mustonen S, Tuorila H, Schaal B (2008) Children's awareness and uses of odor cues in everyday life: a Finland-France comparison. Chemosens Percept 1:190–198

Ferdenzi C, Schirmer A, Roberts SC, Delplanque S, Porcherot C, Cayeux I, Velazco M-I, Sander D, Scherer KR, Grandjean D (2011) Affective dimensions of odor perception: a comparison between Swiss, British, and Singaporean populations. Emotion 11:1168–1181. https://doi.org/10.1037/a0022853

Ferdenzi C, Delplanque S, Barbosa P, Court K, Guinard J-X, Guo T, Roberts SC, Schirmer A, Porcherot C, Cayeux I, Sander D, Grandjean D (2013) Affective semantic space of scents. Towards a universal scale to measure self-reported odor-related feelings. Food Qual Prefer 30:128–138. https://doi.org/10.1016/j.foodqual.2013.04.010

Fialova J, Roberts SC, Havlicek J (2016) Consumption of garlic positively affects hedonic perception of axillary body odour. Appetite 97:8–15. https://doi.org/10.1016/j.appet.2015.11.001

Frumin I, Perl O, Endevelt-Shapira Y, Eisen A, Eshel N, Heller I, Shemesh M, Ravia A, Sela L, Arzi A, Sobel N (2015) A social chemosignaling function for human handshaking. eLife 4. https://doi.org/10.7554/elife.05154

Gelstein S, Yeshurun Y, Rozenkrantz L, Shushan S, Frumin I, Roth Y, Sobel N (2011) Human tears contain a chemosignal. Science 331:226–230. https://doi.org/10.1126/science.1198331

Hall ET (1966) The hidden dimension. Doubleday, New York

Hart R (1980) Human body odor. Nexus 1

Havlicek J, Lenochova P (2006) The effect of meat consumption on body odor attractiveness. Chem Senses 31:747–752. https://doi.org/10.1093/chemse/bjl017

Haze S, Gozu Y, Nakamura S, Kohno Y, Sawano K, Ohta H, Yamazaki K (2001) 2-Nonenal newly found in human body odor tends to increase with aging. J Invest Dermatol 116:520–524. https://doi.org/10.1046/j.0022-202x.2001.01287.x

Herz RS (2009) Aromatherapy facts and fictions: a scientific analysis of olfactory effects on mood, physiology and behavior. Int J Neurosci 119:263–290

Herz RS, Inzlicht M (2002) Sex differences in response to physical and social factors involved in human mate selection: the importance of smell for women. Evol Hum Behav 23:359–364. https://doi.org/10.1016/s1090-5138(02)00095-8

Jaeger SR, McRae JF, Bava CM, Beresford MK, Hunter D, Jia Y, Chheang SL, Jin D, Peng M, Gamble JC, Atkinson KR, Axten LG, Paisley AG, Tooman L, Pineau B, Rouse SA, Newcomb RD (2013) A mendelian trait for olfactory sensitivity affects odor experience and food selection. Curr Biol 23:1601–1605. https://doi.org/10.1016/j.cub.2013.07.030

Karlson P, Lüscher M (1959) Pheromones: a new term for a class of biologically active substances. Nature 183:55–56

Keller A, Zhuang H, Chi Q, Vosshall LB, Matsunami H (2007) Genetic variation in a human odorant receptor alters odour perception. Nature 449:468–472. https://doi.org/10.1038/nature06162

Kuhn F, Natsch A (2009) Body odour of monozygotic human twins: a common pattern of odorant carboxylic acids released by a bacterial aminoacylase from axilla secretions contributing to an inherited body odour type. J R Soc Interface R Soc 6:377–392. https://doi.org/10.1098/rsif.2008.0223

Kuukasjarvi S, Eriksson CJP, Koskela E, Mappes T, Nissinen K, Rantala MJ (2004) Attractiveness of women's body odors over the menstrual cycle: the role of oral contraceptives and receiver sex. Behav Ecol 15:579–584. https://doi.org/10.1093/beheco/arh050
Largey G, Watson R (2006) The sociology of odors. In: Drobnick J (ed) The smell culture reader. Berg, Oxford, UK, pp 29–40
Le Guérer A (2002) Olfaction and cognition: a philosophical and psychoanalytic view. In: Rouby C, Schaal B, Dubois D, Gervais R, Holley A (eds) Olfaction, taste, and cognition. Cambridge University Press, Cambridge, pp 3–15
Leyden JJ, McGinley KJ, Holzle E, Labows JN, Kligman AM (1981) The microbiology of the human axilla and its relationship to axillary odor. J Invest Dermatol 77:413–416
Lundström JN, Olsson MJ (2005) Subthreshold amounts of social odorant affect mood, but not behavior, in heterosexual women when tested by a male, but not a female, experimenter. Biol Psychol 70:197–204
Mainland JD, Keller A, Li YR, Zhou T, Trimmer C, Snyder LL, Moberly AH, Adipietro KA, Liu WLL, Zhuang H, Zhan S, Lee SS, Lin A, Matsunami H (2014) The missense of smell: functional variability in the human odorant receptor repertoire. Nat Neurosci 17:114–120. https://doi.org/10.1038/nn.3598
Marchal S, Bregeras O, Puaux D, Gervais R, Ferry B (2016) Rigorous training of dogs leads to high accuracy in human scent matching-to-sample performance. PLoS ONE 11:e0146963. https://doi.org/10.1371/journal.pone.0146963
Martin A, Saathoff M, Kuhn F, Max H, Terstegen L, Natsch A (2010) A functional ABCC11 allele is essential in the biochemical formation of human axillary odor. J Invest Dermatol 130:529–540
McGann JP (2017) Poor human olfaction is a 19th-century myth. Science 356:eaam7263. https://doi.org/10.1126/science.aam7263
McRae JF, Mainland JD, Jaeger SR, Adipietro KA, Matsunami H, Newcomb RD (2012) Genetic variation in the odorant receptor OR2J3 is associated with the ability to detect the "grassy" smelling odor, cis-3-hexen-1-ol. Chem Senses 37:585–593. https://doi.org/10.1093/chemse/bjs049
Menashe I, Abaffy T, Hasin Y, Goshen S, Yahalom V, Luetje CW, Lancet D (2007) Genetic elucidation of human hyperosmia to isovaleric acid. PLoS Biol 5:e284. https://doi.org/10.1371/journal.pbio.0050284
Mitro S, Gordon AR, Olsson MJ, Lundström JN (2012) The smell of age: perception and discrimination of body odors of different ages. PLoS ONE 7:e38110. https://doi.org/10.1371/journal.pone.0038110
Mujica-Parodi LR, Strey HH, Frederick B, Savoy R, Cox D, Botanov Y, Tolkunov D, Rubin D, Weber J (2009) Chemosensory cues to conspecific emotional stress activate amygdala in humans. PLoS ONE 4:e6415. https://doi.org/10.1371/journal.pone.0006415
Natsch A, Gfeller H, Gygax P, Schmid J (2005) Isolation of a bacterial enzyme releasing axillary malodor and its use as a screening target for novel deodorant formulations. Int J Cosmet Sci 27:115–122. https://doi.org/10.1111/j.1467-2494.2004.00255.x
Olsson MJ, Lundström JN, Kimball BA, Gordon AR, Karshikoff B, Hosseini N, Sorjonen K, Höglund CO, Solares C, Soop A, Axelsson J, Lekander M (2014) The scent of disease: Human body odor contains an early chemosensory cue of sickness. Psychol Sci 25:817–823. https://doi.org/10.1177/0956797613515681
Pangborn RM, Guinard J-X, Davis RG (1988) Regional aroma preferences. Food Qual Prefer 1:11–19. https://doi.org/10.1016/0950-3293(88)90003-1
Parma V, Soares SC (2017) The role of human body odors in ingroup-outgroup relationships. Chem Senses 42:e19

Pause BM, Lübke K, Laudien JH, Ferstl R (2010) Intensified neuronal investment in the processing of chemosensory anxiety signals in non-socially anxious and socially anxious individuals. PLoS ONE 5:e10342. https://doi.org/10.1371/journal.pone.0010342

Penn DJ, Oberzaucher E, Grammer K, Fischer G, Soini HA, Wiesler D, Novotny MV, Dixon SJ, Xu Y, Brereton RG (2007) Individual and gender fingerprints in human body odour. J R Soc Interface 4:331–340. https://doi.org/10.1098/rsif.2006.0182

Porcherot C, Delplanque S, Raviot-Derrien S, Le Calvé B, Chrea C, Gaudreau N, Cayeux I (2010) How do you feel when you smell this? Optimization of a verbal measurement of odor-elicited emotions. Food Qual Prefer 21:938–947

Porter J, Craven B, Khan RM, Chang SJ, Kang I, Judkewitz B, Judkewicz B, Volpe J, Settles G, Sobel N (2007) Mechanisms of scent-tracking in humans. Nat Neurosci 10:27–29

Preti G, Willse A, Labows JN, Leyden JJ, Wahl J, Kwak J (2006) On the definition and measurement of human scent: comments on Curran et al. J Chem Ecol 32:1613–1616. https://doi.org/10.1007/s10886-006-9095-y

Prokop-Prigge KA, Greene K, Varallo L, Wysocki CJ, Preti G (2016) The effect of ethnicity on human axillary odorant production. J Chem Ecol 42:33–39. https://doi.org/10.1007/s10886-015-0657-8

Quignon P, Giraud M, Rimbault M, Lavigne P, Tacher S, Morin E, Retout E, Valin A-S, Lindblad-Toh K, Nicolas J, Galibert F (2005) The dog and rat olfactory receptor repertoires. Genome Biol 6:R83. https://doi.org/10.1186/gb-2005-6-10-r83

Regenbogen C, Axelsson J, Lasselin J, Porada DK, Sundelin T, Peter MG, Lekander M, Lundström JN, Olsson MJ (2017) Behavioral and neural correlates to multisensory detection of sick humans. Proc Natl Acad Sci USA 114:6400–6405. https://doi.org/10.1073/pnas.1617357114

Roberts SC, Gosling LM, Spector TD, Miller P, Penn DJ, Petrie M (2005) Body odor similarity in noncohabiting twins. Chem Senses 30:651–656

Roberts SA, Simpson DM, Armstrong SD, Davidson AJ, Robertson DH, McLean L, Beynon RJ, Hurst JL (2010) Darcin: a male pheromone that stimulates female memory and sexual attraction to an individual male's odour. BMC Biol 8:75. https://doi.org/10.1186/1741-7007-8-75

Santos DV, Reiter ER, DiNardo LJ, Costanzo RM (2004) Hazardous events associated with impaired olfactory function. Arch Otolaryngol Head Neck Surg 130:317–319

Savic I, Berglund H, Gulyas B, Roland P (2001) Smelling of odorous sex hormone-like compounds causes sex-differentiated hypothalamic activations in humans. Neuron 31:661–668. https://doi.org/10.1016/s0896-6273(01)00390-7

Saxton TK, Lyndon A, Little AC, Roberts SC (2008) Evidence that androstadienone, a putative human chemosignal, modulates women's attributions of men's attractiveness. Horm Behav 54:597–601. https://doi.org/10.1016/j.yhbeh.2008.06.001

Schaal B, Porter RH (1991) "Microsmatic humans" revisited: the generation and perception of chemical signals. Adv Study Behav 20:135–199

Schaal B, Soussignan R, Marlier L, Kontar F, Karima IS, Tremblay RE (1997) Variability and invariants in early odour preferences: comparative data from children belonging to three cultures. Chem Senses 22:212

Schaal B, Coureaud G, Langlois D, Ginies C, Semon E, Perrier G (2003) Chemical and behavioural characterization of the rabbit mammary pheromone. Nature 424:68–72

Schleidt M, Hold B, Attili G (1981) A cross-cultural study on the attitude towards personal odors. J Chem Ecol 7:19–31

Schleidt M, Neumann P, Morishita H (1988) Pleasure and disgust: memories and associations of pleasant and unpleasant odours in Germany and Japan. Chem Senses 13:279–293. https://doi.org/10.1093/chemse/13.2.279

Singh D, Bronstad PM (2001) Female body odour is a potential cue to ovulation. Proc R Soc Lond B Biol Sci 268:797–801. https://doi.org/10.1098/rspb.2001.1589
Starkenmann C, Niclass Y, Troccaz M, Clark AJ (2005) Identification of the precursor of (S)-3-methyl-3-sulfanylhexan-1-ol, the sulfury malodour of human axilla sweat. Chem Biodivers 2:705–716. https://doi.org/10.1002/cbdv.200590048
Stevenson RJ (2010) An initial evaluation of the functions of human olfaction. Chem Senses 35:3–20. https://doi.org/10.1093/chemse/bjp083
Stoddart DM (1990) The scented ape: the biology and culture of human odour. Cambridge University Press, Cambridge
Wackermannová M, Pinc L, Jebavý L (2016) Olfactory sensitivity in mammalian species. Physiol Res 65:369–390
Wedekind C, Seebeck T, Bettens F, Paepke AJ (1995) MHC-dependent mate preferences in humans. Proc R Soc Lond B Biol Sci 260:245–249. https://doi.org/10.1098/rspb.1995.0087
Wedekind C, Escher S, Van de Waal M, Frei E (2007) The major histocompatibility complex and perfumers' descriptions of human body odors. Evol Psychol 5:330–343
Wilke K, Martin A, Terstegen L, Biel SS (2007) A short history of sweat gland biology. Int J Cosmet Sci 29:169–179
Wyatt TD (2015) The search for human pheromones: the lost decades and the necessity of returning to first principles. Proc R Soc Lond B Biol Sci 282:20142994. https://doi.org/10.1098/rspb.2014.2994
Wysocki CJ, Pierce JD, Gilbert AN (1991) Geographic, cross-cultural, and individual variation in human olfaction. In: Getchell TV, Doty RL, Bartoshuk LM, Snow JB (eds) Smell and taste in health and disease. Raven Press, New York, pp 287–314

Adaptation of the University of Pennsylvania Smell Identification Test for the Population of Central Russia

Vera V. Voznessenskaya[1(✉)], Maria A. Klyuchnikova[1], Elena I. Rodionova[1,2], and Anna Voznesenskaya[1,2]

[1] A.N. Severtsov Institute of Ecology and Evolution, Russian Academy of Sciences, 33 Leninski Prospect, Moscow 119071, Russia
veravoznessenskaya@gmail.com
[2] Kharkevich Institute for Information Transmission, Russian Academy of Sciences, Moscow, Russia

Abstract. No standardized test is currently used in research or clinical practice in Russia, a country with a total population of ~150 million people. To close this gap, we have tested the original version of the University of Pennsylvania Smell Identification Test (UPSIT) on the population of central Russia including groups of different ages residing in urban areas and in the rural part of the country. Our results demonstrate that the UPSIT is applicable for the evaluation of olfactory function within the population of Central Russia, and allows the detection of age-related differences in human olfactory function. However, several odor samples present within the original UPSIT proved to be unfamiliar to the subjects due to cultural differences. We have therefore tested and identified odor items that may replace poorly recognizable smells of the original UPSIT. Thus, we have developed a culturally adapted version of the UPSIT to be used within the population of the Central Russia.

1 Introduction

Vision and auditory tests have become routine in medical procedures worldwide a long time ago. However, sensory loss in taste and smell received much less medical attention in the past, and although standardized approaches to diagnose impairments in chemosensory function have been developed, they are not well established in medical practice. This is primarily due to the fact that the impacts of a deficiency in a patient's sense of smell on health and well-being are commonly underestimated. Nevertheless, recent research shows that impairments in olfactory function significantly affect the quality of life and oftentimes, overall health (Hummel and Nordin 2005; Keller and Malaspina 2013). A decreased sense of smell compromises the ability to detect the full array of food flavors and, thus, may lead to poor dietary choices resulting in nutrition-related health conditions such as malnutrition and/or obesity (Aschenbrenner et al. 2008). Complete loss of the sense of smell even poses a safety concern because anosmic individuals are unable to detect warning odors such as gas leakage or the smell of spoiled food (Schiffman 1997; Hummel and Nordin 2005; Keller and Malaspina

© Springer Nature Switzerland AG 2019
C. D. Buesching (Ed.): *Chemical Signals in Vertebrates 14*, pp. 153–161, 2019.
https://doi.org/10.1007/978-3-030-17616-7_12

2013). Standardized procedures for diagnostics of smell loss are also important because of the association of olfactory decline and development of neurodegenerative disorders such as Alzheimer's and Parkinson's diseases (Doty et al. 1987; Doty 2007; Djordjevic et al. 2008; Woodward et al. 2017). Decrease in olfactory abilities typically precedes the development of other, more serious symptoms of these disorders and thus, evaluation of the olfactory function may aid in early detection of neurodegenerative disorders when used in combination with other diagnostics procedures.

Several tests aimed to diagnose olfactory impairments within a medical setting have been developed in the last few decades (Amoore and Ollman 1983; Stevens et al. 1988; Kobal et al. 2000; Yousem et al. 2003; Lötsch et al. 2004; Hummel et al. 2007). The University of Pennsylvania Smell Identification Test (UPSIT) is a self-administered smell identification test which utilizes microencapsulated crystals for odorant presentation. The test contains 40 individual odors presented on separate cards gathered in a booklet. An individual being tested has to scratch and sniff a strip containing the odorant and is presented with a multiple choice question regarding the identity of the odor. UPSIT is one of the most widely used odor identification tests with a high test–retest reliability (Doty et al. 1984b; 1989; 1995). However, the ability to recognize, as opposed to simply sense, the test odors largely depends on an individual's familiarity with the smell, which is defined by the everyday sensory environment and thus, is subject to cultural differences.

The UPSIT was primarily developed to assess olfactory function within the US population and is not applicable without prior adaptation to culturally distinct populations. Since its introduction in the early 80s, a number of culturally adapted versions of UPSIT have been developed (Mackay-Sim and Doty 2001; Ogihara et al. 2011; Altundag et al. 2015; Fornazieri et al. 2015; Hsu et al. 2015). However, no standardized test for olfactory evaluation in Russian population exists at the moment.

Thus, in this study, we aimed to evaluate the applicability of the UPSIT for assessment of olfactory function in a population of Central Russia and adapt it for the specific Central Russian cultural context.

2 Materials and Methods

2.1 Participants

Test subjects were 340 healthy volunteers ("participants") of 18–82 years old, men and women, with no apparent olfactory or taste problems. Exclusion criteria were upper respiratory tract infections, allergic rhinitis, history of head trauma, and neurological disorders. Information on medical history was based on self-report. To evaluate the applicability of the original UPSIT to the Russian population, we administered the test to a group of participants aged from 18 to 60 years (mean age 27.4 ± 9.8 years) residing either in Moscow ($n = 143$) or in a rural area ($n = 65$). We included residents of a big city and rural countryside because both everyday life and food culture differ dramatically between these two areas and so does the sensory environment and familiarity with the test odors (Table 1).

Table 1. Study participants

Age (years)	$N = 340$	Living environment
18–60	271	
Females	118	City
	53	Rural
Males	59	City
	41	Rural
>60	69	
Females	15	City
	26	Rural
Males	10	City
	18	Rural

2.2 Test Procedure

We followed the administration procedure recommended by the manufacturer of the University of Pennsylvania Smell Identification Test™, UPSIT (Sensonics, Inc, Haddo Heights, NJ) (Doty 1995). The response alternatives of the UPSIT were translated into Russian. Some of the odorant labels were changed during the translation process to provide descriptors that would be more familiar to the respondents: "wintergreen" was changed to "methyl salicylate containing products/Orbit Wintergreen/warming ointments)", "smoke" to "smoked meat products", and "root beer" to "nutmeg soda." The response alternatives were provided in the written form printed on separate sheets of A4 paper. The instructions were read to each subject by an operator who monitored the test session to ensure that the correct procedure was followed.

2.3 Study Design

In order to verify that UPSIT would allow to detect medically relevant degree of olfactory decline in the Russian population, we administered the test to a group of participants aged over 60 years (n = 44, mean age 70.3 ± 7.4 years), because olfactory deterioration with age is well documented to be present in over half of individuals between the ages of 65 and 80 years (Doty et al. 1984a; Murphy et al. 2002; Doty and Kamath 2014).

To replace poorly identifiable odor items, an additional group of individuals (n = 86) residing either in Moscow (n = 45, 19 men and 26 women, mean age 46.9 ± 24.1 years) or a rural area (n = 41, 18 men and 23 women, mean age 45.3 ± 20.8 years) was tested to recognize 7 potential replacement odors provided by the manufacturer of the UPSIT. Potential replacement odors were provided in the same sniff and scratch form as the odors within the UPSIT and were evaluated using the same procedure as for the original UPSIT items.

2.4 Data Analysis

All data are reported as mean ± SD. UPSIT scores within the study population aged between 18 and 60 years were analyzed with factorial ANOVA including Sex, Smoking, and Geographical area as factors. UPSIT scores within the group of subjects aged over 60 were compared to those within the study group of subjects aged 18–60 years using factorial ANOVA with factors of Age and Sex. The criterion for statistical significance was $p < 0.05$. Statistical analyses were conducted using Dell Statistica software.

The study was conducted according to the principles of the Declaration of Helsinki. Participants provided informed written consent. All procedures were approved by the research ethics committee of A.N. Severtsov Institute of Ecology and Evolution.

3 Results

UPSIT scores for the main study group are provided in Tables 2 and 3. The average number of correct answers within the group of participants aged between 18 and 60 years was 33 ± 3. The number of correct responses was not affected by sex or by the type of the geographical area the participants resided in (Table 2, $F_{1,199} = 1.8, p = 0.19$ and $F_{1,199} = 0.2$, $p = 0.65$, respectively). We also observed only a weak (i.e., non-significant) effect of smoking on the performance scores within this age group (Table 3, $F_{1,199} = 2.7$, $p = 0.10$).

Table 2. UPSIT score in participants aged 18–60 years residing either in a rural or an urban area of the European part of the Russian Federation (all data presented as mean ± SD)

	$N =$	UPSIT score
Rural area	65	33 ± 3
Females	36	33 ± 3
Males	29	34 ± 3
City area	143	34 ± 3
Females	99	33 ± 3
Males	44	34 ± 3
Total	208	33 ± 3

The number of correct answers was significantly smaller in the group of participants aged over 60 years (Table 4, $F_{1,248} = 108$, $p < 0.001$). We also observed a significant negative correlation between a participant's age and performance in the test ($r^2 = -0.21$, $p < 0.01$).

The percentage of correct responses for the identification of individual odors is reported in Table 5. The majority of the samples were correctly identified by more than 75% of participants. However, 6 odor samples were correctly identified by less than 50% of the study participants (highlighted in bold in Table 5). Those were cedar, lime, fruit punch, lilac, cheddar, and grass.

Table 3. UPSIT score in participants aged 18–60 years depending on smoking status (all data presented as ± SD)

	$N =$	UPSIT score
Nonsmokers	152	33 ± 3
Smokers	55	34 ± 3

Table 4. Comparison of UPSIT scores in elderly (>60 years of age) and young (18–60 years of age) participants residing in the European part of the Russian Federation. All data presented as mean ± SD. Groups sharing the same letter did not differ from each other, ***$p < 0.001$, $F_{1,248} = 108$

Age, years	$N =$	UPSIT score
18–60***		
Females	135	33 ± 3[b]
Males	73	33 ± 3[b]
>60***		
Females	28	29 ± 3[a]
Males	16	28 ± 3[a]

Table 5. Identification of individual UPSIT odors. Data are given as the percentage of correct responses provided by respondents of the respective population group

Test odor	Age, years		Living environment		All (%)
	18–60 (%)	>60 (%)	City (%)	Rural (%)	
Pizza	81	**61**	84	**68**	78
Bubble gum	8	**61**	90	76	84
Menthol	84	82	84	83	84
Cherry	88	**52**	90	**69**	81
Motor oil	89	89	86	94	89
Mint	93	**64**	94	79	88
Banana	96	75	92	92	92
Clove	78	82	80	78	79
Leather	99	89	98	96	97
Coconut	94	82	96	86	92
Onion	99	89	99	95	97
Fruit punch	**24**	**18**	**26**	**18**	**23**
Licorice	94	91	94	93	93
Cheddar cheese	**42**	**25**	**50**	**22**	**39**
Cinnamon	80	**68**	80	**75**	78
Gasoline	95	95	93	98	95
Strawberry	82	**64**	81	77	79

(*continued*)

Table 5. (*continued*)

Test odor	Age, years		Living environment		All (%)
	18–60 (%)	>60 (%)	City (%)	Rural (%)	
Cedar	**73**	**64**	**71**	**72**	**71**
Chocolate	77	84	**74**	84	78
Gingerbread	95	77	96	86	92
Lilac	**31**	**30**	**31**	**30**	**30**
Turpentine	89	**61**	91	**74**	84
Peach	97	84	95	93	94
Root beer	90	77	88	87	88
Dill pickle	90	77	88	88	88
Pineapple	90	**73**	85	90	87
Lime	**28**	**5**	**18**	**33**	**24**
Orange	91	77	88	90	89
Wintergreen	96	80	94	92	93
Watermelon	96	95	94	98	96
Paint thinner	94	98	94	95	94
Grass	**52**	**27**	**47**	**49**	**48**
Smoke	100	98	100	99	10
Pineapple	97	84	95	93	94
Grapes	88	**68**	87	80	84
Lemon	97	89	94	98	96
Soap	82	86	79	88	83
Natural gas	96	77	94	91	93
Rose	93	**68**	93	82	89
Peanuts	96	75	94	90	92

The most frequent alternate for "fruit punch" was "soap" (95% of respondents), which allowed us to rename the odorant to "berry soap." There was no consistency in naming the alternates for other poorly recognizable test items.

To replace poorly identifiable odor items, we tested seven potential replacement odors in a new group of participants residing either in a city or in a rural area (Table 6). Five odor items were consistently recognized by over 75% of respondents (highlighted in bold in Table 6) and, thus, were selected to replace unfamiliar items from the original UPSIT version.

Table 6. Identification of seven potential replacement UPSIT odors. All data are given as the percentage of correct responses per group

Test odor	Living environment	
	City (%)	Rural (%)
Apple	51.1	78.1
Raspberry	73.3	82.9
Garlic	**100**	**100**
Baby powder	**93.3**	**80.5**
Grapefruit	**95.6**	**97.6**
Rubber tire	**82.2**	**95.1**
Coffee	**97.8**	**100**

4 Discussion

Here, we have evaluated the applicability of a well established standardized olfactory evaluation procedure, UPSIT, to the population of Central Russia. To the best of our knowledge, no such test is currently used in research or clinical practice in Russia, a country with a total population of ~150 million people.

Our results demonstrate that in a group of healthy individuals under 60 years of age, the UPSIT score was 33 ± 3 out of the original 40 items. Given that 6 items were subsequently excluded or modified because they were unfamiliar to the subjects, the observed average score is comparable to the normosmic range of 34–40 correct answers established for the original version of UPSIT in the US population (Doty 1995).

Although we did not detect any significant differences related to gender or smoking within our study population, the decline in test performance with age could be seen clearly. Olfactory deterioration with age is well documented (Doty et al. 1984a; Murphy et al. 2002; Doty and Kamath 2014); and the magnitude of the observed clinically significant olfactory decline falls within the range of age-related changes rather than the range of gender or health habit-related variations. Thus, our data support potential applicability of the modified UPSIT for clinical practice in the Russian Federation.

It is plausible that gender differences and the effect of smoking behaviour could be detected if all 40 items were familiar to our subjects. For instance, the study of evaluating the Turkish version of the UPSIT demonstrated that sex differences in odor identification were only detected with the culturally modified, but not in the original version of the UPSIT; while the effect of smoking, although detectable with both versions of the UPSIT, was more prominent when the culturally adjusted version was administered (Altundag et al. 2015). We have already identified five odor items that may replace poorly recognizable items of the original UPSIT. The sixth frequently mislabeled odor, originally “fruit punch”, was relabeled as “berry soap” because soap was the alternate chosen by the subjects in 95% of the cases. This misinterpretation of the odor is likely related to the fact that fruit punch is not a common drink in Russia, and that the odor sample smells very similar to one of the most commonly used soap brands with a fruity smell.

Thus, we have developed a culturally adapted version of the UPSIT to be used within the population of the Central Russia.

Acknowledgements. We thank Dr. Richard Doty for kindly providing samples of potential replacement odors for testing. The research was supported by the Russian Science Foundation 16-1510312 to VVV.

References

Altundag A, Tekeli H, Salihoglu M, Cayonu M, Yasar H, Kendirli MT et al (2015) Cross-culturally modified University of Pennsylvania Smell Identification Test for a Turkish population. Am J Rhinol 29:e138–e141

Amoore J, Ollman B (1983) Practical test kits for quantitatively evaluating the sense of smell. Rhinology 21:49–54

Aschenbrenner K, Hummel C, Teszmer K, Krone F, Ishimaru T, Seo HS et al (2008) The influence of olfactory loss on dietary behaviors. Laryngoscope 118:135–144

Djordjevic J, Jones-Gotman M, De Sousa K, Chertkow H (2008) Olfaction in patients with mild cognitive impairment and Alzheimer's disease. Neurobiol Aging 29:693–706

Doty RL (1995) The Smell Identification Test™ administration manual. Sensonics. Inc, Philadelphia, PA

Doty RL (2007) Olfaction in Parkinson's disease. Parkinsonism & related disorders 13:S225–S228

Doty RL, Frye RE, Agrawal U (1989) Internal consistency reliability of the fractionated and whole University of Pennsylvania Smell Identification Test. Percept Psychophys 45:381–384

Doty RL, Kamath V (2014) The influences of age on olfaction: a review. Front Psychol 5:20

Doty RL, McKeown DA, Lee WW, Shaman P (1995) A study of the test-retest reliability of ten olfactory tests. Chem Senses 20:645–656

Doty RL, Reyes PF, Gregor T (1987) Presence of both odor identification and detection deficits in Alzheimer's disease. Brain Res Bull 18:597–600

Doty RL, Shaman P, Applebaum SL, Giberson R, Siksorski L, Rosenberg L (1984a) Smell identification ability: changes with age. Science 226:1441–1443

Doty RL, Shaman P, Dann M (1984b) Development of the University of Pennsylvania Smell Identification Test: a standardized microencapsulated test of olfactory function. Physiol Behav 32:489–502

Fornazieri MA, dos Santos CA, Bezerra TF, Pinna Fde R, Voegels RL, Doty RL (2015) Development of normative data for the Brazilian adaptation of the University of Pennsylvania Smell Identification Test. Chem Senses 40:141–149

Hsu NI, Lai JT, Shen PH (2015) Development of Taiwan Smell Identification Test: a quick office-based smell screening test for Taiwanese. Am J Rhinol 29:e50–e54

Hummel T, Kobal G, Gudziol H, Mackay-Sim A (2007) Normative data for the "Sniffin'Sticks" including tests of odor identification, odor discrimination, and olfactory thresholds: an upgrade. Eur Arch Otorhinolaryngol 264:237–243

Hummel T, Nordin S (2005) Olfactory disorders and their consequences for quality of life. Acta Otolaryngol 125:116–121

Keller A, Malaspina D (2013) Hidden consequences of olfactory dysfunction: a patient report series. BMC Ear Nose Throat Disord 13:8

Kobal G, Klimek L, Wolfensberger M, Gudziol H, Temmel A, Owen C et al (2000) Multicenter investigation of 1,036 subjects using a standardized method for the assessment of olfactory function combining tests of odor identification, odor discrimination, and olfactory thresholds. Eur Arch Otorhinolaryngol 257:205–211

Lötsch J, Lange C, Hummel T (2004) A simple and reliable method for clinical assessment of odor thresholds. Chem Senses 29:311–317

Mackay-Sim A, Doty RL (2001) The University of Pennsylvania Smell Identification Test: normative adjustment for Australian subjects. Aust J Otolaryngol 4:174–177

Murphy C, Schubert CR, Cruickshanks KJ, Klein BE, Klein R, Nondahl DM (2002) Prevalence of olfactory impairment in older adults. JAMA 288:2307–2312

Ogihara H, Kobayashi M, Nishida K, Kitano M, Takeuchi K (2011) Applicability of the cross-culturally modified University of Pennsylvania Smell Identification Test in a Japanese population. Am J Rhinol 25:404–410

Schiffman SS (1997) Taste and smell losses in normal aging and disease. JAMA 278:1357–1362

Stevens JC, Cain WS, Burke RJ (1988) Variability of olfactory thresholds. Chem Senses 13: 643–653

Woodward MR, Amrutkar CV, Shah HC, Benedict RH, Rajakrishnan S, Doody RS et al (2017) Validation of olfactory deficit as a biomarker of Alzheimer disease. Neurol Clin Pract 7:5–14

Yousem D, Sobel N, Johnson B, Mainland J (2003) Functional Neuroimaging of Human Olfaction. In: Doty RL (ed) Handbook of Olfaction and Gustation, 2nd edn. CRC Press, Boca Raton, pp 251–273

Part III

Inter-specific Cues and Signals

House Mouse (*Mus musculus*) Avoidance of Olfactory Cues from Ferrets and Other Mammalian and Reptilian Predators: Preliminary Results

Carlos Grau[1(✉)], Eva Teruel[2], Julius Leclercq[3], and Patrick Pageat[4]

[1] Department of Behavioural and Physiological Mechanisms of Adaptation (DMPCA), Research Institute in Semiochemistry and Applied Ethology (IRSEA), Quartier Salignan, Apt, France
c.grau@group-irsea.com

[2] Statistical Analysis Department, Research Institute in Semiochemistry and Applied Ethology (IRSEA), Quartier Salignan, Apt, France

[3] Animal Experimentation Unit, Research Institute in Semiochemistry and Applied Ethology (IRSEA), Quartier Salignan, Apt, France

[4] Department of Chemicals Identification and Development of Analogues, Research Institute in Semiochemistry and Applied Ethology (IRSEA), Quartier Salignan, Apt, France

Abstract. Like other small rodents, house mice (*Mus musculus*) are at the bottom of vertebrate predator-dominated food chain. After humans, house mice are the most widespread mammal on earth. With their short reproductive cycle and prolificity, they can quickly produce large populations that form a basic source of nutrients for a wide spectrum of predators. As macrosmatic animals, mice use olfaction as a primary tool to avoid predators. However, further research is required to fully understand the main predator olfactory cues and behavioral reactions linked to these stimuli. This study examines the chemical cues used by house mice to detect mammalian and reptilian predators. For this purpose, we carried out a choice test where, during a 10 min period, we measured the total duration that mice remained in either the area closest to or farthest from the predatory stimulus (ferret fur and faeces, snake sheds, fox faeces, dog faeces, and cat urine); mice had no physical access to the stimulus, and both compared areas were identical. Our preliminary results show that mice significantly avoided ferret olfactory stimuli from fur and faeces. The other predator stimuli did not elicit significant avoidance. However, in some cases, this may be due to specific genetic and phenotypic features of the mouse strain tested. Our results are in line with previous work with ferret olfactory stimuli in mice. Nevertheless, further research should examine the role of ferret olfactory cues and semiochemicals as good indicators of their presence that lead to avoidance behaviour of rodents.

© Springer Nature Switzerland AG 2019
C. D. Buesching (Ed.): *Chemical Signals in Vertebrates 14*, pp. 165–181, 2019.
https://doi.org/10.1007/978-3-030-17616-7_13

1 Introduction

1.1 Predation

The detection of predator cues by prey constitutes a valuable tool for survival, making this feature a criterion for selection throughout evolution. Predators and prey run a constant arms race that leads to continuous evolution (Dietl and Kelley 2002), which commensal species have continued in human habitats (Lowry et al. 2013; Bull and Maron 2016). Along with rodents, members of the orders Carnivora and Squamata (lizards and snakes) are macrosmatic, and olfaction and chemical communication play an overarching role in their lives. In the wild, and more recently in anthropogenic environments, rodents and members of the orders Carnivora and Squamata have co-evolved (Abrams 2000) with each having a major influence on the other: the former functioning as an important food source and the latter two posing major predator risks.

House mice are at the bottom of the mammalian food chain, and aside from being the predominant prey of many raptors, also have a number of mustelid, feline, canid and reptilian predators. The main aim of our study was to explore the behavioral reactions of mice to a complete repertoire of olfactory stimuli produced by ferrets, a rodent specialist predator. The stimuli were composed of male and female faeces and fur olfactory cues. We are aware of the loss of specificity with this approach, however, this configuration likely provides a more realistic and complex set of stimuli.

In addition, we tested other predator olfactory stimuli that could play a significant role in the ecology of the house mouse. These species included snakes, the red fox, the domestic cat, and dog.

Our research hypothesis was that these stimuli could act as predator olfactory messages to elicit avoidance in house mice.

1.2 Ferret Olfactory Cues as a Predatory Stimulus

The ferret (*Mustela furo*) is a domestic mustelid (Alexander 1951; Church 2007) whose probable wild ancestor species are the European polecat (*Mustela putorius*) and the Steppe polecat (*M. eversmanni*) (Church 2007) that are both important rodent predators. The main prey of the European polecat are rodents followed by rabbits and anurans, depending on their abundance (Lodé 1997; Santos et al. 2009). The diet of the steppe polecat is similar, with a preference for small rodents (Lanszki and Heltai 2007). Ferrets and polecats are both obligate carnivores, even more so than cats, with a simple digestive tract (Bradley et al. 2006) that obliges them to feed several times per day, a fact which likely conditions their hunting habits.

Scent marking behaviour in ferrets has been described with several behaviors: defecation and urination, anal drag, wiping, body rubbing, and chin rubbing (Clapperton 1985). Some experiments have demonstrated olfactory preference and sex identification in ferrets using olfactory cues. (Clapperton et al. 1989). Species, sex, and age recognition has been confirmed later in wild relative species: *Mustela eversmanni* (Siberian Weasel) and *Mustela sibirica* (Steppe polecat) (Zhang et al. 2002) along with sex and individual recognition in ferrets (Zhang et al. 2005). These olfactory cues could be used by rodents to identify and avoid this predator.

Masini and collaborators found that rats avoided olfactory stimuli from ferret fur (Masini et al. 2005). However, Zimmerling et al. (1994) did not find any effect of anal gland secretion semiochemicals in feral populations of the deer mouse (*Peromyscus manuculatus*). The use of isolated ferret compounds and related species was tested recently by Sievert and Laska (2016). Their results indicated that 2-propylthietane, a chemical identified in the anal gland secretion of several mustelids (Brinck et al. 1983), as well as 3-methyl-1-butanethiol and 2,2-dimethylthietane a chemical identified in several species of skunks (Musteloidea superfamily) decreased general motor activity and elicited avoidance in cd-1 mice (Sievert and Laska 2016).

Nevertheless, there are far fewer publications related to ferret olfactory stimuli as a predator stimulus for rodents than for the most commonly studied species in this area, the cat (*Felis catus*). This can likely be attributed to pragmatic reasons. First, cats outnumber ferrets as companion animals (AVMA 2012), and secondly, pet ferrets have been sold castrated and without anal glands in many countries, which means that their anatomy, physiology, secretions, and behaviour have been altered and they are therefore not suitable as models for chemical signals in the species.

In summary, however, there is some evidence that rodents use ferret cues by associating these stimuli to danger, and volatile putative kairomone compounds have been identified, but no heavier non-volatile compounds such as proteins have been established as anti-predator cues.

1.3 Snake Olfactory Cues as a Predator Stimulus

Olfaction is a primary sense for snakes, and they use it for intraspecific communication and leave chemical messages produced by secretory glands in the skin or in the anal glands (Mason and Parker 2010; Parker and Mason 2011). Terrestrial snakes travel from one point to another by characteristic undulatory movements, with the ventral part in direct contact with the ground. During these movements, they can leave chemical compounds produced in the skin or the anal glands. These trails can be used by other conspecifics to identify their sex and/or mating status. These compounds have been identified mainly as fatty acids (Parker and Mason 2011), with the exception of some rare airborne pheromones, which are detected via direct contact (tongue flicking) with the Vomeronasal Organ (Shine and Mason 2012). These chemical cues could potentially be good kairomone candidates as they would indicate the relatively recent presence of snakes and where they have traveled, and rodents widely use sniffing behaviour to detect heavy molecules like the Major Urinary Proteins (MUP).

Rodents and snakes have a long co-evolutionary history together, probably longer than with any other predator. Snakes and their clade Serpentes diverged approximately 35 million years before rodents (Graphodatsky et al. 2011; Reyes-Velasco et al. 2015); rodents have therefore shared habitats with snakes throughout their entire evolutionary history of 75 million years. Mechanisms and chemicals used to detect these predators probably evolved before speciation of actual species of rodents (Boursot et al. 1993) and snakes, where the order Squamata is the largest order within the Class Reptilia with 10265 living species, and its suborder Serpentes accounts for approximately 3600 species (www.reptiledatabase.org 2017). Therefore, allopatric species (i.e., species not sharing the same geographic habitat) could demonstrate anti-predator behaviors based on ancient mechanisms originating from this long co-evolutionary history. Chemical compounds used to detect predators can come from compounds that already have a chemical meaning in intra-specific communication or between competitor species (Banks et al. 2016). Indeed, some studies have provided evidence of anti-predator behaviors to snake olfactory stimuli in rodents or lizards, and in some cases, the species used were allopatric. However, this information is incomplete and sometimes contradictory (Papes et al. 2010; de Oliveira Crisanto et al. 2015).

In this present study, we decided to test olfactory stimuli that could have an ecological meaning for rodents in avoiding these predators. We tested stimuli from three species. *Rinechis scalaris* and *Vipera aspis* are terrestrial medium-sized species found in southern Europe with a diet composed largely of small mammals such as rodents from the genus *Mus*, *Rattus* or *Apodemus* (Saviozzi and Zuffi 1997; Pleguezuelos et al. 2007), and *Trimesurus albolabris* is a semi-arboreal species from southeast Asia that feeds partially on rodents, but probably in a smaller proportion than the other two species (Coborn 1991). We used skin sheds samples as they probably contain the chemical compounds left by the snakes as they travel and are easily sampled and transported. In addition, skin sheds have already shown anti-predator effects in rodents (Papes et al. 2010).

1.4 Fox and Dog Olfactory Cues as a Predator Stimulus

The dog (*Canis familiaris*) is a domesticated carnivorous mammal species descended from the wolf (*Canis lupus*). Along with the red fox *(Vulpes Vulpes),* both are considered generalist predators (Hanski et al. 1991). However, rodents represent a higher proportion of the diet of the red fox (*Vulpes vulpes*) (Leckie et al. 1998; Díaz-Ruiz et al. 2013) than of the wolf, whose diet also includes rodents, but who prefers larger prey such as ungulates (Capitani et al. 2004; Wagner et al. 2012). Here, we tested if rodents differed in their predator avoidance behaviour against faeces from these two carnivore species according to their prey preference.

1.5 Cat Urine Olfactory Cues as a Predator Stimulus

Cats are proteinuric, excreting large quantities of proteins (0.5–1.0 mg/mL) in their urine, where ca. 90% consists of cauxin, a carboxylesterase-like protein (Spotte 2014). Cauxin is directly correlated with L-Felinine, an amino acid excreted in cat urine (Miyazaki et al. 2006a, b). Cat urine and the amino acid L-Felinine have shown some negative effect on rodent reproduction (Vasilieva et al. 2000; Voznessenskaya 2014). We thus hypothesize that this protein or other chemical compounds in cat urine could be identified by house mice as chemical cues indicating danger, affecting mouse exploratory behaviour and/or locomotor activity. Although the role of cauxin has not been clearly identified (Spotte 2014), we know that its production depends on the sex of the animal, with males secreting higher amounts than females (Miyazaki et al. 2006a), as is also the case for Major Urinary Proteins in rodents, and therefore we used a non-castrated male in this experiment.

2 Material and Methods

2.1 Animals

Forty-eight (24 males and 24 females) RjOrl: Swiss 16–22 week old mice (Janvier Labs, France) were kept at the facilities of the Research Centre in Semiochemistry and Applied Ethology (IRSEA) under the supervision of a veterinarian specializing in laboratory animals according to the requirements of French and European Law (2010/63/EU). All protocols and techniques described in this paper were approved by the IRSEA ethics committee (approval number AFCE20150501).

The breeding room was kept at a temperature of 22 ± 2 °C and 60 ± 20% humidity. A 12–12 h inversed (light: dark) light cycle regimen was used with the light cycle beginning at 12:00 PM (lights off). All procedures were conducted between 12:00 PM and 5:00 PM, as the beginning of the dark cycle is one of the most active periods in mice (McLennan and Taylor-Jeffs 2004).

Housing cages were Eurostandard type IIL (Tecnipast, Italy), (369 × 156 × 132 mm), with a total floor surface of 435 cm^2. Animals were group housed (except during the tests and the habituations) to minimize stress due to isolation. As mice are a social species, each cage housed three animals. Food was available ad libitum, with 2014 global rodent diet (Envigo, UK). The lignocel 3–4 (Envigo, UK) bedding was changed weekly. As enrichment material, each cage was equipped with a red plastic tube and craft paper and white paper as nesting material (Genobios, Laval, France), as mice prefer complex nests with more than one material (Hess et al. 2008).

2.2 Apparatus

Rectangular arenas with a 4 mm thick 50 × 30 cm glass base covered with a transparent plastic top were used for the replicates. The treatment was applied on a piece of medical gauze (4 × 4 cm) and placed on one of the two sides. The square of glass was marked underneath with electric tape to distinguish two laterals and a central area in the arena. Lateral areas were separated by a 1 mm thick opaque plastic PVC barrier measuring 24 × 30 cm, which was attached to the top of the arena. A small square (4.5 × 4.5 cm) was cut out in the centre to allow the tested mouse to move freely (Fig. 13.1).

2.3 Treatments and Treatment Application

The animals were raised in the laboratory and were naive to the tested stimuli having had no previous contact with any of them (Table 13.1). Treatments were poured over a 4 × 4 cm piece of medical gauze, which was placed inside a metallic drilled tea ball; the tea ball was placed on a circular recipient of the same diameter with a flat base to avoid rolling and set on a square of glass (8 × 8 cm, 3 mm thick) to diminish contact with the arena. Another empty tea ball was placed on the other area in the same conditions. Treatment for each animal was chosen according to a randomized procedure. Treatment position was balanced, the treatment was placed on each side for half

Fig. 13.1. Experimental device used for measuring avoidance. **a** On the left side we see the treated area (1), central area (2) and starting point, right side: untreated area (3). **b** Close-up photography for the treatment containers, one of both sides contained the treatment and the other remained always empty as control. This device was used as a modified open-field in order to measure the house mouse's exploratory and avoidance behaviour, as well as specific anti-predatory and fearful responses to predator stimuli. The vertical plastic divisions had the role of reducing the passage of volatile compounds to adjacent areas and acted as a physical visual barrier

of the replicates to avoid bias. Treatment was considered as the only independent variable; 6 groups of 8 animals were tested for each treatment.

Ferret (*Mustela furo*) odour: Samples were generously donated from two anonymous owners of pet ferrets in the south of France. Samples were recovered from two ferrets, one male and one female. The skin/fur olfactory cue sample was recovered by means of two cotton towels which remained in the cage of the male or the female for one month. The faeces/anal gland stimuli were recovered by taking fresh faeces from the ferret cages and were stored at −20 °C until the tests. The treatment used as a complex general ferret olfactory cue was composed of 2 pieces of tissue (2.5 × 2.5 cm) which remained in the cage for 1 month, 1 piece with the male; 1 piece with the female + 1.5 g of faeces from the male and 1.5 g from the female.

Snake odour: Ecdysis or shedding is the change of the most external keratinized part of the epidermis in snakes; snakes must regularly renew this layer to keep pace with continuous growth throughout their life span. Shedding also plays a role in sanitation and eliminating skin diseases such as mycosis (Tu et al. 2002; Jacobson

Table 13.1. Predator species tested as olfactory stimuli of the house mouse (*Mus musculus*). [1]Stimuli tested together as snake olfactory cue; [2]not applicable; [3]stimuli tested together as ferret olfactory cue

Species	Vernacular name	Breed/subspecies	Sex	Olfactory cue/s	Amount of sample
Canis familiaris	Dog	Beagle	Females	Faeces	3 g
Felis cattus	Cat	European	Males	Urine	1 mL
Rinechis scalaris1	Ladder snake	N/A[2]	Unknown	Snake skin shed	1 piece 5 × 5 cm
Vipera aspis1	Aspic viper		Both sexes		1 piece 5 × 5 cm
Trimesurus albolabris[1]	White-lipped pit viper		Female		1 piece 5 × 5 cm
Mustela putorius	Domestic ferret	Furo	Both sexes	Cotton tissue in contact with the animals for 1 month + faeces[3]	2 pieces of towel 2.5 cm × 2.5 cm (1♂, 1♀) 3 g faeces (1.5♂, 1.5♀)
Vulpes vulpes	Red fox	N/A[2]	Unknown	Faeces	3 g

2007), and is the first barrier between snakes and their environment. Snake sheds are the most external part of these reptiles, and therefore probably contain the chemicals found in snake trails. For this reason, and because of the ease of sampling and transporting the material without the need to manipulate or stress the animals, we decided to use it as a snake olfactory predatory cue. Snake sheds were generously donated by the Centre d'Études Biologiques de Chizé (France). The sheds were recovered during the month of May after hibernation. They were sent to IRSEA between two and three weeks after being deposited and stored at −20 °C until their use for the behavioral tests. Three 5 × 5 cm pieces of snake shed, one for each species, *Rinechis scalaris (*ladder snake, unknown sex*)*, *Vipera aspis* (aspic viper, both sexes) and *Trimeresurus albolabris* (white-lipped pit viper, female) were used all together as a single predator stimulus. All three species were fed with mice.

Red fox (*Vulpes vulpus*) odour: Faeces of *Vulpes vulpes* were taken from fresh droppings of wild animals in the south of France. The sex of the animals was unknown. Faecal samples were stored at −20 °C until the beginning of the experiments. 3 g of faeces were used as a predator stimulus capable of eliciting avoidance according to previous research (Hacquemand et al. 2010).

Dog (*Canis familiaris*) odour: Faeces of un-spayed female beagles (CEDS, centre d'elevage du domaine des souches, Mezilles, France) were sampled in the facilities of the IRSEA. The animals were 8 years old when the samples were collected. 3 g of female faeces were used as a generalist predator stimulus; this amount was based on previous studies with another member of the Canidae family (Hacquemand et al. 2010).

Cat (*Felis domesticus*) odour: Cat urine from a European breed uncastrated male was collected in the facilities of the IRSEA from a non-absorbent bedding (Katkor, Rein Vet products, The Netherlands). Urine was frozen in aliquots after sampling and unfrozen the morning of the tests at room temperature. 1 mL of urine was used as a cat predator olfactory stimulus during the tests.

Blank/Control odour: A clean medical gauze was used as a control treatment as it was the physical support for all treatments.

During experimental manipulation, blinding was not possible as all conditions were easily visually recognizable to the human observer before being hidden by the tea balls.

2.4 Behavioral Test

All animals were habituated to the device for a period of 10 min on 2 days during the 2 weeks leading up to the experiment. No treatment was applied during these sessions, only the empty tea balls were present.

On the day of the experiment, animals were transported from the holding cage to the testing room, placed in the arena using red PVC tubes in order to decrease stress from tail manipulation (Hurst and West 2010), and video-recorded for 10 min. The treatment was applied and its position in the arena was randomized for all the replicates.

Each treatment and control group were composed of four males and four females. Animals were not euthanized at the end of the experiments. The tests were conducted in the experimental room between 12:00 PM and 5:00 PM, with temperatures in the range of 21 ± 2 °C, and humidity 50 ± 20%. The same operator manipulated the mice throughout all tests.

The glass base, the transparent cover, and the PVC separations were cleaned between replicates with Vigor surpuissant® disinfectant cleaner (eau écarlate, Ste Geneviève des Bois, France); they were then cleaned with white paper towels dampened with water, and finally dried with clean white paper towels. Two identical arenas were rotated between replicates in order to dissipate possible volatile traces of cleaner product. Each tea ball was used only for one treatment with its own glass square to diminish the risk of cross contamination. The external part of the tea ball and the metallic base were cleaned between animals with white paper wetted with alcohol.

2.5 Measures and Video Analysis

Each replicate was video-recorded with a video-camera placed 1 m over the arena (JVC HD Everio 1920x1080 full HD model GZ-HM446), located at a 90° angle viewing position to the arena. This perspective allowed a complete analysis of avoidance behaviour and locomotor activity. Video analysis was performed by two independent observers (CG and JL).

Video analyzes were carried out blind. The manipulators knew the treatment when conducting the tests but had no notion of the experimental condition during the video analysis.

Avoidance behaviour was measured using the dependent variables: treated area total duration, untreated area total duration. Avoidance behaviour was defined as animals significantly increasing the time spent in the "untreated area," or decreasing the time spent in the "treatment area."

2.6 Statistical Analysis

Data analysis was carried out using SAS 9.4 software Copyright (c) 2002–2012 by SAS Institute Inc., Cary, NC, USA. Bilateral situation; the significance threshold was fixed at 5%. Student t tests were used with the t test procedure as normality (analyzed with the UNIVARIATE procedure) was established.

3 Results

No difference was observed for the medical gauze alone, used as our control odour ($P = 0.51$), with an average of 201.62 s in the treated area and 222.12 s in the untreated area. This means that the medical gauze did not have any significant attractant or repulsive olfactory effect.

Mice spent significantly less time in the treated area than in the untreated area when they were exposed to the ferret's olfactory stimuli ($P = 0.0474$), with a mean of 168.25 s spent in the treatment area and 221.65 s in the area farthest from the treatment (untreated area).

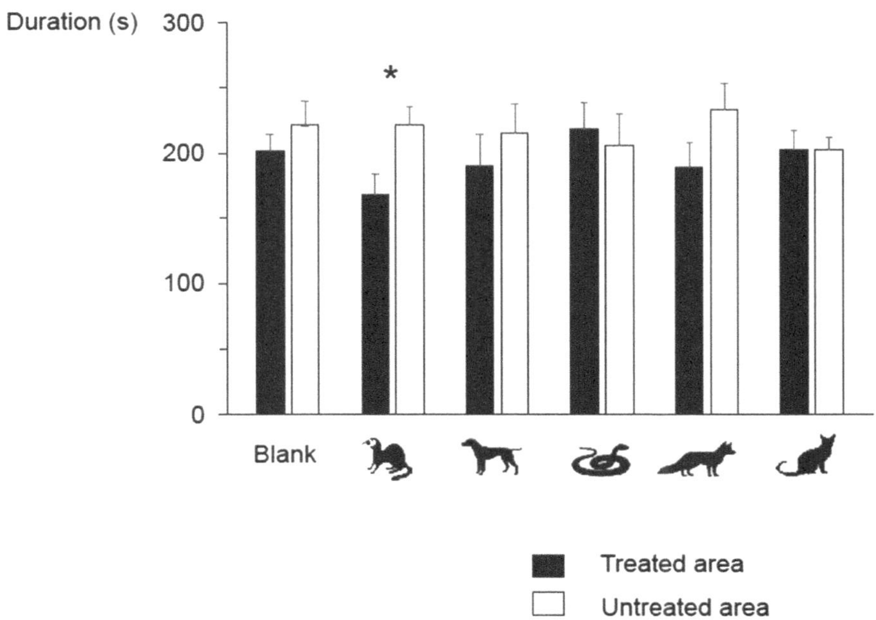

Fig. 13.2. Total duration of time remaining in the treated area and the untreated area. The multiple comparisons have been computed using the *T-test*. Significance threshold was fixed at 5%. $*P \leq 0.05$. Data is shown as the mean ± standard error, as parametric conditions where accomplished

None of the other treatments showed any significant differences or any discriminative tendency between the treated and the untreated area (Fig. 13.2). However, regarding the descriptive data, animals exposed to fox faeces showed higher average times in the untreated area (233.25 s) and the second lowest average time in the treated area (189.12 s) when compared to all other treatment conditions. However, the *t*-test results are far from a significant ($P = 0.25$) when comparing the times, the animals remained in both areas for the fox faeces condition.

The treatment with dog faeces showed no significant difference between the two areas ($P = 0.59$), with an average of 190 s in the treated area and 215 s in the untreated area.

Regarding the ophidian treatment, the snake sheds, mice showed no significant difference between the two areas ($P = 0.77$), with an average of 218.5 s in the treated area and 206.25 in the untreated area.

Finally, the treatment with cat urine showed no significant difference between the two areas ($P = 0.97$), with an average of 203.43 s in the treated area and 202.62 s in the untreated area.

4 Discussion

Our results show that the outbreed strain of mice used here, avoided a complex olfactory mustelid stimulus from *Mustela furo* (Ferret), spending significantly more time in the farther end of the device than close to this predatory stimulus. No differences were observed for any of the other predator stimuli, including cat, dog, fox, and snake odors. Furthermore, no differences were observed between the control and the blank, which means that the medical gauze did not have an attractant or repulsive olfactory effect.

Our results on avoidance behaviour to ferret olfactory stimuli appear to agree with previous studies. Masini et al. (2005) found similar avoidance behaviors in Sprague-Dawley rats, where ferret fur stimuli elicited anti-predator behaviours and increased ACTH and plasma corticosterone. Fur also elicited the expression of c-fos in brain areas related to stress (Masini et al. 2005), and did not produce habituation to this predator stimulus in Sprague-Dawley-rats (Masini et al. 2006). Although these authors did not find anti-predator responses to ferret faeces, urine or anal gland secretions (Masini et al. 2005), these responses have been observed in other studies (Apfelbach et al. 2005).

Apfelbach et al. (2001) found that ferret urine influenced the estrus cycle of female Campbell's hamsters (*Phodopus cambelli*), delaying or inhibiting ovulation. In a similar study (Apfelbach et al. 2015), the same authors found that Dwarf hamsters could distinguish ferret urine from animals fed with hamsters versus mice or chickens (as controls), where latency to approach the olfactory stimulus, the number of visits, and the total duration close to the stimulus was decreased when ferrets were fed with hamsters. Ferret urine also decreased counter-marking in high marking male mice (Roberts et al. 2001), where the authors hypothesized that the hamsters were able to distinguish between the urines because of the different ratios of chemical compounds based on the diet composition of the ferrets, rather than the presence of new compounds. Specifically, pyrazines could play a role in the odour sensing and caution expressed in the presence of the ferret urine (Apfelbach et al. 2015), as it has been shown in prey species such as mice and deer (Osada et al. 2015).

As our ferret predator stimulus contained faecal material, the avoidance observed during our tests could be due to chemical compounds such as the sulfurous compound 2,2-dimethylthietane, which has already been identified as a putative kairomone (Sievert and Laska 2016). However, this compound can be considered as highly volatile and our samples were collected over a relatively long period. The most volatile compounds had most likely already volatilized, which could mean that this effect was produced by a heavier, long-lasting molecule. The heaviest and most resistant molecules in chemical communication are proteins (Wyatt 2014).

Lipocalins are the only protein family for which a kairomone effect has been identified in mammals; these include the major rat urinary protein MUP13, and the cat's lipocalin Fel d 4 (Papes et al. 2010). The ferret has a protein allergen whose structure and origin has not been identified (Díaz-Perales et al. 2013). However, its size (17 kDa) is identical to mouse or rat lipocalins (Konradsen et al. 2015) and similar to other lipocalins, such as the cat allergen Fel d 4 (19.7 kDa). From these notions, we can hypothesize that one explanation for avoidance of olfactory stimuli from ferrets in mice could be the presence of a lipocalin protein found in ferret fur, saliva and urine (Díaz-Perales et al. 2013). This protein would share properties and homology with other kairomone lipocalins, which would simplify the recognition of these different stimuli.

One possible critique of this hypothesis lies in the fact that, in our experiments, mice had no direct contact with the compounds (the drills in the metallic mesh measured 0.5 mm which prevented any contact); this may impede heavier molecules such as peptides or proteins from reaching the Vomeronasal Organ, where molecules are typically transferred to by direct physical contact with their nostrils of scent marks (Kimoto et al. 2005; Roberts et al. 2010). However, the metallic balls were perforated, and the mouse's nose could approach the stimuli at a very short distance (0.16 mm, the diameter of the wires), which may be close enough to absorb this small protein/peptide with the air stream created by sniffing.

Our results showed no avoidance behaviour toward snake sheds in an outbreed laboratory mouse strain; several factors may explain this result. The snake samples may not have been fresh enough to send a significant message of risk and this could diminish antipredator responses such as avoidance (Bytheway et al. 2013). The period of ecdysis can be especially rich in the production of pheromones in snakes and the skin shed has been shown to contain a large quantity of pheromones at this particular moment (Parker and Mason 2011). The sheds were transported at room temperature and the periods between some of the sampling and delivery were from 2 to 3 weeks, so there is some possibility that the lighter chemical compounds had evaporated and the heaviest had degraded (Bytheway et al. 2013). Regarding the relevance of the species used, *Rinechis scalaris* and *Vipera aspis* have a high proportion of small mammals in their diets, mainly rodents from the genus Mus, Rattus or Apodemus (Saviozzi and Zuffi 1997; Pleguezuelos et al. 2007). *Trimesurus albolabris* is a semi-arboreal species and rodents are part of its diet, but probably in a smaller proportion than for the other two species (Coborn 1991). The use of predator cues from animals that specialize in hunting the species being tested is likely preferable but not mandatory for predator cue experiments. We find generalized prey responses to similar predatory species even if they are not the prey's own direct predator (Webb et al. 2010).

Snake sheds have already been tested with different results in mammals or other species, eliciting fear behaviors and avoidance in some studies and having no effect in others. Papes et al. (2010) found that snake sheds increased corticosterone in blood, elicited avoidance and specific antipredator behaviors including freezing and risk assessment, and produced a significant activation of the vomeronasal organ sensory neurons. The species used in this case was not specified in the paper, it was only identified as a "snake pet" and the sample was fresh. Pillay et al. (2003) found that striped mice (*Rhadomys pumilio*) avoided the feces of a predator snake (*Hemachatus haemachatus*), and the response was bigger when the snake had fed on striped mice. In an older study, Weldon et al. (1987) found responses only in female mice: an increased number of fecal boli and decreased consumption of food. Regarding studies indicating negative or low responses in rodents to snake sheds as a predator stimulus, Wasko et al. (2014) found no responses to faeces, and very limited responses to live snakes in three species of rodents.

Some studies have been performed in reptiles. For example, Sullivan et al. (2002) observed anti-predator responses in red-backed salamanders (*Plethodon cinereus*) to a distilled-water rinse of garter snakes (*Thamnophis sirtalis)* previously fed with this species of salamander. In another study, Webb and collaborators (2010) found anti-predator responses to snake chemical cues in velvet geckos using "scented" cardboard which remained inside the cages of the snakes for 2 days; the snake species used to obtain the chemical cues were not common predators of the geckos.

Negative results regarding mouse avoidance to dog feces can be explained by the fact that the diets of dogs and their wild ancestors don't contain rodents as a primary food source. Rodent consumption increased as dogs became domesticated due to their cohabitation in ecologically impoverished human environments. Unlike dogs, foxes are considered generalist predators that consume rodents in large quantities, and olfactory cues, such as the fox feces compound TMT, have been extensively proven to elicit avoidance (Rosen et al. 2015). However, the TMT doses presented in feces are lower than those used in almost every publication (Buron et al. 2007), which could explain the smaller responses seen in our study.

Our results showed no avoidance in mice to male cat urine. The existing literature provides no clear conclusions about rodent avoidance of cat urine. The traditional hypothesis states that predator mammalian urines could elicit anti-predator responses in prey due to the sulfurous compounds derived from meat metabolism (Nolte et al. 1994). 2-phenylethylamine (PEA) is a chemical compound found in carnivorous mammalian species urines that elicits avoidance and fires fear hard-wires in mice and rats, however its presence is low in cats compared with other carnivorous species (Ferrero et al. 2011). Pyrazines and compound analogues that have been found in wolf urine also elicited anti-predator responses and firing of fear pathways (Osada et al. 2015). In another study, urine from several felines and canids induced defensive behaviors, but those of cats and herbivores showed no influence (Fendt 2006). Taken together, these results seem to validate the idea that urine from canids and felines other than cats containing compounds such as PEA and pyrazines elicit anti-predator behaviors in rodents. The effect is not seen in response to cats because these compounds are absent or present in only small amounts in cat urine. Bramley and collaborators (2002) found unclear responses to cat urine in rats. In a first study with Norway rats, only one of two island populations of wild rats showed avoidance. In a second study with the ship rat (*Rattus rattus*) and Polynesian rats (*R.exulans),* no responses to the predator stimuli were observed (Bramley and Waas 2001). On the other hand, as we stated previously, there is some evidence that cat urine and the amino acid Felinine itself are capable of altering reproductive parameters in rodents (Vasilieva et al. 2000; Voznessenskaya 2014). As a methodological critique, our urine samples were taken from only one cat, and ideally the sample tested should pool urine from several animals, so we cannot say that it is fully representative.

The mouse strain used for the experiments, RjOrl: Swiss, is a laboratory mouse outbreed strain, with a wider genetic repertoire than other laboratory strains. Nevertheless, it has a high degree of inbreeding and consanguinity (Wahlsten and Crabbe 2007). This implies some specific phenotypes that could affect predator chemical cue detection (Dell'Omo et al. 1994). Differences in vomeronasal organ receptors have been identified in other strains due to the process of inbreeding from the original wild species. This may explain differences in responses between strains and between lab and wild animals (Stempel et al. 2016). In addition, laboratory mice strains have been selected to be tame and easily handled, and are less reactive in general to aversive stimuli, so this could also decrease the behavioral response to predator stimuli (Goto et al. 2013).

Finally, we must consider the statistical power and the number of animals used; in our study, we observed a considerable degree of variability. So, the use of more sensitive animals, as is the case in wild animals or other strains probably would increase the behavioral response (but also the variability). Increasing the number of animals per group would augment the chances of observing statistical differences in further research.

5 Conclusion

Our results showed that mice significantly avoided ferret olfactory stimuli from fur and feces. Additional research should further explore ferret olfactory cues as they likely present biologically meaningful messages for mice. The lack of avoidance behavior to the other stimuli tested may be attributed in some cases to phenotypic characteristics of the laboratory strain, statistical power, or aging of the samples. In our opinion, snakes remain a good candidate for finding the first reptile kairomone for rodents due to their ecological importance as predators and their long co-evolutionary history with rodents.

Acknowledgements. We acknowledge X. Duchemin and X. Bonnet for the snake sheds, and E. Landen for English proofreading.

References

Abrams PA (2000) The evolution of predator-prey interactions: theory and evidence. Annu Rev Ecol Syst 31:79–105

Alexander PDT (1951) A history of the ferret. J Hist Med 6:471–480

Apfelbach R, Blanchard CD, Blanchard RJ, Hayes RA, McGregor IS, (2005) The effects of predator odors in mammalian prey species: A review of field and laboratory studies. Neurosci Biobehav Rev 29(8):1123–1144

Apfelbach R, Soini HA, Vasilieva NY, Novotny MV (2015) Behavioral responses of predator-naïve dwarf hamsters (*Phodopus campbelli*) to odor cues of the European ferret fed with different prey species. Physiol Behav 146:57–66

Apfelbach R, Wiest H, Vasilieva NY (2001) Ferret (*Mustela putorius* f. furo) odor affects the estrous cycle in Campbell' s hamster females (*Phodopus campbelli*). Wiss Mitt Niederösterr Landesmuseum 14:147–152

AVMA (2012) U.S. pet ownership & demographics sourcebook

Banks PB, Daly A, Bytheway JP (2016) Predator odours attract other predators, creating an olfactory web of information. Biol Lett 12:20151053

Boursot P, Auffray JC, Britton-Davidian J, Bonhomme F (1993) The evolution of house mice. Annu Rev Ecol Syst 24:119–152

Bradley T, Lightfoot TL, Mayer J (2006) Exotic pet behaviour: birds, reptiles and small mammals, 1st edn. Saunders Elsevier, St. Louis

Bramley GN, Waas JR (2001) Laboratory and field evaluation of predator odors as repellents for kiore (*Rattus exulans*) and ship rats (*R. rattus*). J Chem Ecol 27:1029–1047

Bramley GN, Waas JR, Henderson HV (2000) Responses of wild Norway rats (*Rattus norvegicus*) to predator odors. J Chem Ecol 26:705–719

Bull JW, Maron M (2016) How humans drive speciation as well as extinction. Proc R Soc B Biol Sci 283:20160600

Buron G, Hacquemand R, Pourie G et al (2007) Comparative behavioral effects between synthetic 2,4,5-trimethylthiazoline (TMT) and the odor of natural fox (*Vulpes vulpes*) feces in mice. Behav Neurosci 121:1063–1072

Bytheway JP, Carthey AJR, Banks PB (2013) Risk vs. reward: how predators and prey respond to aging olfactory cues. Behav Ecol Sociobiol 67:715–725

Brinck C, Erlinge S, Sandell M (1983) Anal secretion in mustelids. J Chem Ecol 9:727–745

Capitani C, Bertelli I, Varuzza P et al (2004) A comparative analysis of wolf (*Canis lupus*) diet in three different Italian ecosystems. Mamm Biol—Zeitschrift für Säugetierkd 69:1–10

Church B (2007) Ferret-polecat domestication: genetic,taxonomic and phylogenetic relationships. In: Lewington JH (ed) Ferret husbandry, medicine and surgery, second. Saunders Elsevier, Philadelphia, pp 122–150

Clapperton BK (1985) Olfactory communication in the ferret (*Mustela furo* L.) and its application in wildlife management. Massey University

Clapperton BK, Minot EO, Crump DR (1989) Scent lures from anal sac secretions of the ferret *Mustela furo* L. J Chem Ecol 15:291–308

Coborn J (1991) The atlas of snakes of the world, 1st edn. T.F.H Publications

de Oliveira Crisanto K, de Andrade WMG, de Azevedo Silva KD et al (2015) The differential mice response to cat and snake odor. Physiol Behav 152:272–279

Dell'Omo G, Fiore M, Alleva E (1994) Strain differences in mouse response to odours of predators. Behav Processes 32:105–116

Díaz-Perales A, González de Olano D, Pérez-Gordo M, Pastor-Vargas C (2013) Allergy to uncommon pets: new allergies but the same allergens. Front Immunol 4:3–8

Díaz-Ruiz F, Delibes-Mateos M, García-Moreno JL et al (2013) Biogeographical patterns in the diet of an opportunistic predator: the red fox *Vulpes vulpes* in the Iberian Peninsula. Mamm. Rev. 43:59–70

Dietl G, Kelley P (2002) The fossil record of predator-prey arms races: coevolution and escalation hypotheses. Paleontol Soc Pap 8:353–374

Fendt M (2006) Exposure to urine of canids and felids, but not of herbivores, induces defensive behavior in laboratory rats. J Chem Ecol 32:2617–2627

Ferrero DM, Lemon JK, Fluegge D et al (2011) Detection and avoidance of a carnivore odor by prey. Proc Natl Acad Sci U S A 108:11235–11240

Goto T, Tanave A, Moriwaki K et al (2013) Selection for reluctance to avoid humans during the domestication of mice. Genes Brain Behav 12:760–770

Graphodatsky AS, Trifonov VA, Stanyon R (2011) The genome diversity and karyotype evolution of mammals. Mol Cytogenet 4:22

Hacquemand R, Jacquot L, Brand G (2010) Comparative fear-related behaviors to predator odors (TMT and natural fox feces) before and after Intranasal $ZnSO_4$ treatment in mice. Front Behav Neurosci 4:188

Hanski I, Hansson L, Henttonen H (1991) Specialist predators, generalist predators, and the microtine rodent cycle. J Anim Ecol 60:353–367

Hess SE, Rohr S, Dufour BD et al (2008) Home improvement: C57BL/6J mice given more naturalistic nesting materials build better nests. J Am Assoc Lab Anim Sci 47:25–31

Hurst JL, West RS (2010) Taming anxiety in laboratory mice. Nat Methods 7:825–826

Jacobson ER (2007) Infectious diseases and pathology of reptiles, first. Taylor & Francis, London

Kimoto H, Haga S, Sato K, Touhara K (2005) Sex-specific peptides from exocrine glands stimulate mouse vomeronasal sensory neurons. Nature 437:898–901

Konradsen JR, Fujisawa T, van Hage M et al (2015) Allergy to furry animals: new insights, diagnostic approaches, and challenges. J Allergy Clin Immunol 135:616–625

Lanszki J, Heltai M (2007) Diet of the European polecat and the steppe polecat in Hungary. Mamm Biol 72:49–53

Leckie FM, Thirgood SJ, May R, Redpath SM (1998) Variation in the diet of red foxes on Scottish moorland in relation to prey abundance. Ecography (Cop) 21:599–604

Lodé T (1997) Trophic status and feeding habits of the European Polecat *Mustela putorius* L. 1758. Mamm Rev 27:177–184

Lowry H, Lill A, Wong BBM (2013) Behavioural responses of wildlife to urban environments. Biol Rev Camb Philos Soc 88:537–549

Masini CV, Sauer S, Campeau S (2005) Ferret odor as a processive stress model in rats: neurochemical, behavioral, and endocrine evidence. Behav Neurosci 119:280–292

Masini CV, Sauer S, White J et al (2006) Non-associative defensive responses of rats to ferret odor. Physiol Behav 87:72–81

Mason RT, Parker MR (2010) Social behavior and pheromonal communication in reptiles. J Comp Physiol Neuroethol SensY, Neural, Behav Physiol 196:729–749

McLennan IS, Taylor-Jeffs J (2004) The use of sodium lamps to brightly illuminate mouse houses during their dark phases. Lab Anim 38:384–392

Miyazaki M, Yamashita T, Hosokawa M, et al (2006a) Species-, sex-, and age-dependent urinary excretion of cauxin, a mammalian carboxylesterase. Comp Biochem Physiol B Biochem Mol Biol 145:270–277

Miyazaki M, Yamashita T, Suzuki Y, et al (2006b) A major urinary protein of the domestic cat regulates the production of felinine, a putative pheromone precursor. Chem Biol 13:1071–1079

Nolte DL, Mason JR, Epple G et al (1994) Why are predator urines aversive to prey? J Chem Ecol 20:1505–1516

Osada K, Miyazono S, Kashiwayanagi M (2015) The scent of wolves: pyrazine analogs induce avoidance and vigilance behaviors in prey. Front Neurosci 9:363

Papes F, Logan DW, Stowers L (2010) The vomeronasal organ mediates interspecies defensive behaviour through detection of protein pheromone homologs. Cell 141:692–703

Parker MR, Mason RT (2011) Pheromones in snakes: history, patterns and future research directions. Reprod Biol Phylogeny Snakes 552–571

Pillay N, Alexander GJ, Lazenby SL (2003) Responses of striped mice, *Rhabdomys pumilio*, to faeces of a predatory snake. Behaviour 140:125–135

Pleguezuelos JM, Fernández-Cardenete JR, Honrubia S et al (2007) Correlates between morphology, diet and foraging mode in the ladder snake *Rhinechis scalaris* (Schinz, 1822). Contrib to Zool 76:179–186

Reyes-Velasco J, Card DC, Andrew AL et al (2015) Expression of venom gene homologs in diverse python tissues suggests a new model for the evolution of snake venom. Mol Biol Evol 32:173–183

Roberts S a, Simpson DM, Armstrong SD, et al (2010) Darcin: a male pheromone that stimulates female memory and sexual attraction to an individual male's odour. BMC Biol 8:75

Roberts SC, Gosling LM, Thornton E a, Mcclung J (2001) Scent-marking by male mice under the risk of predation. Behav Ecol 12:698–705

Rosen JB, Asok A, Chakraborty T (2015) The smell of fear: innate threat of 2,5-dihydro-2,4,5-trimethylthiazoline, a single molecule component of a predator odor. Front. Neurosci. 9:292

Santos MJ, Matos HM, Baltazar C et al (2009) Is polecat (*Mustela putorius*) diet affected by "mediterraneity"? Mamm Biol 74:448–455

Saviozzi P, Zuffi MAL (1997) An integrated approach to the study of diet of *Vipera aspis*. Herpetol Rev 28:23–24

Shine R, Mason RT (2012) An airborne sex pheromone in snakes. Biol Lett 8:183–185

Sievert T, Laska M (2016) Behavioral responses of CD-1 mice to six predator odor components. Chem Senses 00:bjw015

Spotte S (2014) Free-ranging cats. Behav Ecol Manag

Stempel H, Jung M, Pérez-Gómez A et al (2016) Strain-specific loss of formyl peptide receptor 3 in the murine vomeronasal and immune systems. J Biol Chem 291:9762–9775

Sullivan AM, Maerz JC, Madison DM (2002) Anti-predator response of red-backed salamanders (*Plethodon cinereus*) to chemical cues from garter snakes (*Thamnophis sirtalis*): laboratory and field experiments. Behav Ecol Sociobiol 51:227–233

Tu MC, Lillywhite HB, Menon JG, Menon GK (2002) Postnatal ecdysis establishes the permeability barrier in snake skin: new insights into barrier lipid structures. J Exp Biol 205:3019–3030

Uetz P, Hallermann J, Hosek J (2017) www.reptiledatabase.org

Vasilieva NY, Cherepanova EV, von Holst D, Apfelbach R (2000) Predator odour and its impact on male fertility and reproduction in *Phodopus campbelli* hamsters. Naturwissenschaften 87:312–314

Voznessenskaya VV (2014) Influence of cat odor on reproductive behavior and physiology in the house mouse: *Mus musculus*. CRC Press/Taylor & Francis

Wagner C, Holzapfel M, Kluth G et al (2012) Wolf (*Canis lupus*) feeding habits during the first eight years of its occurrence in Germany. Mamm Biol 77:196–203

Wahlsten D, Crabbe JC (2007) Chapter 15 behavioral testing. In: The mouse in biomedical research

Wasko DK, Bonilla F, Sasa M (2014) Behavioral responses to snake cues by three species of Neotropical rodents. J Zool 292:142–150

Webb JK, Du W, Pike D, Shine R (2010) Generalization of predator recognition: velvet geckos display anti-predator behaviours in response to chemicals from non-dangerous elapid snakes. Curr Zool 56:337–342

Weldon PJ, Divita FM, Middendorf GA (1987) Responses to snake odors by laboratory mice. Behav Process 14:137–146

Wyatt TD (2014) Proteins and peptides as pheromone signals and chemical signatures. Anim Behav 97:273–280

Zhang J-X, Soini HA, Bruce KE et al (2005) Putative chemosignals of the ferret (*Mustela furo*) associated with individual and gender recognition. Chem Senses 30:727–737

Zhang J-X, Sun L, Bin Zhang Z et al (2002) Volatile compounds in anal gland of Siberian weasels (*Mustela sibirica*) and steppe polecats (*M. eversmanni*). J Chem Ecol 28:1287–1297

Zimmerling LM, Sullivan TP (1994) Influence of mustelid semiochemicals on population dynamics of the deer mouse (*Peromyscos maniculatus*). J Chem Ecol 20:667–689

Do Carnivores Have a World Wide Web of Interspecific Scent Signals?

Peter Apps[1(✉)], Kasim Rafiq[1,2], and J. Weldon McNutt[1]

[1] Botswana Predator Conservation Trust, Private Bag 13, Maun, Botswana
peterjapps@gmail.com
[2] School of Natural Science and Psychology, Liverpool John Moores University, Liverpool, UK

Abstract. The orthodox view of mammalian chemical communication is that individuals exchange information with members of their own species, but there is now a growing body of evidence that carnivores also send messages to members of different species. A camera trapping study in northern Botswana captured video of 21 species of carnivore, of which 18 scent marked a total of 105 times. Seventeen marks countermarked scent from the same species, and 24 marks countermarked scent from a different species. Of the 39 species of carnivore in southern Africa, 24 are involved in some kind of interspecific scent marking, and smaller webs of interspecific scent marking are found among carnivores in Europe, Australia, north and central America, and the Himalayas. Interspecific scent communication has been missed in the past due to an almost exclusive focus on single-species studies, and because the tool that is essential to study it; camera trapping, has come into widespread use only in the past decade. Now we know that interspecific scent marking is common and widespread, we need to take a closer look at it to understand its role in carnivore guild ecology.

1 Introduction

Mammal semiochemistry has been concerned almost exclusively with communication between individuals of the same species, with the small fraction of the work that has addressed interspecific scent communication focussed sharply on eavesdropping, most commonly by prey on predator odors (reviews Apfelbach et al. 2005; Jones et al. 2016; Parsons et al. 2017). When eavesdroppers sniff scent marks but do not leave marks of their own, information flows only one way, but there is a possibility of two-way communication when individuals of different species countermark each other's marks. Such interspecific marking has hardly ever been examined; the first results were published in 1991 when Paquet (1991) reported that wolves (*Canis lupus*) and coyotes (*Canis latrans*) countermark each other's urine, and it was not until 19 years later that McTavish and Gibeau (2010) published a short popular article on multiple species' use of communal marking sites. The pace of publication accelerated from 2013 when Li et al. (2013) recorded multispecies use of a snow leopard (*Panthera uncia*) marking site, followed by Allen et al. (2015) on multiple species' use of puma (*Puma concolor*) scent marking sites, King et al. (2016) on ocelot (*Leopardus pardalis)* latrines in central America, Wikenros et al. (2017) on red fox (*Vulpes vulpes*) responses to lynx (*Lynx*

© Springer Nature Switzerland AG 2019
C. D. Buesching (Ed.): *Chemical Signals in Vertebrates 14*, pp. 182–202, 2019.
https://doi.org/10.1007/978-3-030-17616-7_14

lynx) faeces in Europe, and Apps et al. (2017) on interspecific marking by African carnivores. It now looks as though interspecific scent marking may be common and widespread, and that a new field in mammalian semiochemistry is waiting to be explored.

Although the study of interspecific scent communication is growing rapidly, it has a very short history and a very small literature that we add to here with the results of a straightforward quantitative comparison of the rates of inter- and intra-specific countermarking in the southern African predator guild. We discuss the limited background information and argue that interspecific marking will turn out to be ecologically important, because scent communication between members of different species could have a role in ecological processes such as intra-guild competition, with important implications for conserving threatened populations and managing pests. At the very least, we need to take it into account when we study chemical communication in the field. The focus is exclusively on carnivores because they have been the subjects of all previously published studies.

2 Methods

Camera trapping records of scent marking and countermarking were collected in northern Botswana as "bycatch" from a study of the repellent effects of 3-mercapto-3-methyl butanol on leopards (*Panthera pardus*).

2.1 Study Area

The work was carried out in the Botswana Predator conservation trust study area in the wildlife management areas to the east of the Moremi Game Reserve, in northern Botswana. The vegetation is mixed tree savanna and the area supports populations of all its naturally occurring mammal species, including 24 species of carnivore ranging in size from lions (*Panthera leo*) to dwarf mongooses (*Helogale rufula*).

2.2 Camera Trapping

Reconyx Ultrafire XR6 (Reconyx, Holmen, WI) camera traps were deployed in four sets of four, and one set of three. The sets of four each had two cameras 30 m apart on opposite sides of a vehicle track, facing each other along the track, mounted 78 cm off the ground, and two cameras 10 m apart on opposite sides of the track, mounted 58 cm off the ground and aimed diagonally across the track towards two slow-release scent dispensers giving out tomcat thiol; 3-mercapto-3-methylbutanol, being tested for its repellent effects on leopards (for methods see Apps and McNutt 2018). All cameras were positioned 1 m from the edge of the track (Fig. 1). The set of three had two cameras 30 m apart and one camera 5 m from the edge of the track monitoring both scent dispensers. These multiple camera arrays overcame shortcomings in camera performance and provided video surveillance of approximately 28 m of track. The cameras were set on high sensitivity, with a 1 s trigger delay (the minimum), and captured 30 s of video on each trigger. Videos taken at night were illuminated by long

Fig. 1 A four-camera trapping array. The view is from a camera trap 15 m from the scent dispensers (circled) and shows the three other cameras (arrowed); two at 5 m from the dispensers and the other one at 15 m

wave (940 nm) infrared floodlights in the cameras. Cameras were set up on the 9th and 10th of January 2017 and operated until June 8-26, 2017 depending on site. Videos were captured over 3586 camera days. Damage by elephants, and flat batteries inactivated some cameras for varying lengths of time, but this will not have led to bias in favor of either intra- or inter-specific countermarking. The scent dispensers were initially empty to provide a scentless control condition, and were loaded with tomcat thiol on 26 March 2017. Since sample sizes for most species are small, data with and without scent in the dispensers have been pooled.

2.3 Data Processing

For each video that showed a carnivore scent marking, the species and number of individuals, the time they spent sniffing and/or scent marking, and what and where they sniffed or marked were extracted to spread sheets. Defecation, urination, body and cheek rubbing, and anogenital rubbing were all counted as scent marking. A mark was considered to be a countermark if it was in the same place as an earlier mark, with no limit on the time since the first mark was deposited, and the number of countermarks by animals of the same species was compared with countermarking by a different species. The relative number of intra- and interspecific countermarks is robust to differences in the times for which cameras were in action; camera failure would have reduced the number of marks recorded, and missed markings would have led to an undercount of marks that were countermarked, but there are no plausible grounds for supposing that camera failure could be biased toward inter- or intra-specific countermarking.

Except for lions and known individual leopards it was rare to be able to sex predators in videos and the data have not been separated according to sex.

3 Results

Among 2376 video records in which carnivores were in view for a total of 24,554 s (nearly 7 h), there were 105 records of scent marking and rubbing from carnivores of 18 species (Table 1). Sixty-four marks were not subsequently countermarked within the recording period, and of the 41 marks that were countermarked 17 were countermarked directly by the same species as deposited the mark, and 24 were countermarked directly by a different species. Four marks were indirect countermarks of the same species; they were where the same species had previously marked and had then been countermarked by a different species, and 17 countermarks were where a different species had marked and had subsequently been countermarked. The maximum number of marks on one spot was six, deposited successively by a banded mongoose, a slender mongoose, two cape foxes (*Vulpes chama*), a slender mongoose (*Galerella sanguinea*) and a Cape fox, in that order. Nine spots had three or more marks, for a total of 38 marks. The maximum number of species that used one spot was three, which occurred at five spots. Sixty-four marks were on "clean" surfaces (in other words surfaces with no previous recorded scent marks, they may have been marked before the cameras were set). Whether or not a mark was countermarked was not significantly affected by its being on a clean surface (Fishers exact test, 2-tailed, $P = 0.23$). The 22 countermarks of marks on clean surfaces were equally divided between inter- and intra-specific, but there was a bias toward subsequent countermarks (i.e., countermarks of countermarks; three or more marks on one spot) being interspecific; of the 19 direct countermarks of interspecific countermarks only four were by the same species as the previous mark (Fisher's $P = 0.10$), and of the 21 countermarks that indirectly overlaid another countermark only four were intraspecific.

Of the species that countermarked, only two, Selous' mongoose (*Paracynictis selousi*) and bat-eared fox (*Otocyon megalotis*) with one countermark each, countermarked only their own species. Another two; large-spotted genet (*Genetta maculata*) and yellow mongoose *(Cynictis penicillata)* did not countermark, but were countermarked. Animals from 6 species; banded mongoose (*Mungos mungo*), slender mongoose, civet (*Civettictis civetta*), cheetah (*Acinonyx jubatus*), caracal (*Caracal caracal*) and leopard directly countermarked only the marks of other species but did not countermark scent from conspecifics. Animals from three species; brown hyaena (*Hyaena brunnea*), African wild dog (*Lycaon pictus*) and black-backed jackal (*Canis mesomelas*) did not countermark at all, and were also not countermarked by other individuals of the same or any other species. Honey badgers (*Mellivora capensis*), spotted hyaenas (*Crocuta crocuta*) and dwarf mongooses were never recorded to scent-mark.

4 Discussion

4.1 *How Much Interspecific Marking Goes on?*

Interspecific scent marking is widespread and common. In addition to the results of this study it has also been recorded in other circumstances in our study area; African wild

Table 1 Summary of scent marking and countermarking by carnivores in northern Botswana

species that deposited the first mark	species that deposited the overmark																					
	small-spotted genet	large-spotted genet	serval	yellow mongoose	wildcat	lion	banded mongoose	slender mongoose	Cape fox	Selous mongoose	bat-eared fox	civet	cheetah	caracal	leopard	brown hyaena	African wild dog	black-backed jackal	total deposited	not countermarked	overmarked by same sp.	overmarked by other sp.
clean	3	4	5	2	25	2	1	4	0	6	1	1	0	4	1	3	1	1	64	42		22
small-spotted genet	4		1																9	4	4	1
large-spotted genet					1														4	3	0	1
serval	1		3		1							2		1					11	3	3	5
yellow mongoose					1														2	1	0	1
wildcat	1		1		6		1	2						1	1				36	23	6	7
lion						1													4	3	1	0
banded mongoose								1	1										2	0	0	2
slender mongoose									1										8	7	0	1
Cape fox								1	1										3	1	1	1
Selous mongoose										1									7	6	1	0
bat-eared fox											1								2	1	1	0
civet													1						3	2	0	1
cheetah																			1	1	0	0

Table 1 (continued)

caracal			2		1														6	3	0	3
leopard						1													2	1	0	1
brown hyaena																			3	3	0	0
African wild dog																			1	1	0	0
black-backed jackal																			1	1	0	0
column total direct countermarks intra-specific	4	0	3	0	6	1	0	0	1	1	1	0	0	0	0	0	0	0			17	
column total direct countermarks inter-specific	2	0	4	0	4	1	1	4	2	0	0	2	1	2	1	0	0	0				24
total countermarks per species	6	0	7	0	10	2	1	4	3	1	1	2	1	2	1	0	0	0	105	64	17	24

dogs, black-backed jackals, civets, and leopards scent mark at spotted hyaena latrines (Table 2) (Vitale 2017), and slender mongooses overmarked small-spotted genets (*Genetta genetta*) in a pilot of the present study (Table 2) (Apps et al. 2017). In addition, there are several passing mentions of interspecific scent marking in the literature, and records from other sources (Tables 3 and 4), and adding these to the Botswana results shows a dense web of connections among southern African carnivores (Fig. 2); of the 39 species of terrestrial carnivore in southern Africa, 24 are involved in some form of interspecific scent marking.

In other carnivore guilds on other continents, there are similar webs on a smaller scale (Table 2). Wolves and coyotes overmark one another's urine marks more than marks from their own species (Paquet 1991). "Communication trees" occur widely in Canada and are scent marked by wolves, coyotes, red foxes, lynx (*Lynx canadensis*), pumas, grizzly bears (*Ursus arctos*), black bears (*Ursus americanus*), wolverines (*Gulo gulo*), martens (*Martes americana*), and domestic dogs (*Canis familiaris*) (McTavish and Gibeau 2010). In the Santa Cruz Mountains, California, "community scrapes" are areas of soil 3-4 m across raked up and scent-marked by pumas, where coyotes, bobcats (*Lynx rufus*), grey foxes (*Urocyon cinereoargenteus*), and striped skunks (*Mephitis mephitis*) also scent mark (Allen et al. 2015). Scrapes are visited by grey foxes more often than by pumas, and foxes urinate on 38% of visits, a higher rate than female pumas, and cheek rub on puma urine marks on 10% of visits (Allen et al. 2017). Bobcats make scent scrapes during 13% of visits, spray urine during 12% and rub during 0.4% (Allen et al. 2015). Ocelot latrines in Costa Rica are visited by 14 species of mammals (King et al. 2016), of which four; tayras (*Eira barbara*), ocelots, white-nosed coatis (*Nasua narica),* and common opossums (*Didelphis marsupialis*) scent mark or rub. Tayras visit 40% more frequently than ocelots. In Poland, translocated scats from European lynx were visited by red foxes, European badgers (*Meles meles*), martens (*Martes* sp), and racoon dogs (*Nyctereutes procyonoides*), but not by lynx. Both martens and foxes countermarked the lynx scat; the foxes countermarking with urine on 38 out of 75 visits (Wikenros et al. 2017). In central China, seven species; snow leopards, wolves, dogs, brown bears, red foxes, Tibetan foxes (*Vulpes ferrilata*) and Pallas' cat (*Otocolobus manul*) marked a shared site at the foot of a cliff where two major valleys meet (Li et al. 2013).

Multispecies use of marking sites and latrines, and short mentions of interspecific countermarking are scattered through the literature (Table 3), other instances may still be hidden there, but finding them depends on luck or exhaustive trawling; interspecific scent marking is not used as a key search phrase. Personal communication with carnivore researchers revealed further, unpublished incidents of shared marking sites (Table 4). There are also examples of interspecific countermarking in videos from recreational camera trapping posted on Facebook and YouTube, but the hurdle of searchability is even higher than it is in the scientific literature.

https://www.youtube.com/watch?v=LK21zlz1sMg.

https://www.youtube.com/watch?v=5pn9yJx_tlI.

http://northernwoodlands.org/game-camera/article/marking-behavior.

Most of these records are single incidents or small-scale studies, but they are numerous and widespread enough to suggest that interspecific scent marking merits a closer look.

Table 2 Quantitative comparisons of inter- and intraspecific scent marking from the literature

Locations	Species A	Species B	Countermarks A to A	Countermarks B to B	Countermarks A to B	Countermarks B to A	References
Canada	Wolves	Coyotes	37/92	56/79	58/79	58/92	Paquet (1991)
Belize	Jaguars	Pumas	14	6	2	2	Harmsen et al. (2016)
Botswana	Small-spotted genets	Slender mongooses				9/9	Apps et al. (2017)
California USA	Pumas	Grey foxes	Pumas visited puma scrapes 762 times and marked 471 times, grey foxes visited 903 times and marked 557 times, bobcats visited 496 times and marked 131 times				Allen et al. (2014, 2015, 2017)
Costa Rica	Ocelots	Tayras, white-nosed coatis, opossums	Ocelots visited latrines 63 times, marked or rubbed 84 times, tayras visited 88 times, marked or rubbed 178 times, white-nosed coatis visited 44 times, marked or rubbed 20 times, opossums visited 50 times, rubbed 23 times				King et al. (2016)
Poland	Lynx	Red foxes	Red foxes visited lynx scats 75 times, no-scat controls 17 times; urinated on lynx scats 38 times, on controls once. Lynx did not visit				Wikenros et al. (2017)
Botswana	Spotted hyaenas	African wild dogs, civets, black-backed jackals, leopards	Hyaenas visited hyaena latrines 249 times, scent marked 50 times, wild dogs visited 72 times, marked 15 times, jackals visited 89 times, marked 14 times, civets visited 35 times, marked 4 times, leopards visited 47 times, marked once				Vitale (2017)
China	Snow leopards	Wolves, dogs, brown bears, red foxes, Tibetan foxes, Pallas' cats	Snow leopard 145 visits 86 marks, wolf 42 visits 5 marks, dog 30 visits 3 marks, brown bear 12 visits 7 marks, red fox 81 visits 11 marks, Tibetan fox 44 visits 3 marks, Pallas's cat 16 visits 1 mark				Li et al. (2013)

Clearly, the number of records from this study is too small to be any more than indicative, but the outcome is both clear and surprising; interspecific countermarking was more frequent than intraspecific countermarking, both overall and in most of the species that countermarked. This may simply have been because all the animals were exposed to more allospecific than conspecific marks, and were countermarking according to availability (Fig. 3). Even the most common marks, 36 from African wildcats (*Felos Silvestre's capra*), were outnumbered nearly 2:1 by other marks (Table 1), and there was also a weak trend for the number of times that a species' marks were countermarked to rise with the number of marks deposited by that species. Nevertheless, since mammals routinely make subtle olfactory discriminations, this tendency cannot be ascribed to an inability to discriminate conspecific from allospecific marks. In other carnivore guilds heterospecific marking being more frequent than conspecific marking is the rule rather than the exception (Table 2), which implies that interspecific countermarking is as significant to carnivore ecology as intraspecific countermarking is to carnivore social behavior.

4.2 What Took Us So Long?

Although incidents of interspecific scent marking are widespread, and it has been studied in some detail in six predator guilds on four continents, there is still a huge body of work on mammal scent communication in which it is never mentioned, including authoritative reviews from the early years (Macdonald 1980; Gorman and Trowbridge 1989), extensive reviews of eavesdropping (Kats and Dill 1998; Apfelbach et al. 2005; Jones et al. 2016; Parsons et al. 2017), a detailed review on overmarking (Ferkin and Pierce 2007), and a review of trends in the first 13 CSiV conferences (Müller-Schwarze 2016). If interspecific scent marking is as common and widespread as it now appears to be, why did it take so long for it to be recognized; or conversely, since records of interspecific marking are so scarce, is it really of any significance?. We believe that interspecific marking has been almost invisible due to the way that chemical communication has been studied, and the tools that have been available to study it.

The holy grail of semiochemistry is the chemical characterization and practical application of pheromones, and the definition of a pheromone is explicitly intraspecific; "…substances which are secreted to the outside by an individual and received by a second individual **of the same species**, in which they release a specific reaction,…." (my emphasis) (Karlson and Lüscher 1959).

Clearly, interspecific marking is impossible in the single-species populations of laboratory animals where the large majority of studies of mammalian scent communication have been conducted. Even in the field, research projects have nearly always focussed on single species, and have had a strong bias toward the immediate responses to scent marks by group members in the small minority of mammals that are conspicuously social (e.g., Jordan et al. 2011a, b). A researcher working on a species among the solitary majority, who saw an individual of species X scent mark in a particular spot will not have been watching when an individual of species Y scent marked the spot the day before, or when Y sniffed and marked again at the same spot, but may well have been following species X when it marked again at the site of its

Table 3 Pointers to potential interspecific scent marking from the literature, found opportunistically and by trawling; there are no useful specific search terms

Species A	Species B	Circumstances	References
Aardwolves	Black-backed jackals, Cape foxes, bat-eared foxes	Namibia, scats from each of the other species found in one of 30 aardwolf middens	Nel and Bothma (1983)
Cheetahs	Civets	South Africa, civet(s) repeatedly visited and one scent marked at a cheetah marking tree	Marnewick et al. (2006)
Lynx	Martens	Poland, martens overmark translocated lynx scats	Wikenros et al. (2017)
Tiger (*Panthera tigris*)	Tapir (*Tapirus indicus*)	Louisville Zoo, when a tapir was moved into an enclosure vacated by a tiger it overmarked the tiger's urine	Coe (2004)
Dog, cat (*Felis silvestris catus*)	Red fox	Australia, red foxes pay more attention to dog and cat faeces than to fox faeces, and may defecate near them	Banks et al. (2016)
Lions	Black-backed jackals	South Africa, black-backed jackals defecate on lion scat	Hayward and Hayward (2010)
Brown hyaenas	Honey badgers	Southern Kalahari, honey badgers and brown hyaenas make latrines under the same trees	Begg et al. (2003)
Wolves, coyotes	Domestic dogs	Alaska, feral dogs overmark wolves and coyotes	Gipson (1983), cited by Lenth et al. (2006)
Kit foxes (*Vulpes macrotis*)	Coyotes	California, 10 of 61 kit fox latrines contained coyote scats	Ralls and Smith (2004)
Swift foxes (*Vulpes velox*)	Coyotes	Colorado, some swift fox latrines contain coyote scats	Darden et al. (2008)
River otters	Mink, dogs	Chile, invasive mink and domestic dogs use river otter latrines	Sepúlveda et al. (2014)
Stone martens (*Martes fiona*), small-spotted genets	Pine martens	In captivity pine martens overmark stone marten and genet marks, but at the same rate as clean surfaces	De Monte and Roeder (1990)
European lynx urine	Wolf, small spotted genet, European badger, red fox	In captivity, rubbing in lure tests	Monterroso et al. (2011)
Stone marten scents	Wolf, genet		
Red fox urine	Wolf, genet		
European lynx urine	Red fox, stone marten, small-spotted genet	Foxes rubbed and urinated, genets and martens defecated	

Table 4 Unpublished pointers to potential interspecific scent marking

Species	Notes	Sources
Small-spotted genets, Cape genets *Genetta tigrina*	South Africa both species defecate in the same middens	Emmanuel Do Linh San, University of Fort Hare
Large-spotted genet, leopard	South Africa, a leopard urinated over a genet anal-gland mark on a tree	
Slender mongoose, yellow mongoose	Botswana, lip marking on the same twig	pers obs
Yellow mongoose, Selous mongoose	Botswana, anal marking the same object	pers obs
Bobcat, Andean cats (*Leopardus jacobita*),	South America, bobcat urine used as a camera trap lure was countermarked with urine, faeces, face rubs	Rodrigo Villalobos, Andean Cat Alliance
Cape clawless otters (*Aonyx capensis*), spotted-necked otters (*Lutra maculicollis*), water mongooses (*Atilax palidinosus*)	South Africa, defecate in shared latrines	Michael Somers, University of Pretoria
Cheetahs, slender mongooses, small-spotted genets	Botswana, slender mongooses and small-spotted genets mark at cheetah scent-marking trees	Jane Horgan, Cheetah Conservation Botswana
Brown hyaenas, black-backed jackals	Namibia, on the Namib coast, jackals defecate in brown hyaena latrines	Ingrid Wiesel, Brown Hyaena Research Project, Namibia

original mark. The countermark will then have been interpreted as a response to the original mark from species X, and not to the unrecorded marks from species Y.

The only countermarking that is likely to be detected after the marks have been deposited is when both marks are faeces and/or visible, discrete glandular secretions whose appearance differs between species or individuals. This is how multispecies use of latrines has been recognized (Tables 2, 3, 4) (e.g., Nel and Bothma 1983; Stewart et al. 2002; Ralls and Smith 2004; Vitale 2017), but the large majority of marks with urine and thin smears of secretions, including all of those captured on video in this study, will not be detected post hoc. When researchers have found scats from another species in their focal species' latrines it was typically reported (if at all) only briefly (e.g., Nel and Bothma 1983; Ralls and Smith 2004).

The recognition of interspecific scent marking has also been held back by refutation bias; Harmsen et al. (2016) recorded scent marking by jaguars or pumas at 12 of 19 camera sites used for population monitoring. Jaguars (*Panthera onca*) marked more than once at five sites, pumas marked more than once at three sites, and both pumas and jaguars marked at two sites. In other words, the rate of interspecific marking was 40% of the rate of intraspecific jaguar marking and 67% of the rate of puma-puma marking.

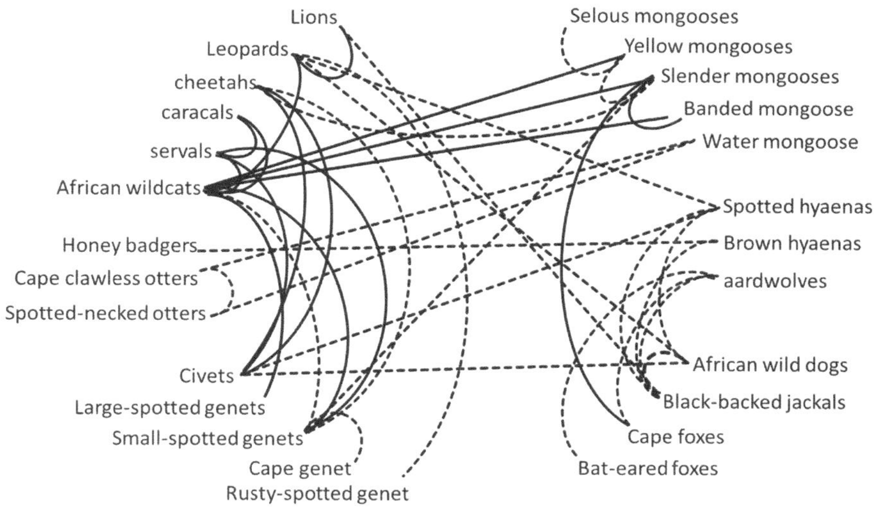

Fig. 2 Connections between southern African carnivores that countermark one another's scent marks. Solid lines are from this study, dashed lines are from the literature, personal communications, and personal observations. Sources are in Tables 2, 3, and 4

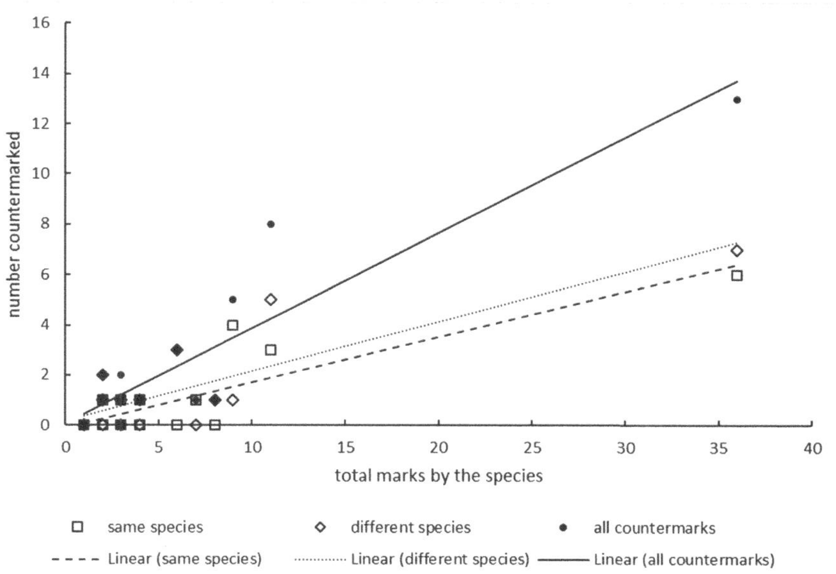

Fig. 3 The number of scent marks from each species that were countermarked increased with the number of marks that species deposited

Nevertheless, they state; "We found no evidence of interspecific response or countermarking between jaguars and pumas."

An important, almost critical, tool for recording interspecific scent marking did not come into routine use until after about 2000. Digital camera traps overcome many of the shortcomings of the usual approaches to work on scent communication in the field; they monitor scent marks instead of following animals, they record all species more or less indiscriminately, they work continuously, and have perfect spatial memory over periods of weeks or months so that countermarking can be recognized. Paquet's (1991) snow tracking study of wolves and coyotes is the only detailed work on interspecific marking that did not use camera trapping, and the pace of publication has accelerated sharply since fast-triggering cameras with good quality video have been available.

There is now enough evidence to conclude that interspecific scent marking has been going on behind the scenes while we have been using study designs and methods that could not have detected it even if we had been searching for it.

4.3 Is Interspecific Scent Marking Communication?

From current evidence, countermarking is widespread, and in some predator guilds it happens frequently, but it does not necessarily follow that messages are transmitted between the countermarkers. When individuals of the same species produce both marks, countermarking is taken as good evidence that both the original mark and the countermark are scent marks (Jordan et al. 2013), but when different species are involved this may not apply, and there are other possible reasons why an animal would put a scent mark close to or on top of one from another species. Putting a scent mark near another one may be the olfactory equivalent of putting it in a visually prominent position to make it easier to detect (e.g., Peters and Mech 1975; Barja et al. 2005; Monclús et al. 2009; Hayward and Hayward 2010). If several species favor the same kind of rare sites for scent marking, countermarking will be a consequence of their using the same sites, even if they take no notice of one another's marks. This may have occurred to some extent in the current study; the cameras and scent dispensers were popular marking sites; 38 of the 105 marks were on camera poles or boxes, and 15 on scent dispensers; 23 (62%) of the marks on cameras were countermarked and 8 (53%) of the marks on dispensers, compared to the 10 out of 52 (19%) marks on natural surfaces that were countermarked. Shared use of a favored site may account for multispecies use of a marking site in the Himalayas (Table 2) (Li et al. 2013). On the other hand, sniffing followed by precise overmarking when similar sites were abundantly available; for example, the lion that cheek rubbed a tree after it had been urine-marked by a leopard (http://www.momo-p.com/showdetail-e.php?movieid=momo181217pp01b), the African wildcat precisely overmarking a large-spotted genet mark on flattened grass (http://www.momo-p.com/showdetail-e.php?movieid=momo181217fs02b) and an African wildcat precisely overmarking a caracal mark on a vehicle track (http://www.momo-p.com/showdetail-e.php?movieid=momo181217fs01b), show that interspecific countermarking was not confined to rare sites. Mammals of several species scent mark when they encounter new or clean objects (e.g., Roeder 1980; De Monte and Roeder 1990; Hoffman et al. 2010), and if novelty is detected by scent, a new mark will stimulate countermarking and the countermark will stimulate another mark and so on, even if none of the markers get any information from the marks. Captive pine martens (*Martes martes*) mark both con- and heterospecific marks at the same rate as clean surfaces (De Monte and Roeder 1990), but since in this study the marked spots were hugely outnumbered by unmarked spots, a similar lack of discrimination by the Botswana carnivores would have resulted in hardly any overmarking at all.

Definitive evidence that information flows in both directions between marker and countermarker will need more work than has been possible so far, but although some incidents of interspecific countermarking are probably due to marking prominent, rare or new spots, the evidence so far favors an exchange of information as the reason why carnivores countermark instead of simply eavesdropping.

4.4 What is Interspecific Scent Marking for?

With so little background information and a restricted sample size, much of the discussion has to be speculative. With the exception of a few specialist insectivores and herbivores, most carnivores' diets overlap with others' (Caro and Stoner 2003; Hunter and Caro 2008), and they compete for these shared resources and possibly others such as den sites. Competing predators reduce both exploitation and interference competition by avoiding one another in space and time (Hunter and Caro 2008; De Satgé et al. 2017), sometimes with surprising specificity (Vanak et al. 2013; Grassel et al. 2015; Kushata et al. 2018), and this needs a way for individuals of one species to detect the presence of individuals of the other, and to forecast the probability that they will be an area. Prey avoid predators by eavesdropping on their scent (Apfelbach et al. 2005; Jones et al. 2016; Parsons et al. 2017), and eavesdropping on scent marks (or other scent traces) provides a plausible explanation for how carnivores of one species know that another has been in an area, and predict future movements based on movements in the recent past. Nevertheless, eavesdropping does not need countermarking, so why do different species of carnivore countermark one another? If interspecific scent marking is communication, what messages are being transmitted? And what advantages accrue to the countermarking individual? Intriguingly, except for reciprocal eavesdropping on alarm calls, bidirectional communication between different species of mammal is very rare (Westrip and Bell 2015), and so there is not much guidance to be had from work on physical channels of communication.

In principle, only two of Ferkin and Pierce's (2007) hypotheses about overmarking —numbers 3 and 4 which deal with attracting and guarding mates—do not apply when one species countermarks another. Most of the potential functions of interspecific marking do not exclude one another, and it is almost certain that different species pairs allospecific for different reasons. Conspecific top marks sometimes cancel the marks that they overlie (Ferkin and Pierce 2007), and, similarly, allospecific overmarks may disrupt a competitor's messages to its conspecifics. If interspecific countermarking is an interspecific equivalent of intraspecific territorial marking, whether or not a countermark is subsequently countermarked back again will distinguish between transients, residents, and invaders in a similar way to bulletin boards (Ferkin and Pierce 2007) and scent wars (Hurst and Beynon 2004). If individuals of different species have an armed truce based on individual recognition or familiarity in an interspecific version of dear enemies (Palphramand and White 2007) or pragmatic resource relationships

(Tinnesand et al. 2015), it may be to each individual's advantage to signal its continued presence in an area. In a way similar to the "book-keeping" marking of emptied caches (Henry 1977) countermarking allospecific marks may tick them off a list of items that need attending to.

Picking up a superior competitor's scent by rubbing on it may sometimes be a defence against predation (Allen et al. 2017) but this is very unlikely to apply to countermarking among small carnivores such as genets and slender mongooses whose contrasting activity cycles mean that they never meet in the wild, and it cannot explain why a 3 kg nocturnal Cape fox rubbed on the scent of a 700 g diurnal slender mongoose (http://www.momo-p.com/showdetail-e.php?movieid=momo181217vc01b), or a lion rubbed on leopard urine (http://www.momo-p.com/showdetail-e.php?movieid=momo181217pp01b).

The big question is; does interspecific scent marking mediate intra-guild interactions and contribute to guild structures? Interspecific interference competition's effects (Grether et al. 2017) have important implications for carnivore conservation and pest control (Linnell and Strand 2000; Prugh et al. 2009) and we need to understand interspecific scent communication's role in it.

4.5 Implications for the Study of Mammal Chemical Communication

Interspecific scent marking, even if it is not always two-way olfactory communication, calls for some changes in the way that we study chemical communication, especially in the field. Some of the changes are purely practical, while others affect the strategy of study design. For all mammals we know even less about interspecific marking than we do about intraspecific marking; "who marks and who responds, how, where, when and under what circumstances?" will need answers before we can decipher what messages are being sent. Studies of intraspecific scent communication will need to take into account mammals' responses to scent marks that were not deposited by their own species.

Peters and Mech (1975) presciently wrote "… attempts must be made to watch the animals actually responding to the marks of their neighbors." and although they were writing about intraspecific marking, their remarks apply also to interspecific marking; to detect multispecies marking we have to monitor scent marks, not animals. Camera trapping is the tool of choice, especially as better cameras allow increasing use of video. There is still plenty of room for improvements in cameras, and we need to develop and apply study designs and camera setups specifically to investigate scent marking and responses to marks. In particular, because cameras are often popular marking sites there have to be cameras watching cameras, and cameras will need to be deployed for several months at a time in order to capture longer series of scent marks from different species. The bycatch of camera trapping that was not aimed at scent marking (or not at interspecific marking per se) will still be a valuable (and low cost) source of data if it can be accessed, but currently, the repositories for camera trap data accept only still images. In addition to the immediate reactions to marks, such as overmarking or rubbing recorded by camera traps, we also need to investigate deferred reactions and long-term effects on, e.g., ranging behaviour (Paquet 1991) or activity cycles, using remote surveillance with GPS and accelerometers in tracking collars.

A better understanding of interspecific scent marking might provide a rational foundation for the development of "super-lures" for mammal pest control (Linklater et al. 2013). Interspecific responses may underlie idiosyncratic responses to perfumes (Braczkowski and Watson 2013), which classically contain odorous animal secretions, or lately their synthetic substitutes, and the use of mammal scent as lures (Clapperton et al. 1994; Monterroso et al. 2011).

Investigations of choice of marking sites and surfaces (e.g., Ruiz-Olmo et al. 2013; Espirito-Santo et al. 2007; Barja et al. 2005) have focussed on scent mark encounterability, detectability, persistence, and detection range. If there is a cost to being countermarked then marks may be put where other species cannot countermark them even if doing so compromises their activity; urine sprayed onto a prickly bush is not likely to be rubbed or countermarked with anal-sac secretion (Apps et al. 2017) for example.

Interspecific overmarking presents field researchers with some specific practical problems. If scent marks need to be moved around to test animal's reactions, or collected for chemical analysis, how will we ever be sure that a scent-mark collected in the wild is not contaminated by an earlier one, which on the evidence so far would more likely than not be from a different species?. We may need to put out clean marking targets and monitor them with camera traps to check that only the target species has marked them. Experimental scents and odorless controls are likely to be marked by somebody sooner or later, and all subsequent reactions, even from animals of a species different from the marker's might then be a response to the adventitious mark. If the experimental scent or response to controls is the exclusive focus of the study (rather than countermarking) checking camera traps very frequently, reviewing videos immediately and replacing any hardware that had been scent-marked will be daily necessities.

If interspecific marking is common, the naive view that the active components of scent marks are species-specific, as expressed by Apps et al. (2013); "Our criteria for a compound to be considered a good candidate for a semiochemical role are as follows: …not present in the scent marks, faeces or urine of other species from the same area, since territorial signals must be species-specific." will lead us astray. If scent signals and responses to them are species-specific, the compounds that elicit, say, urine overmarking in African wildcats will be specific to wildcats, and a different set of chemicals would elicit overmarking in caracals. To find the active ingredients we would look for compounds that differ between the two species. Now we know that wildcats overmark both wildcat and caracal urine we should look also at the compounds that are found in both species and that do not occur in species which caracals and wildcats do not overmark. This means that we have to know which species do and do not elicit overmarking, and analyze marks from a larger number of species. More optimistically, we can speculate that interspecific communication is carried on in a sort of semiochemical pidgin with simplified grammar and vocabulary that might be easier to unravel than the complexities of intraspecific communication with its signature mixtures (Wyatt 2010) and signals of genetic quality (Charpentier et al. 2010).

The foundational methods of mammal semiochemistry and five decades of experience with intraspecific olfactory communication can be drawn on to guide future research into interspecific communication. Although the full complexity and subtlety of scent communication in social behaviour and ecology can appear only in natural communities, there is no reason not to use experimental methods with captive, tame and trained animals to investigate specific aspects of interspecific communication under controlled conditions (Gorman 1976; Müller-Schwarze 1977; Macdonald 1979; De Monte and Roeder 1990; Monterroso et al. 2011).

4.6 Implications Beyond Mammal Chemical Communication

Animals' attraction to scent marks, and their countermarking responses have implications that extend beyond semiochemistry.

If interspecific countermarking is communication then it markedly increases the "numbers of differentiated relationships in which individual animals are involved" and might account for those aspects of the evolution of carnivore brain size and intelligence that cannot be explained by the size of social groups (Holekamp and Benson-Amram 2017).

If carnivores commonly rub themselves on one another's urine, faeces, and secretions, disease transmission is likely to be faster and more widespread than if they confine their attentions to their own species. In Chile, invasive American mink (*Neovison vison*), which use river otter (*Lontra provocax*) latrines are a potential bridge host for transmission of canine distemper virus from domestic dogs to the otters (Sepúlveda et al. 2014). In Botswana, Cape foxes and slender mongooses overmarked banded mongooses, which suffer from a tuberculosis pathogen that transmits via anal gland secretions (Alexander et al. 2016). Prey animals infected with *Toxoplasma gondii* are attracted by the scent of potential predators (Berdoy et al. 2000; Poirotte et al. 2016), and something similar may be affecting the responses of competing predators to one another's scents.

Attraction to scent marks near or on camera traps (Wagnon and Serfass 2016), and cascades of interspecific countermarking, are likely to bias image captures for all the species involved and violate the assumptions of the models that are used to generate the estimates of population size and density that inform conservation management decisions for whole landscapes and ecosystems (MacKenzie et al. 2002; Rowcliffe et al. 2013; Wearn and Glover-Kapfer 2017). Increased detectability due to the short-range attraction and responses to heterospecific marks on or near cameras that we recorded here may account for apparent large-scale positive associations in carnivore distributions in our study area (Rich et al. 2017). In the interests of making population estimates fit for purpose, we need to understand the effects of scent marks and countermarking on camera trap detections.

5 Conclusion

Interspecific communication among carnivores has been found wherever it has been searched for with methods capable of detecting it, and its existence opens up a new field of research into mammalian semiochemistry. On present evidence, it appears to be widespread and common enough to be a significant contributor to ecosystem processes and structures, which will give it a role in management and conservation. We need to take a closer look at it.

References

Alexander KA, Sanderson CE, Larsen MH, Robbe-Austerman S, Williams MC, Palmer MV (2016) Emerging tuberculosis pathogen hijacks social communication behavior in the group-living banded mongoose (*Mungos mungo*). MBio 7(3):e00281–e002816

Allen ML, Gunther MS, Wilmers CC (2017) The scent of your enemy is my friend? The acquisition of large carnivore scent by a smaller carnivore. J Ethol 35:13–19

Allen ML, Wallace CF, Wilmers CC (2015) Patterns in bobcat (*Lynx rufus*) scent marking and communication behaviors. J Ethol 33:9–14

Allen ML, Wittmer HU, Wilmers CC (2014) Puma communication behaviours: understanding functional use and variation among sex and age classes. Behaviour 151:819–840

Apfelbach R, Blanchard CD, Blanchard RJ, Hayes RA, McGregor IS (2005) The effects of predator odors in mammalian prey species: a review of field and laboratory studies. Neurosci Biobehav Res 29:1123–1144

Apps PJ, McNutt JW (2018) Are camera traps fit for purpose? A rigorous, reproducible and realistic test of camera trap performance. Afr J Ecol 56:710–720

Apps P, Mmualefe L, McNutt JW (2013) A reverse engineering approach to identifying which compounds to bioassay for signalling activity in the scent marks of African wild dogs (*Lycaon pictus*). In: East ML, Denhard M (eds) Chemical signals in vertebrates XII. Springer, Berlin, pp 417–432

Apps PJ, Claase M, Yexley B, McNutt JW (2017) Interspecific responses of wild African carnivores to odour of 3-mercapto-3-methylbutanol, a component of wildcat and leopard urine. J Ethol 35:153–159

Banks PB, Daly A, Bytheway JP (2016) Predator odours attract other predators, creating an olfactory web of information. Biol Lett 12:20151053

Barja I, de Miguel FJ, Bárcena F (2005) Faecal marking behaviour of Iberian wolf in different zones of their territory. Folia Zool 54:21–29

Begg CM, Begg KS, Du Toit JT, Mills MGL (2003) Scent-marking behaviour of the honey badger, *Mellivora capensis* (Mustelidae), in the southern Kalahari. Anim Behav 66:917–929

Berdoy M, Webster JP, Macdonald DW (2000) Fatal attraction in Toxoplasma-infected rats: a case of parasite manipulation of its mammalian host. Proc R Soc B 267:1591–1594

Braczkowski A, Watson L (2013) Observations of leopard and caracal responses to novel scents in South Africa. Cat News 58:13–15

Caro TM, Stoner CJ (2003) The potential for interspecific competition among African carnivores. Biol Conserv 110:67–75

Charpentier MJ, Crawford JC, Boulet M, Drea CM (2010) Message 'scent': lemurs detect the genetic relatedness and quality of conspecifics via olfactory cues. Anim Behav 80:101–108

Clapperton BK, Eason CT, Weston RJ, Woolhouse AD, Morgan DR (1994) Development and testing of attractants for feral cats, *Felis catus* L. Wildl Res 21:389–399

Coe J (2004) Mixed species rotation exhibits. In: 2004 ARAZPA conference proceedings, Australia

Darden SK, Steffensen LK, Dabelsteen T (2008) Information transfer among widely spaced individuals: latrines as a basis for communication networks in the swift fox? Anim Behav 75:425–432

De Monte M, Roeder JJ (1990) Responses to inter-and intraspecific scent marks in pine martens (*Martes martes*). J Chem Ecol 16:611–618

De Satgé J, Teichman K, Cristescu B (2017) Competition and coexistence in a small carnivore guild. Oecologia 184:873–884

Espírito-Santo C, Rosalino LM, Santos-Reis M (2007) Factors affecting the placement of common genet latrine sites in a Mediterranean landscape in Portugal. J Mammal 88:201–207
Ferkin MH, Pierce MA (2007) Perspectives on over-marking: is it good to be on top? J Ethol 25:107–116
Gipson PS (1983) Evaluation and control implications of behavior of feral dogs in interior Alaska. In Kaukemen DE (ed) Vertebrate pest control and management materials: fourth symposium. ASTM STP 817. American Society for Testing and Materials: Philadelphia, Pennsylvania, pp 285–294
Gorman ML (1976) A mechanism for individual recognition by odour in *Herpestes auropunctatus* (Carnivora: Viverridae). Anim Behav 24:141–145
Gorman ML, Trowbridge BJ (1989) The role of odor in the social lives of carnivores. In Gittleman JL (ed) Carnivore behavior, ecology, and evolution. Springer, New York, pp. 57–88
Grassel SM, Rachlow JL, Williams CJ (2015) Spatial interactions between sympatric carnivores: asymmetric avoidance of an intraguild predator. Ecol Evol 5:2762–2773
Grether GF, Peiman KS, Tobias JA, Robinson BW (2017) Causes and consequences of behavioral interference between species. Trends Ecol Evolut. http://dx.doi.org/10.1016/j.tree.2017.07.004
Harmsen BJ, Sanchez E, Foster RJ (2016) Differential marking behaviour by sympatric felids in a neotropical forest. CATnews 64:8–12
Hayward MW, Hayward GJ (2010) Potential amplification of territorial advertisement markings by black-backed jackals (*Canis mesomelas*). Behaviour 147:979–992
Henry JD (1977) The use of urine marking in the scavenging behavior of the red fox (*Vulpes vulpes*). Behaviour 61:82–105
Hoffman KL, Decasa DH, Ruiz MB, González-Mariscal G (2010) Scent marking by the male domestic rabbit (*Oryctolagus cuniculus*) is stimulated by an object's novelty and its specific visual or tactile characteristics. Behav Brain Res 207:360–367
Holekamp KE, Benson-Amram S (2017) The evolution of intelligence in mammalian carnivores. Interf Focus 7:20160108. https://doi.org/10.1098/rsfs.2016.0108
Hunter J, Caro T (2008) Interspecific competition and predation in American carnivore families. Ethol Ecol Evol 20:295–324
Hurst JL, Beynon RJ (2004) Scent wars: the chemobiology of competitive signalling in mice. BioEssays 26:1288–1298
Jones ME, Apfelbach R, Banks PB, Cameron EZ, Dickman CR, Frank A, McLean S, McGregor IS, Müller-Schwarze D, Parsons MH, Sparrow E, Blumstein DT (2016) A nose for death: integrating trophic and informational networks for conservation and management. Front Ecol Evol 4:124. https://doi.org/10.3389/fevo.2016.00124
Jordan NR, Golabek KA, Apps PJ, Gilfillan GD, McNutt JW (2013) Scent-mark identification and scent-marking behaviour in African Wild Dogs (*Lycaon pictus*). Ethology 119:644–652
Jordan NR, Manser MB, Mwanguhya F, Kyabulima S, Rüedi P, Cant MA (2011a) Scent marking in wild banded mongooses: 1. Sex-specific scents and overmarking. Anim Behav 81:31–42
Jordan NR, Mwanguhya F, Kyabulima S, Rüedi P, Hodge SJ, Cant MA (2011b) Scent marking in wild banded mongooses: 3. Intrasexual overmarking in females. Anim Behav 81:51–60
Karlson P, Lüscher M (1959) 'Pheromones': a new term for a class of biologically active substances. Nature 183:55–56
Kats LB, Dill LM (1998) The scent of death: chemosensory assessment of predation risk by prey animals. Ecoscience 5:361–394
King TW, Salom-Pérez R, Shipley LA, Quigley HB, Thornton DH (2016) Ocelot latrines: communication centers for Neotropical mammals. J Mammal 98:106–113

Kushata JNT, Périquet S, Tarakini T, Muzamba M, Mafuwa B, Loveridge AJ, MacDonald DW, Fritz H, Valeix M (2018) Drivers of diurnal rest site selection by spotted hyaenas. J Zool 304:132–140

Lenth BE, Knight RL, Brennan ME (2006) The effects of dogs on wildlife communities. Nat Area J 28:218–227

Li J, Schaller GB, McCarthy TM, Wang D, Jiagong Z, Cai P, Basang L, Lu Z (2013) A communal sign post of snow leopards (*Panthera uncia*) and other species on the Tibetan Plateau, China. Int J Biodiversity 2013

Linklater WL, Greenwood D, Keyzers RA, Duckworth JA, Banks P, MacKay JWB, Stockum C (2013) Pied-pipers wanted: the search for super-lures of New Zealand mammal pests. New Zealand Science Review 70:31–36

Linnell JD, Strand O (2000) Interference interactions, co-existence and conservation of mammalian carnivores. Divers Distrib 6:169–176

Macdonald DW (1979) Some observations and field experiments on the urine marking behaviour of the red fox, *Vulpes vulpes* L. Ethology 51:1–22

Macdonald DW (1980) Patterns of scent marking with urine and feces amongst carnivore communities. Symp Zool Soc Lond 45:107–139

MacKenzie DI, Nichols JD, Lachman GB, Droege S, Royle JA, Langtimm CA (2002) Estimating site occupancy rates when detection probabilities are less than one. Ecology 83:2248–2255

Marnewick KA, Bothma JDP, Verdoorn GH (2006) Using camera-trapping to investigate the use of a tree as a scent-marking post by cheetahs in the Thabazimbi district. S Afr J Wildl Res 36:139–145

McLean S, Müller-Schwarze D, Sparrow EE (2017) Biologically meaningful scents: a framework for understanding predator–prey research across disciplines. Biol Rev 93:98–114

McTavish C, Gibeau M (2010) How animals send "tree mail". Remote imaging reveals forest communication. The Wildlife Professional Summer 2010:66–68

Monclús R, Arroyo M, Valencia A, De Miguel FJ (2009) Red foxes (*Vulpes vulpes*) use rabbit (*Oryctolagus cuniculus*) scent marks as territorial marking sites. J Ethol 27:153–156

Monterroso P, Alves PC, Ferreras P (2011) Evaluation of attractants for non-invasive studies of Iberian carnivore communities. Wildl Res 38:446–454

Müller-Schwarze D (1977) Complex mammalian behavior and pheromone bioassay in the field. In: Müller-Schwarze D, Mozell MM (eds) Chemical Signals in Vertebrates. Springer, Boston, pp 413–433

Müller-Schwarze D (2016) Chemical Signals in Vertebrates 13: where do we stand and what might be next. In Schulte BA, Goodwin TE, Ferkin MH (eds) Chemical signals in vertbrates 13. Springer, New York, pp 11–16

Nel JAJ, Bothma J (1983) Scent marking and midden use by aardwolves (*Proteles cristatus*) in the Namib Desert. Afr J Ecol 21:25–39

Palphramand KL, White PC (2007) Badgers, *Meles meles*, discriminate between neighbour, alien and self scent. Anim Behav 74:429–436

Paquet PC (1991) Scent-marking behavior of sympatric wolves (*Canis lupus*) and coyotes (*C. latrans*) in Riding Mountain National Park. Can J Zool 69:1721–1727

Parsons MH, Apfelbach R, Banks PB, Cameron EZ, Dickman CR, Frank AS, Jones ME, McGregor IS, McLean S, Müller-Schwarze D, Sparrow EE (2017) Biologically meaningful scents: a framework for understanding predator–prey research across disciplines. Biol Rev 93:98–114

Peters RP, Mech LD (1975) Scent-marking in wolves: radio-tracking of wolf packs has provided definite evidence that olfactory sign is used for territory maintenance and may serve for other forms of communication within the pack as well. Am Sci 63:628–637

Poirotte C, Kappeler PM, Ngoubangoye B, Bourgeois S, Moussodji M, Charpentier MJ (2016) Morbid attraction to leopard urine in Toxoplasma-infected chimpanzees. Curr Biol 26:R98–R99

Prugh LR, Stoner CJ, Epps CW, Bean WT, Ripple WJ, Laliberte AS, Brashares JS (2009) The rise of the mesopredator. Bioscience 59:779–791

Ralls K, Smith DA (2004) Latrine use by San Joaquin kit foxes (*Vulpes macrotis mutica*) and coyotes (*Canis latrans*). West N Am Nat 64:544–547

Rich LN, Miller DAW, Robinson HS, McNutt JW, Kelly MJ (2017) Carnivore distributions in Botswana are shaped by resource availability and intraguild species. J Zool 303:90–98

Roeder J (1980) Marking behaviour and olfactory recognition in genets (*Genetta genetta* L., Carnivora-Viverridae). Behaviour 72:200–210

Rowcliffe JM, Kays R, Carbone C, Jansen PA (2013) Clarifying assumptions behind the estimation of animal density from camera trap rates. J Wildl Manag 77:876

Ruiz-Olmo J, Such-Sanz A, Piñol C (2013) Substrate selection for urine spraying in captive wildcats. J Zool 290:143–150

Sepúlveda MA, Singer RS, Silva-Rodríguez EA, Eguren A, Stowhas P, Pelican K (2014) Invasive American mink: linking pathogen risk between domestic and endangered carnivores. EcoHealth 11:409–419

Stewart PD, MacDonald DW, Newman C, Tattersall FH (2002) Behavioural mechanisms of information transmission and reception by badgers, *Meles meles*, at latrines. Anim Behav 63:999–1007

Tinnesand HV, Buesching CD, Noonan MJ, Newman C, Zedrosser A, Rosell F, Macdonald DW (2015) Will trespassers be prosecuted or assessed according to their merits? A consilient interpretation of territoriality in a group-living carnivore, the European badger (*Meles meles*). PLoS ONE 10(7):e0132432. https://doi.org/10.1371/journal.pone.0132432

Vanak AB, Fortin D, Thaker M, Ogden M, Owen C, Greatwood S, Slotow R (2013) Moving to stay in place: behavioral mechanisms for coexistence of African large carnivores. Ecology 94:2619–2631

Vitale J (2017) The olfactory behaviour of spotted hyaenas (*Crocuta crocuta*) and sympatric mammals in the Okavango Delta, Botswana. PhD thesis, University of Nottingham

Wagnon CJ, Serfass TL (2016) Camera traps at northern river otter latrines enhance carnivore detectability along riparian areas in eastern North America. Global Ecol Conserv 8:138–143

Wearn OR, Glover-Kapfer P (2017) Camera-trapping for conservation: a guide to best-practices. WWF conservation technology series, vol 1, no 1. WWF-UK, Woking, UK

Westrip JR, Bell MB (2015) Breaking down the species boundaries: selective pressures behind interspecific communication in vertebrates. Ethology 121:725–732

Wikenros C, Jarnemo A, Frisén M, Kuijper DP, Schmidt K (2017) Mesopredator behavioral response to olfactory signals of an apex predator. J Ethol 35:161–168

Wyatt TD (2010) Pheromones and signature mixtures: defining species-wide signals and variable cues for identity in both invertebrates and vertebrates. J Comp Physiol A 196:685–700

Part IV

Semio-chemistry and Evolution

Chemistry Between Salamanders: Evolution of the SPF Courtship Pheromone System in Salamandridae

Franky Bossuyt(✉), Margo Maex, Dag Treer, Lisa M. Schulte, Ines Van Bocxlaer, and Sunita Janssenswillen

Amphibian Evolution Lab, Biology Department, Vrije Universiteit Brussel, Pleinlaan 2, B-1050 Brussels, Belgium
Franky.Bossuyt@vub.be

Abstract. Chemical communication is an important aspect of courtship and reproduction in salamanders (*Urodela*), where males secrete several protein pheromones from various sexually dimorphic glands. The most widely used sex pheromone system consists of proteins of the Sodefrin Precursor-like Factor (SPF) family. This protein family already started diversifying through gene duplications in the Late Palaeozoic and continued to do so in various urodelan lineages. As a result, males of multiple extant salamander species secrete different sets or ratios of SPF protein pheromones which, as shown in behavioral tests, can evoke various female responses during the courtship process, such as following behavior, cloacal gaping, or an acceleration of courtship in general. Still, all observable effects essentially are a consequence of female receptivity enhancement, indicating a preserved role for SPF pheromones. Salamandridae is a large family with currently 119 described species that show a considerable variety in courtship behaviour. In this family, different lineages evolved various modes of pheromone transfer, ranging from direct application during amplexus on land, to an indirect transfer in which males abandoned physical contact and instead tail-fan pheromones underwater toward a nearby female. In one genus, an additional decapeptide pheromone, sodefrin, originated by cleavage from a precursor of the SPF family and was co-opted alongside uncleaved SPF protein pheromones. Here we discuss our current understanding and knowledge gaps on the use of the SPF pheromone system in salamandrids, and we define some testable hypotheses on its evolution in relation to changes in mating habitat and courtship modes.

1 Introduction

Throughout the animal kingdom, sex pheromones play an essential role in eliciting a variety of responses during reproductive interactions between males and females. In amphibians, the role of chemical signals during courtship is especially well documented in salamanders (*Urodela*). Most salamanders have internal fertilization, meaning that eggs and sperm merge inside the female body. Since salamanders do not have a copulatory organ, females of species in the internally fertilizing clade (Fig. 1) have to pick up the male spermatophore with their cloaca from the environment to

© Springer Nature Switzerland AG 2019
C. D. Buesching (Ed.): *Chemical Signals in Vertebrates 14*, pp. 205–220, 2019.
https://doi.org/10.1007/978-3-030-17616-7_15

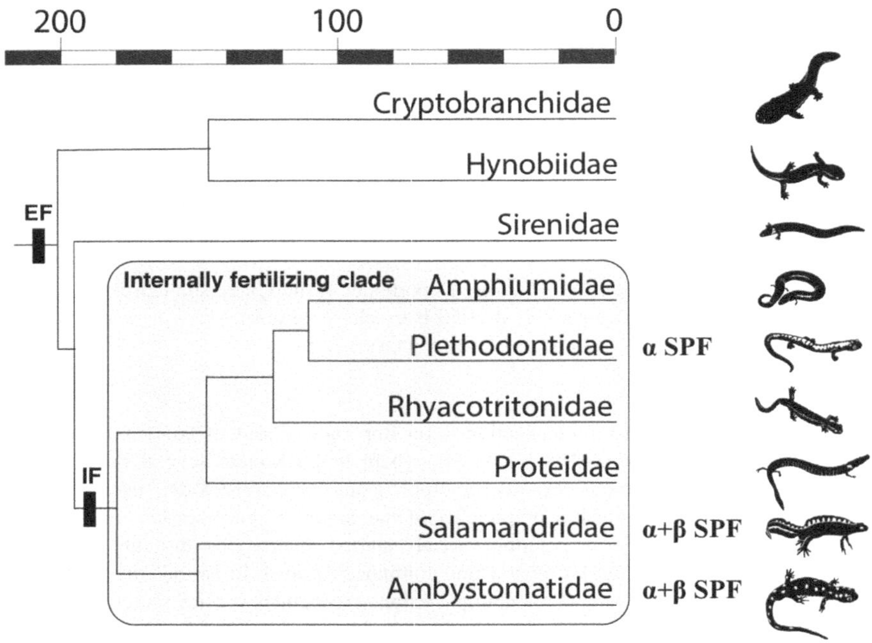

Fig. 1. Known SPF pheromone use in salamanders. The phylogeny and timescale are based on TimeTree (www.timetree.org). With the exception of terrestrial representatives of the families Plethodontidae, Ambystomatidae, and Salamandridae, the remaining salamanders mate in water. The urodelan ancestor had external fertilization (EF), while internal fertilization (IF) originated in the ancestor of all Salamandroidea. The confirmed use of SPF pheromones in three extant families suggests the potential of pheromone use in all internally fertilizing salamanders, while the use of both alpha and beta SPF (products of a gene duplication that happened before the diversification of extant salamanders) in Salamandridae and Ambystomatidae suggests extension of the potential use of SPF pheromones to all (i.e., including externally fertilizing) salamanders

complete courtship and reproduction successfully. The use of pheromones is an essential component in coordinating the courtship process that precedes fertilization. To this aim, males often possess sexually dimorphic glands, which hypertrophy during the breeding season and have been demonstrated or postulated to produce courtship pheromones (Sever 1991; Sever 2003; Van Bocxlaer et al. 2015). Some well-known examples are the mental glands on the chin in plethodontids (Houck et al. 2008) and the highly specialized dorsal glands in the cloaca and abdominal cavity of newts (Kikuyama et al. 1995; Van Bocxlaer et al. 2015). The chemicals secreted from these glands are transmitted from male to female either by direct contact (e.g., transfer of courtship pheromones from mental glands to nares or skin-abrasions in terrestrially mating plethodontids: Arnold 1977; Houck et al. 1998; Rollmann et al. 2003; Houck et al. 2008) or, provided that both sexes are in close proximity, by transmission through water (e.g., transfer of pheromones from the cloaca by tail-fanning in aquatically mating newts; Halliday 1974, 1977, 1990; Arntzen and Sparreboom 1989; Wells 2007; Treer et al. 2013; Van Bocxlaer et al. 2015).

Salamandridae forms a diverse group of 119 species in three subfamilies, Pleurodelinae (newts, 99 species), Salamandrinae (true salamanders, 18 species) and Salamandrininae (2 species) (AmphibiaWeb. 2018. https://amphibiaweb.org University of California, Berkeley, CA, USA. Accessed 17 January 2018). This family evolved an interesting diversity in mating strategies, with different courtship modes both on land and in water, the presence of various forms of contact during courtship, and different modes of sperm transfer (Halliday 1977, 1990; Houck and Arnold 2003; Sparreboom 2014). Since salamandrids only started diversifying in the Late Cretaceous (Zhang et al. 2008; Kieren et al. 2018), they form an interesting group to study pheromone evolution in relation to different courtship strategies over a relatively short evolutionary time.

One of the best-known modes of underwater pheromone transfer in salamandrids is tail-fanning behaviour in Pleurodelinae, in which males use their tail to create a water current that carries courtship pheromones from the cloaca toward the female's snout. This behavior has already been described and illustrated in the eighteenth (Spallanzani 1780) and early nineteenth centuries (Rusconi 1821), and detailed courtship sequences have been outlined in multiple species since then (e.g., Tinbergen and ter Pelkwijk 1938; Nelson 1960; Halliday 1977, 1990; Duellman and Trueb 1994). The female responds by essentially following the male path and taking up his spermatophore. In search for molecules secreted by tail-fanning males, the decapeptide sodefrin, which was isolated from the male dorsal glands of the Japanese fire belly newt *Cynops pyrrhogaster*, was the first salamander pheromone to be identified (Kikuyama et al. 1995). Sodefrin was shown to function as a close-range attractant, and variants were detected in different populations of this species, as well as in its sister species *C. ensicauda* (Yamamoto et al. 2000; Iwata et al. 2004, 2005; Nakada et al. 2007). Subsequent studies have shown that proteins of about 20 kDa act as courtship pheromones in several salamander families (Houck et al. 2008; Van Bocxlaer et al. 2015; Maex et al. 2016). Although these pheromones do not share any similarity with sodefrin at amino acid level, their transcripts and precursors are phylogenetically related with those of sodefrin, and they were therefore termed Sodefrin Precursor-like Factors (SPF).

Other pheromones have been described in salamanders (Rollmann et al. 1999; Houck et al. 2007; Nakada et al. 2017; Maex et al. 2018), but SPF proteins are the most widely distributed male courtship pheromones among urodeles. Here we review our current understanding on the use of pheromones of the SPF family (i.e., sodefrin and SPF) in relation to the wide variety of courtship interactions in salamandrids (Kikuyama et al. 1995; Yamamoto et al. 2000; Janssenswillen et al. 2014, 2015; Van Bocxlaer et al. 2015, 2016).

2 Female Courtship Responses and the Biological Roles of SPF

Several behavioral experiments have been developed to demonstrate the presence of chemical communication between sexes in salamanders. For example, linear as well as Y-shaped olfactometers were used to perform choice tests with different species (e.g., Dawley 1984; Malacarne and Vellano 1987; Secondi et al. 2005; Poschadel et al.

2007). In other setups, the time that a newt spent in the vicinity of sponges containing potential pheromones was compared to that of sponges containing only water (e.g., Toyoda et al. 1994; Kikuyama et al. 1995; Yamamoto et al. 2000). These experiments all compared time that a female spent in olfactometer compartments with different stimuli and were able to indicate the presence of chemical communication. However, because such experiments are less suited for understanding the biological roles of these chemical stimuli, new tests were developed that allowed females to show more natural courtship responses in the presence of another individual. In plethodontids, the duration of courtship after application of putative pheromones on the female nares was measured to assay pheromone activity in a setup in which a female was coupled with a male whose pheromone gland (i.e., mental gland; Rollmann et al. 1999) was removed. In newts with internal pheromone glands that have courtship without contact, the problem of a pheromone-releasing male was circumvented by developing a two-female test, in which female responses were compared in water with and without male courtship pheromones (Treer et al. 2013). The latter test allows females to show the same courtship responses, including following behavior, as in the presence of a courting male. In newts with amplexus, i.e., where the male grasps the female, a two-female test was developed in which cloacal gaping in response to the presence of male pheromones was measured (Janssenswillen and Bossuyt 2016). Since opening the cloaca is a required response for insemination, this test also allows a more straightforward interpretation of the biological role of pheromones than general choice experiments. The development of all these tests enabled studying the effects of SPF pheromones in a more realistic biological context. Although SPF pheromones are able to attract females in olfactometers (Kikuyama et al. 1995), these proteins do not play a role in attracting females from a long distance, since males only start administering their pheromones when the female is already in close vicinity. Instead, an important role of SPF pheromones is to enhance female receptivity during close-range courtship, which at several instants during the courtship process results in observable changes, such as following behaviour, cloacal gaping, or an acceleration of the courtship process in general.

Despite the evident effects that male SPF courtship pheromones have on female receptivity in experimental setups, the exact impact of these chemical cues in the totality of all available cues during natural courtship in newts is not yet entirely understood. When breeding newts meet underwater, they can deliver short-range communication signals of different modalities at the same time, such as visual, chemical and tactile cues. On the one hand, courtship pheromones are able to induce courtship responses in female newts in the absence of a male, obviating the need for male visual and tactile cues to elicit female mating responses (Treer et al. 2013). On the other hand, while pheromones are beneficial for urodelan courtship, they do not always seem to be a necessary prerequisite for successful reproduction. For example, in the Spanish ribbed newt (*Pleurodeles waltl*), a successful mating without obvious delivery of male courtship pheromones to the female could be observed in about 5% of the cases (Janssenswillen and Bossuyt 2016). Similar behaviour was observed in plethodontid salamanders, which engaged in mating even when the male courtship pheromone-producing mental gland was ablated (Houck et al. 1998; Rollmann et al. 1999), although it must be noted that the availability of other pheromone-producing glands could not be excluded in these animals (Sever 1994, 2003; Rollins and Staub 2017). Altogether, behavioural experiments indicate a key role for SPF pheromones in influencing female courtship responses (Van Bocxlaer et al. 2015).

3 The SPF pheromone System

Phylogenetic analyses have shown that all transcripts of the SPF family are derived from genes that belong to the three-finger protein (TFP) superfamily (Janssenswillen et al. 2014), which is present in most vertebrates and has a wide variety of biological functions (e.g., toxic, cholinergic signalling, and immune regulation; Tsetlin 1999; Loughner et al. 2016). Three-finger proteins are characterized by a distinct tertiary structure of three beta-strand loop fingers, shaped by a conserved cysteine pattern (Fleming et al. 1993; Casey et al. 1994; Kieffer et al. 1994; Palmer et al. 2007; Kini and Doley 2010). SPFs are two-domain three-finger proteins (2D-TFPs), often consisting of either an anterior 8-cysteine pattern and a posterior 6-cysteine pattern (resulting in the loss of one 'finger'), or an anterior 10-cysteine pattern and a posterior 8-cysteine pattern (Fig. 2), within mature proteins of around 20 kDa (Janssenswillen et al. 2014). Little is known about the functions of 2D-TFPs in vertebrates. The *Anolis carolinensis* genome contains 19 2D-TFP genes, and several of these are highly expressed in male testes and dewlap, and in female ovaries (NCBI Protein Database. 2004. https://www.ncbi.nlm.nih.gov/protein/, accessed on 21 January 2018), suggesting a function in reproduction (Baeckens et al. 2016). In snakes, 2D-TFPs have been postulated to function as a self-protecting agent against phospholipase A_2 from their own venom, which could enter the blood circulation (Dunn and Broady 2001; Fortes-Dias et al. 2017), but nonvenomous snakes secrete them as well (Okomura et al. 1999). In the genome of the frog *Silurana tropicalis*, 26 2D-TFPs were identified, but the function of most of these is unknown. A reconstruction of vertebrate 2D-TFP evolution suggests that amphibian 2D-TFPs (including SPF pheromones) form a monophyletic group, suggesting that the diversification of these genes in Amphibia started after their divergence from Amniota (Janssenswillen et al. 2014). Other than SPF genes that are mainly expressed in male sexually dimorphic glands, genes of this clade are found to be expressed in the gastrula embryo stage in frogs (Amaya et al. 2006; Clark et al. 2016), and in the eyes and regenerative tissues of salamanders (Maki et al. 2010; Campbell et al. 2011), but the exact biological function in these tissues remains unknown.

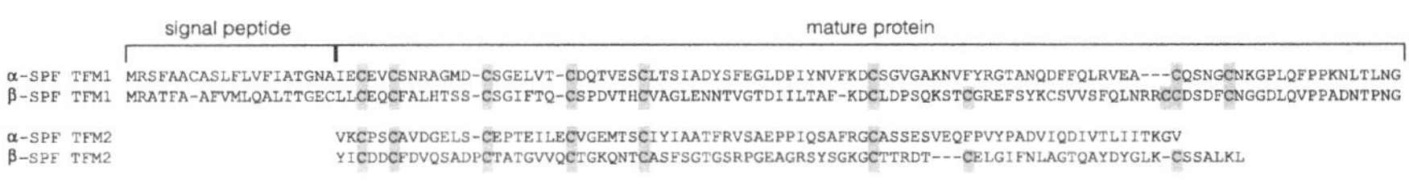

Fig. 2. Conserved cysteine patterns in alpha and beta SPF proteins. The shown sequences are from *Cynops pyrrhogaster* (GenBank-numbers KU213615 and KU213626). SPF proteins have two three-finger motifs (TFM1 and TFM2), containing 8 and 6 cysteines (alpha SPF), or 10 and 8 cysteines (beta SPF). Cysteines are indicated in gray. TFM = three-finger motif

The SPF gene repertoire has diversified substantially through gene duplications that started already during the early amphibian evolution. Phylogenetic analyses have shown that several extant species of salamandrids and ambystomatids secrete pheromone proteins, termed SPF alpha and beta, whose divergence can be traced back to a gene duplication that happened in the late Palaeozoic, before the diversification of extant salamanders (Van Bocxlaer et al. 2015; Maex et al. 2016). Each of these gene lineages further diversified through additional duplications during more recent urodelan evolution, and several extant salamander species now secrete highly divergent isoforms of both alpha and beta SPF proteins alongside each other during courtship. The genus *Cynops* has an additional set of transcripts, making up about 15% of the total SPF family blend (Van Bocxlaer et al. 2016), that leads to the secretion of another pheromone, the decapeptide sodefrin (Kikuyama et al. 1995).

4 The Origin of Sodefrin

Sodefrin and SPF proteins are products from the same gene family, and a large part of their precursors is homologous at the nucleotide level. However, sodefrin transcripts contain an insertion of 62 base pairs that causes a frameshift in protein translation. This frameshift results in a different amino acid sequence with new cleavage sites, which allows generating the decapeptide sodefrin (Fig. 3). SPF transcripts do not have this insert and are secreted as full-length proteins (Janssenswillen et al. 2014). Because of the frameshift, the C-terminus of the sodefrin and SPF amino acid precursor sequences are nonhomologous, and as a consequence sodefrin shows no resemblance to intact SPF proteins. Precursors containing the insert were only found in the genus *Cynops* (Janssenswillen et al. 2014), indicating that a sodefrin-like decapeptide originated and

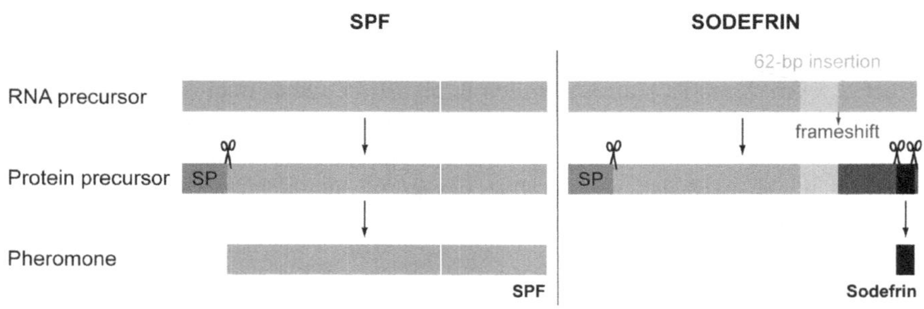

Fig. 3. Comparison of SPF and sodefrin precursors. SPF and sodefrin pheromones originate from the same gene family, and are largely homologs at the nucleotide level (RNA precursors are indicated in the same tint of gray). SPF protein precursors become mature pheromones after cleavage (indicated with scissors) of the signal peptide (SP) (shown on the left). Sodefrin RNA precursors differ from SPF RNA precursors by an insertion of 62 base pairs that, compared to SPF, causes a translational frameshift (shown on the right). This leads to the origin of a protein precursor that shows no amino acid sequence similarity with the SPF precursor at the C-terminal region (indicated in dark gray). The new amino acid sequence contains additional cleavage sites that allow cleavage of the decapeptide sodefrin (indicated in black)

obtained a pheromone function in the shadow of uncleaved SPF proteins shortly after the divergence of *Cynops* from its sister clade (Fig. 4) (Janssenswillen et al. 2014; Van Bocxlaer et al. 2016). The unique combination of an ancient pheromone blend (SPF) with a new pheromone (sodefrin) sheds light on how a random molecule can become a pheromone, even though it shows no similarity to existing signals. Because most SPF genes continued to encode for uncleaved SPF pheromones, the insertion in one of the genes likely did not have a large effect on the overall pheromone function of the altered

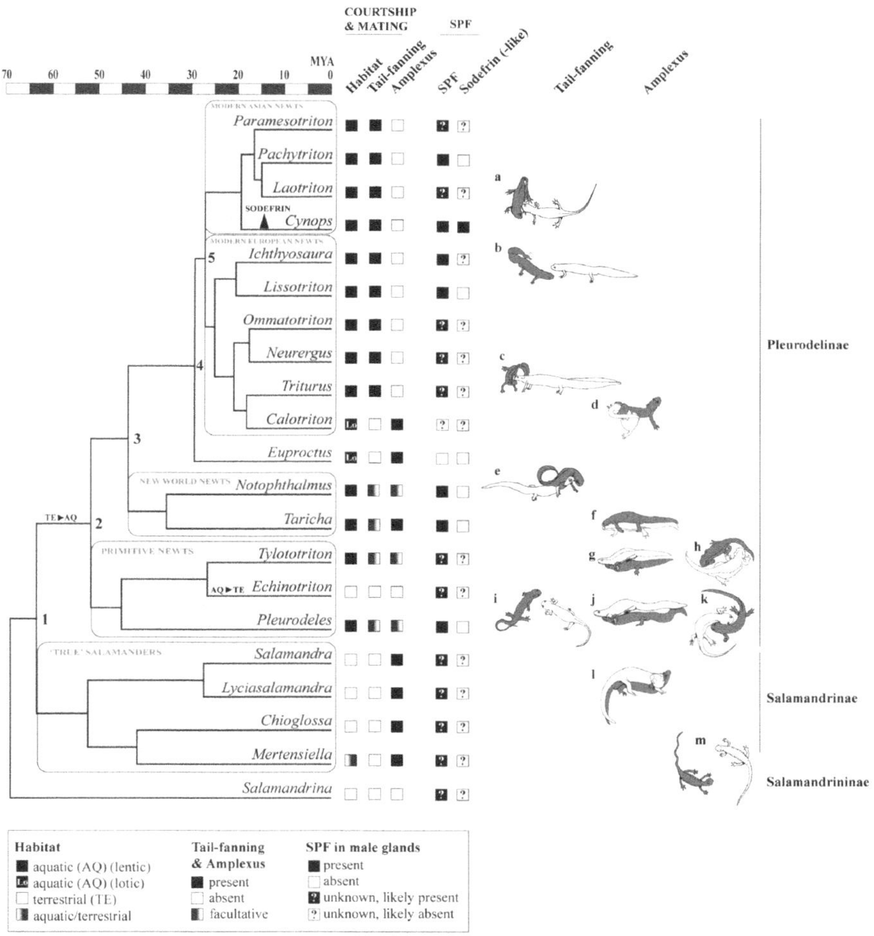

Fig. 4. The evolution of courtship in Salamandridae. The figure gives an overview of courtship and mating in salamandrids, with an indication of known use of SPF and sodefrin-like decapeptide pheromones. Squares with question marks indicate probably present (black) or probably absent (white), based on presence or absence in studied species and their phylogenetic relationships. The timescale follows Zhang et al. 2008 and is only given as indicative measure. Males are drawn in gray, females in white. The decapeptide pheromone sodefrin (and its variants) evolved relatively recently in the lineage to *Cynops*

protein blend. However, because sodefrin was secreted alongside functional SPF proteins, females may have been able to associate the new scent with existing pheromones, making it easier for the new decapeptide to evolve a sex pheromone function (Janssenswillen et al. 2014).

Based on behavioural experiments in which responses of single females were assessed in the absence of other individuals (Toyoda et al. 1994; Kikuyama et al. 1995; Yamamoto et al. 2000; Nakada et al. 2007), sodefrin and its variants have been described as attractants. Different components of the *Cynops* pheromone blend thus would serve distinct but complementary functions, in which the decapeptide functions as an attractant, while full-length SPF proteins are used to enhance female receptivity. However, the decapeptides are produced together with uncleaved SPF pheromones in the male dorsal glands, and both are simultaneously secreted via the cloaca during male tail-fanning at close distance (Kikuyama and Toyoda 1999; Van Bocxlaer et al. 2016). Additionally, SPF protein pheromones from modern European newts were also shown to attract females from a short distance in a Y-maze setup that includes a visual newt stimulus (Treer and Bossuyt 2010). We, therefore, consider it possible that decapeptides and SPF proteins have highly similar functions, but this hypothesis remains to be tested. If proteins and decapeptides would show comparable effects in both behavioural setups (i.e., two-female test and attractant test), the observed short-range attraction for both pheromones could be recognized as a logical consequence of the enhanced female receptivity.

5 Species-Specificity of the SPF Pheromone System

The expression of multiple, highly divergent SPF proteins in a single species prompts the question of how the underlying genes evolved to produce species-specific courtship signals throughout salamander evolution. Interspecific experiments have disclosed varying degrees of behavioural responses to courtship pheromones of the same *versus* other species. In both *Cynops pyrrhogaster* and *Cynops ensicauda*, females are attracted by the scent of their own species- and population-specific decapeptide pheromones (Kikuyama et al. 1995; Yamamoto et al. 2000; Iwata et al. 2004, 2005; Nakada et al. 2007). For species using SPF pheromones, females of one species (*Lissotriton helveticus*) showed responses to male courtship pheromones of a related species (*L. vulgaris*) in a two-female test, but they were much less pronounced than for proteins of their own species, indicating a preference for conspecific signals (Van Bocxlaer et al. 2015). Furthermore, the same species showed no responses to pheromones of a more distantly related newt (*Ichthyosaura alpestris*), suggesting that after speciation, cross-reaction to allospecific courtship pheromones gradually dilutes over evolutionary time as species diverge (Treer et al. 2013). One mechanism through which protein courtship pheromones can diverge to establish species-specificity was disclosed in an in-depth transcriptome study on the multicomponent SPF pheromone system of these species (Treer et al. 2017). Despite the vast diversity of available SPF gene copies, *I. alpestris* and *L. helveticus* predominantly expressed the same subset of orthologs in their pheromone blend. Neither gene loss nor co-option of distinct SPF genes has left a clear imprint on the courtship pheromones of both species, and the absence of major differences in SPF expression suggests that sequence divergence has been the predominant mode of pheromone diversification in these species. The observed species-specificity seems to be a consequence of speciation rather than the cause since cross-reaction between the two congeneric *Lissotriton* species still occurs. However, despite the fact that courtship pheromones do not necessarily prevent mating between closely related species or populations (Rollmann et al. 2003; Van Bocxlaer et al. 2015), they may still contribute to speciation by reducing gene flow through assortative mating (Kozak 2003).

6 Pheromone Glands in Salamandrids

Male salamandrids possess different sexually dimorphic glands and nuptial excrescences, which often are closely associated with specific courtship behavior. The most widely occurring pheromone glands are located in the male cloaca. In salamandrids, these glands are known as dorsal glands because they are situated in the dorsal region of the cloaca (Sever et al. 1990; Sever 1992). They are highly specialized skin glands that secrete via papillae situated at the posterior end of the cloaca and are homologous to vent glands of males of other salamander families (Sever 1991, 2003). In some newt genera, these dorsal glands are quite large and extend from the cloaca into the so-called abdominal cavity, and are therefore often called abdominal glands (Sacerdote 1956; Kikuyama et al. 1995). However, the term 'abdominal glands' has been deemed inappropriate because of the lack of a true abdomen in salamanders (Sever 2003; Osikowski and Cierniak-Zuzia 2013). Pheromones produced in the dorsal glands are released into the environment via the cloaca. Amplecting species (see below) can, in addition to tail-fanning, use cloacal imposition to deliver these pheromones more directly to the female nares (Janssenswillen and Bossuyt 2016). For most other salamandrid male glands and excrescences, it is still unknown whether they are involved in pheromone communication. For example, some species have nuptial pads that assist in holding the female during amplexus (e.g., *Pleurodeles* and *Notophthalmus*), but also could play a role in pheromone application (Mitchell and Gibbons 2010; Sparreboom 2014; Gong et al. 2018). Additionally, males of several species independently evolved various head glands (Sever 2003) which are rubbed against the female nares while amplexing (e.g., *Taricha*; Davis and Twitty 1964; Arnold 1972), during tail-fanning (e.g., *Cynops pyrrhogaster*, Fig. 4a, Kawamura and Sawada 1959; and *Lissotriton boscai*, Rafinski and Pecio 1992), or in an amplexus-tail-fanning combination (e.g., *Notophthalmus*, Fig. 4e; Pool and Dent 1977; Verrell 1982; Janssenswillen et al. 2015). At this point, *Notophthalmus* is the only salamandrid species for which it was shown that non-cloacal glands—in this species the cheek glands—contain an evolutionary diverse set of SPF pheromones (Janssenswillen et al. 2015).

7 SPF and the Evolution of Courtship Behaviour

The salamandrid radiation holds terrestrially as well as aquatically courting species (Fig. 4), and is characterized by various modes and degrees of physical contact during courtship. The basal splits in the salamandrid phylogeny lead to a single evolutionary lineage with no contact (*Salamandrina*, Fig. 4m), and several lineages with amplexus, in which the male holds the female in various ways during courtship. In true salamanders (e.g., *Lyciasalamandra*, Fig. 4l) and primitive newts (e.g., *Tylototriton* and *Pleurodeles*, Fig. 4g, j), the male grabs the female from beneath, which results in a ventral amplexus (Arnold 1972, 1977; Sparreboom 2014). New World newts are largely characterized by some kind of dorsal amplexus (Fig. 4e, f), in which *Notophthalmus* differs from *Taricha* by the fact that the female is constrained by the male's hind limbs around her neck (Fig. 4e; Gallien 1952; Verrell 1982; Petranka 2010). Two genera that independently colonized lotic water (*Euproctus* and *Calotriton*) have copulatory-like contact during courtship and use a form of restraint of the female with the tail or the mouth (Fig. 4d) (Ahrenfeldt 1955; Salthe 1967; Griffiths 1996; Guillaume 1999; Houck and Arnold 2003; Carranza and Amat 2005). Males of Eurasian modern newts largely rely on tail-fanning their pheromones to the female under water, with no or limited physical contact with the female (Fig. 4b, c). Given that the courtship strategies of males define their opportunities for pheromone application, an intriguing question is how evolutionary transitions between courtship habitats or modes on one hand, and pheromone use and the overall evolution of the SPF pheromone system on the other hand, have influenced each other in this family.

Phylogenetic reconstruction of the mating habitat shows that the early evolution of salamandrids is characterized by the split of terrestrially courting lineages (*Salamandrina* and the clade of "true" salamanders, with the exception of *Mertensiella* that has both aquatic and terrestrial mating) from otherwise aquatically courting newts (with the exception of *Echinotriton*) (Fig. 4). This indicates that salamandrid evolution is characterized by one transition from a terrestrial to an aquatic mating habitat (Fig. 4, in the ancestor of newts, between nodes 1 and 2), and one reversal (in *Echinotriton*). Use of SPF pheromones until now only has been demonstrated in aquatically courting species, but given the early origin of the SPF pheromone system (Fig. 1) and the presence of these pheromones in terrestrially courting plethodontids, we hypothesize that SPF proteins are also used by terrestrially mating salamandrids.

The conflicting phylogenetic relationships among major salamandrid lineages in different studies (for an overview, see Frost 2019) currently do not allow to reliably reconstruct transitions between different modes and levels of physical contact. Additionally, ancestral state reconstructions where different amplectic modes are coded as distinct states find absence of contact as the most probable character state for ancestors along the backbone of the salamandrid tree (Fig. 4, nodes 1 to 4) (Kieren et al. 2018), while coding amplexus as a present/absent state in a maximum parsimony framework shows presence of amplexus for these nodes, with loss in the ancestor of all modern newts (Fig. 4, between nodes 4 and 5). While reconstruction is inherently uncertain, the presence of facultative amplexus in multiple extant species (*Notophthalmus*, *Pleurodeles*, *Tylototriton*) is indicative for the plasticity of contact during courtship. For example, a study on the natural courtship behaviour of the Spanish ribbed newt (*Pleurodeles waltl*) showed that amplexus and associated pheromone use are context-dependent (Janssenswillen and Bossuyt 2016). Successful insemination in this species results from pinwheel behaviour, i.e., the couple going into a circular movement that ultimately leads to spermatophore deposition by the male and subsequent uptake through the cloaca by the female (Fig. 4k). Most often, this behaviour starts after the male has taken the female in a firm amplexus for an extended period of time, during which he releases one arm of his grip and regularly transfers pheromones by imposing his cloaca directly on the female nose (cloacal imposition). However, attentive observations on the natural courtship behaviour of these newts showed variations on this theme, in which males also use tail-fanning without contact to deliver their pheromones (Fig. 4i), or in which females engage in a pinwheel without a preceding amplexus, i.e., without a phase of male pheromone application (comparable to courtship in *Tylototriton*, Fig. 4h). Males that adapt their courtship behaviour in the function of the female's responsiveness (facultative amplexus) have also been found in the red-spotted newt (*Notophthalmus viridescens*) (Verrell 1982). Based on this plasticity in both primitive and New World newts, we hypothesize that the backbone of the salamandrid tree may have been equally characterized by ancestors in which amplexus varied across populations or was context-dependent. If true, complete loss of amplexus may have been achieved only in the ancestors of modern Eurasian newts (Fig. 4, between nodes 4 and 5) and could be correlated with the development of more pronounced dorsal glands (Sacerdote 1956; Osikowski and Cierniak-Zuzia 2013; Van Bocxlaer et al. 2015,2016; Treer et al. 2017; Kikuyama et al. 1995; Sever 1992) and increased tail-fanning behaviour, which in turn may have affected pheromone evolution in this clade.

8 Future Perspectives

Our current understanding of the SPF pheromone system will facilitate future studies on the evolution of chemical communication in relation to courtship behaviour in salamanders. To get better insights in pheromone use across evolutionary transitions (aquatic *versus* terrestrial mating, amplexus *versus* no amplexus), integrative approaches that combine pheromone expression studies of the complete spectrum of secreted components with behavioural assays in a broader range of species with diverse courtship strategies would be very valuable. Examining both total pheromone blends and (combinations of) isolated or recombinantly obtained individual SPF proteins will provide insights in whether total quantity or rather genetic and posttranslational (e.g., glycosylation) diversity of specific SPF proteins determine efficiency of a pheromone blend. Furthermore, studies at the level of female chemoreceptors will be essential to understand how pheromones and their receptors have co-evolved and define the variation in behavioural courtship strategies we see today.

The use of SPF proteins by males to enhance female receptivity suggests selective advantages to both males and females in view of better coordination during courtship and shorter mating interval (Houck et al. 2008; Treer et al. 2013). Until now, behavioural tests were developed to investigate the ability of candidate pheromones to enhance female receptivity, but these tests did not examine to what extent females may prefer pheromone mixtures of some males above others (sexual selection). Still, it is plausible that females are choosy regarding the quantities (total SPF amount emitted) or diversity (SPF proteins originating from multiple gene copies) that a male produces. A recent study in the amplecting species *Pleurodeles waltl* (Janssenswillen and Bossuyt 2016) provided an interesting clue in this regard. The study showed that in about one on four natural courtships, the female avoided the stage of pinwheel behaviour (the stage necessary for successful reproduction in this species), either by thanatosis (i.e., feigning death) and/or by escaping from amplexus (Janssenswillen and Bossuyt 2016). This indicates that, even when amplected, females have a choice whether or not they want to engage in reproduction. Enhanced responsiveness to specific male pheromone mixtures could indicate an evolved female preference, for example, if pheromones would directly or indirectly convey information about male quality. Along the same line, the seasonal development of enlarged, conspicuous crests and tail fins, as well as visually conspicuous coloration patterns, suggest that elaborate sexually dimorphic male visual traits play a role in sexual selection as well. Especially in species that do not restrain a female (amplexus) during courtship, males seem to use a combination of visual and chemical stimuli to persuade a female into courtship. Only an integration of all these aspects in overarching studies will allow to fully understand how sexual communication selection has shaped the salamandrid radiation.

Acknowledgements. This work was supported by the European Research Council (ERC 204509 StG grant to FB), Fonds voor Wetenschappelijk Onderzoek-Vlaanderen (FWO research project G026715 N), and the Strategic Research Program of the Vrije Universiteit Brussel (SRP30). We are very grateful to Kim Roelants for providing drawings.

References

Ahrenfeldt RH (1955) Mating behaviour of *Euproctus asper* in captivity. Br J Herpetol 2:194–197

Amaya E, Ashurst JL, Bonfield JK, Croning MDR, Chen C-K, Davies RM, Francis MD, Garrett N, Gilchrist MJ, Grafham DV, McLaren SR, Papalopulu N, Rogers J, Smith JC, Taylor RG, Voigt J, Zorn AM (2006) Direct submission: nucleotide [Internet]. Bethesda (MD): National Library of Medicine (US), National Center for Biotechnology Information; [04 October 2006]—Accession No. CR761728.2, *Xenopus tropicalis* finished cDNA, clone TGas100l21 [cited 2018 January 14]. Available from: https://www.ncbi.nlm.nih.gov/nuccore/CR761728.2

Arnold SJ (1972) The evolution of courtship behaviour in salamanders (Doctoral dissertation). University of Michigan, Ann Arbor, Michigan, p 570

Arnold SJ (1977) The evolution of courtship behavior in new world salamanders with some comments on old world salamandrids. In: Taylor DH, Guttman SI (eds) The reproductive biology of amphibians. Plenum Press, New York, USA, pp 141–183

Arntzen JW, Sparreboom M (1989) A phylogeny for the old world newts, genus *Triturus*: biochemical and behavioural data. J Zool 219:645–664

Baeckens S, Driessens T, Van Damme R (2016) Intersexual chemo-sensation in a "visually-oriented" lizard. Anolis sagrei. PeerJ 4:e1874

Campbell LJ, Suárez-Castillo EC, Ortiz-Zuazaga H, Knapp D, Tanaka EM, Crews CM (2011) Gene expression profile of the regeneration epithelium during axolotl limb regeneration. Dev Dyn 240(7):1826–1840

Clark K, Karsch-Mizrachi I, Lipman DJ, Ostell J, Sayers EW (2016) GenBank. Nucleic Acids Res 44(D1):D67–D72. https://doi.org/10.1093/nar/gkv1276

Carranza S, Amat F (2005) Taxonomy, biogeography and evolution of *Euproctus* (Amphibia: Salamandridae), with the resurrection of the genus *Calotriton* and the description of a new endemic species from the Iberian peninsula. Zool J Linn Soc 145(4):555–582

Casey JR, Petranka JG, Kottra J, Fleenor DE, Rosse W (1994) The structure of the urokinase-type plasminogen activator receptor gene. Blood 84:1151–1156

Davis WC, Twitty VC (1964) Courtship behavior and reproductive isolation in the species of *Taricha* (Amphibia, Caudata). Copeia 1964:601–610

Dawley EM (1984) Recognition of individual, sex and species odours by salamanders of the *Plethodon glutinosus-P. jordani* complex. Anim Behav 32(2):353–361

Duellman WE, Trueb L (1994) Biology of amphibians. Johns Hopkins University Press, USA, p 670

Dunn RD, Broady KW (2001) Snake inhibitors of phospholipase A_2 enzymes. Biochim Biophy Acta (BBA)—Mol. Cell Biol. Lipids 1533(1):29–37

Fleming TJ, O'hUigin C, Malek TR (1993) Characterization of two novel Ly-6 genes. Protein sequence and potential structural similarity to alpha-bungarotoxin and other neurotoxins. J Immunol 150(12):5379–5390

Fortes-Dias CL, Campos PC, Fernandes CAH, Fontes MRM (2017) Phospholipase A2 inhibitors from snake blood (sbPLIs). In: Gopalakrishnakone P, Inagaki H, Mukherjee A, Rahmy T, Vogel CW (eds) Snake venoms. Toxinology. Springer, Dordrecht, pp 105–122

Frost DR (2019) Amphibian species of the world, an online reference. Version 6.0. Accessed on 10 January 2019. Electronic database accessible at http://research.amnh.org/herpetology/amphibia/index.html. American Museum of Natural History, New York, USA

Gallien L (1952) Elevage et comportement du *Pleurodèle* au laboratoire. Bulletin de Société Zoologique de France 77:456–461

Gong Y, Shu G, Huang F, He L, Li C, Xie F (2018) Courtship behaviour and male sexual competition of the Taliang crocodile newt, *Liangshantriton taliangensis*. Amphibia-Reptilia 39(3):275–288

Griffiths RA (1996) Newts and salamanders of Europe. Poyser Natural History, London, UK, p 188

Guillaume O (1999) Does the Pyrenean salamander *Euproctus asper* use chemical cues for sex identification and mating behaviour? Behav Proc 46(1):57–62

Halliday TR (1974) Sexual behaviour of the smooth newt *Triturus vulgaris* (Urodela, Salamandridae). J Herpetol 8(4):277–292

Halliday TR (1977) The courtship of European newts: an evolutionary perspective. In: Taylor DH, Guttman SI (eds) The reproductive biology of amphibians. Plenum Press, New York, USA, pp 185–232

Halliday TR (1990) The evolution of courtship behavior in newts and salamanders. Adv Study Behav 19:137–169

Houck LD, Arnold SJ (2003) Courtship and mating behaviour. In: Sever DM (ed) Phylogeny and reproductive biology of Urodela (Amphibia). Science Publishers, Enfield, New Hampshire, USA, pp 383–424

Houck LD, Bell AM, Reagan-Wallin NL, Feldhoff RC (1998) Effects of experimental delivery of male courtship pheromones on the timing of courtship in a terrestrial salamander, *Plethodon jordani* (Caudata: Plethodontidae). Copeia 1998:214–219

Houck LD, Palmer CA, Watts RA, Arnold SJ, Feldhoff PW, Feldhoff RC (2007) A new vertebrate courtship pheromone, PMF, affects female receptivity in a terrestrial salamander. Anim Behav 73(2):315–320

Houck LD, Watts RA, Mead LM, Palmer CA, Arnold SJ, Feldhoff PW, Feldhoff RC (2008) A candidate vertebrate pheromone, SPF, increases female receptivity in a salamander. In: Hurst JL, Beynon RJ, Roberts SC, Wyatt TD (eds) Chemical signals in vertebrates 11. Springer-Verlag Inc., New York, USA, pp 213–221

Iwata T, Conlon JM, Nakada T, Toyoda F, Yamamoto K, Kikuyama S (2004) Processing of multiple forms of preprosodefrin in the abdominal gland of the red-bellied newt *Cynops pyrrhogaster*: regional and individual differences in preprosodefrin gene expression. Peptides 25(9):1537–1543

Iwata T, Ishizuka Y, Nakada T, Toyoda F, Yamamoto K, Conlon JM, Kikuyama S (2005) Regionally specific occurrence of an active sodefrin variant in the red-bellied newt. Ann NY Acad Sci 1040(1):351–353

Janssenswillen S, Bossuyt F (2016) Male courtship pheromones induce cloacal gaping in female newts (Salamandridae). PLoS ONE 11(1):e0144985

Janssenswillen S, Vandebergh W, Treer D, Willaert B, Maex M, Van Bocxlaer I, Bossuyt F (2014) Origin and diversification of a salamander sex pheromone system. Mol Biol Evol 32 (2):472–480

Janssenswillen S, Willaert B, Treer D, Vandebergh W, Bossuyt F, Van Bocxlaer I (2015) High pheromone diversity in the male cheek gland of the red-spotted newt *Notophthalmus viridescens* (Salamandridae). BMC Evol Biol 15:54

Kawamura T, Sawada S (1959) On the sexual isolation among different species and local races of Japanese newts. J Sci Hiroshima Univ Ser B 18:17–31

Kieffer B, Driscoll PC, Campbell ID, Willis AC, van der Merwe PA, Davis SJ (1994) Three-dimensional solution structure of the extracellular region of the complement regulatory protein CD59, a new cell-surface protein domain related to snake venom neurotoxins. Biochemistry 33(15):4471–4482

Kieren S, Sparreboom M, Hochkirch A, Veith M (2018) A biogeographic and ecological perspective to the evolution of reproductive behaviour in the family Salamandridae. Mol Phylogenet Evol 121:98–109

Kikuyama S, Toyoda F, Ohmiya Y, Matsuda K, Tanaka S, Hayashi H (1995) Sodefrin: a female-attracting peptide pheromone in newt cloacal glands. Science 267:1643–1645

Kikuyama S, Toyoda F (1999) Sodefrin: a novel sex pheromone in a newt. Rev Reprod 4(1):1–4

Kini RM, Doley R (2010) Structure, function and evolution of three-finger toxins: mini proteins with multiple targets. Toxicon 56(6):855–867

Kozak KH (2003) Sexual isolation and courtship behavior in salamanders of the *Eurycea bislineata* species complex, with comments on the evolution of the mental gland and pheromone delivery behavior in the Plethodontidae. Southeast Nat 2(2):281–293

Loughner CL, Bruford EA, McAndrews MS, Delp EE, Swamynathan S, Swamynathan SK (2016) Organization, evolution and functions of the human and mouse Ly6/uPAR family genes. Hum Genomics 10(1):10

Maex M, Van Bocxlaer I, Mortier A, Proost P, Bossuyt F (2016) Courtship pheromone use in a model urodele, the Mexican axolotl (*Ambystoma mexicanum*). Sci Rep 6:20184

Maex M, Treer D, De Greve H, Proost P, Van Bocxlaer I, Bossuyt F (2018) Exaptation as a mechanism for functional reinforcement of an animal pheromone system. Curr Biol 28 (18):2955–2960.e5

Maki N, Martinson J, Nishimura O, Tarui H, Meller J, Tsonis PA, Agata K (2010) Expression profiles during dedifferentiation in newt lens regeneration revealed by expressed sequence tags. Mol Vision 16:72–78

Malacarne G, Vellano C (1987) Behavioral evidence of a courtship pheromone in the crested newt, *Triturus cristatus carnifex* Laurenti. Copeia 1987(1):245–247

Mitchell J, Gibbons W (2010) Salamanders of the Southeast. University of Georgia Press, Georgia, USA, p 324

Nakada T, Toyoda F, Iwata T, Yamamoto K, Conlon JM, Kato T, Kikuyama S (2007) Isolation, characterization and bioactivity of a region-specific pheromone, [Val^8] sodefrin from the newt *Cynops pyrrhogaster*. Peptides 28(4):774–780

Nakada T, Toyoda F, Matsuda K, Nakakura T, Hasunuma I, Yamamoto K, Onoue S, Yokosuka M, Kikuyama S (2017) Imorin: a sexual attractiveness pheromone in female red-bellied newts (*Cynops pyrrhogaster*). Sci Rep 7:41334

Nelson J (1960) Courtship display in the Italian crested newt (*Triturus cristatus carnifex*). Biol J 1:6–9

Okomura K, Masui K, Inoue S, Ikeda K, Hayashi K (1999) Purification, characterization and cDNA cloning of a phospholipase A2 inhibitor from the serum of the non-venomous snake *Elaphe quadrivirgata*. Biochem J 341(1):165–171

Osikowski A, Cierniak-Zuzia K (2013) Cloacal anatomy of the male Carpathian newt, *Lissotriton montandoni* (Amphibia, Salamandridae), in the breeding season. Zool Sci 30(9):748–753

Palmer CA, Watts RA, Houck LD, Picard AL, Arnold SJ (2007) Evolutionary replacement of components in a salamander pheromone signaling complex: more evidence for phenotypic-molecular decoupling. Evolution 61(1):202–215

Petranka J (2010) Salamanders od the United States and Canada. Smithsonian books, Washington, USA, p 592

Pool TB, Dent JN (1977) The ultrastructure and the hormonal control of product synthesis in the hedonic glands of the red-spotted newt *Notopthalmus viridescens*. J Exp Zool 201(2):177–201

Poschadel JR, Rudolph A, Plath M (2007) Nonvisual mate choice in the Pyrenean mountain newt (*Euproctus asper*): females prefer small males. Acta Ethologica 10(1):35–40

Protein [Internet]. Bethesda (MD): National Library of Medicine (US), National Center for Biotechnology Information; 2004—[cited 2018 January 21]. Available from: https://www.ncbi.nlm.nih.gov/protein/

Rafinski J, Pecio A (1992) The courtship behaviour of the Bosca's newt, *Triturus boscai* (Amphibia: Salamandridae). Folia Biol 40:155–165

Rollins RE, Staub NL (2017) The Presence of caudal courtship-like glands in male and female ouachita dusky salamanders (*Desmognathus brimleyorum*). Herpetologica 73(4):277–282

Rollmann SM, Houck LD, Feldhoff RC (1999) Proteinaceous pheromone affecting female receptivity in a terrestrial salamander. Science 285(5435):1907–1909

Rollmann SM, Houck LD, Feldhoff RC (2003) Conspecific and heterospecific pheromone effects on female receptivity. Anim Behav 66:857–861

Rusconi M (1821) Amours des salamandres aquatiques et developpement du tetard de ces salamandres depuis l'oeuf jusqu'a l'animal parfait. Paolo Emilio Giusti, Milano, Italy, p 91

Sacerdote M (1956) Per una migliore conoscenza della topografia delle ghiandole annesse alla cloaca nel tritone crestato. Bolletino di zoologia 23(1):33–50

Salthe SN (1967) Courtship patterns and the phylogeny of the urodeles. Copeia 1967:100–117

Secondi J, Haerty W, Lodé T (2005) Female attraction to conspecific chemical cues in the palmate newt *Triturus helveticus*. Ethology 111(8):726–735

Sever DM, Verrell PA, Halliday TR, Griffiths M, Waights V (1990) The cloaca and cloacal glands of the male smooth newt, *Triturus vulgaris vulgaris* (Linnaeus), with especial emphasis on the dorsal gland. Herpetologica 46(2):160–168

Sever DM (1991) Comparative anatomy and phylogeny of the cloacae of salamanders (Amphibia: Caudata). I. Evolution at the family level. Herpetologica 47(2):165–193

Sever DM (1992) Comparative anatomy and phylogeny of the cloacae of salamanders (Amphibia: Caudata) IV. Salamandridae. Anat Rec 233(2):229–244

Sever DM (1994) Comparative anatomy and phylogeny of the cloacae of salamanders (Amphibia: Caudata). VII. Plethodontidae. Herpetol Monogr 8:276–337

Sever DM (2003) Courtship and mating glands. In: Sever DM (ed) Phylogeny and reproductive biology of Urodela (Amphibia). Science Publishers, Enfield, New Hampshire, USA, pp 323–381

Spallanzani L (1780) Dissertazioni di fisica animale e vegetabile. Presso La Societa Tipografica, Modena, Italy, p 361

Sparreboom M (2014) Salamanders of the old world. KNNV Publishing, Zeist, The Netherlands, p 431

Tinbergen N, Ter Pelkwijk JJ (1938) De kleine watersalamander. De Levende Natuur 43(8):232–237

Toyoda F, Tanaka S, Matsuda K, Kikuyama S (1994) Hormonal control of response to and secretion of sex attractants in Japanese newts. Physiol Behav 55(3):569–576

Treer D, Maex M, Van Bocxlaer I, Proost P, Bossuyt F (2017) Divergence of species-specific protein sex pheromone blends in two related, nonhybridizing newts (Salamandridae). Mol Ecol 27(2):508–519

Treer D, Van Bocxlaer I, Matthijs S, Du Four D, Janssenswillen S, Willaert B, Bossuyt F (2013) Love is blind: indiscriminate female mating responses to male courtship pheromones in newts (Salamandridae). PLoS ONE 8(2):e56538

Treer D, Bossuyt F (2010) Chemical 'versus' visual cues in newts (Salamandridae): experiments in total darkness and in a Y-maze. In: 17th Benelux Congress of Zoology, Ghent, Belgium. Conference abstract, p 87

Tsetlin V (1999) Snake venom α-neurotoxins and other 'three-finger'proteins. Eur J Biochem 264(2):281–286

Van Bocxlaer I, Maex M, Treer D, Janssenswillen S, Janssens R, Vandebergh W, Proost P, Bossuyt F (2016) Beyond sodefrin: evidence for a multi-component pheromone system in the model newt *Cynops pyrrhogaster* (Salamandridae). Sci Rep 6:21880

Van Bocxlaer I, Treer D, Maex M, Vandebergh W, Janssenswillen S, Stegen G, Kok P, Willaert B, Matthijs S, Martens E, Mortier A, de Greve H, Proost P, Bossuyt F (2015) Side-by-side secretion of late palaeozoic diverged courtship pheromones in an aquatic salamander. Proc R Soc Lond B Biol Sci 282:20142960

Verrell P (1982) The sexual behaviour of the red-spotted newt, *Notophthalmus viridescens* (Amphibia: Urodela: Salamandridae). Anim Behav 30(4):1224–1236

Wells KD (2007) The ecology and behavior of amphibians. The University of Chicago Press, Chicago, USA, p 1148

Yamamoto K, Kawai Y, Hayashi T, Ohe Y, Hayashi H, Toyoda F, Kawahara G, Iwata T, Kikuyama S (2000) Silefrin, a sodefrin-like pheromone in the abdominal gland of the sword-tailed newt, *Cynops ensicauda*. FEBS Lett 472:267–270

Zhang P, Papenfuss TJ, Wake MH, Qu L, Wake DB (2008) Phylogeny and biogeography of the family Salamandridae (Amphibia: Caudata) inferred from complete mitochondrial genomes. Mol Phylogenet Evol 49(2):586–597

Comparative Structural Modeling of Bovine Vomeronasal Type-1 Receptor 1 (VN1R1) and Elucidation of Molecular Interactions with Pheromones Using *in silico* Approaches

Rajesh Durairaj[1(✉)], Cécile Bienboire-Frosini[1], and Patrick Pageat[2]

[1] Department of Behavioural and Physiological Mechanisms of Adaptation (D-MPCA), Research Institute in Semiochemistry and Applied Ethology (IRSEA), Quartier Salignan, 84400 Apt, France
r.durairaj@group-irsea.com

[2] Department of Chemicals Identification and Development of Analogues, Research Institute in Semiochemistry and Applied Ethology (IRSEA), Quartier Salignan, Apt, France

Abstract. The perception of pheromone signals is mediated by the axon of chemosensory neurons in the vomeronasal organ (VNO) that project into the accessory olfactory bulb (AOB) in the nasal cavity. Specifically, the vomeronasal type-1 receptor 1 (VN1R1) is a member of the G protein-coupled receptor (GPCR) superfamily of proteins and interconnecting large olfactory sensory neurons that is directed by a multigene superfamily in most species. In bovines, the VN1R1 is crucially involved in mate choice, reproduction and pheromone transport activity during chemical communication. The VNO morphology, genome assembly, sequence alignment and evolutionary divergence of VN1R were described previously. However, the structural modeling and elucidation of bovine VN1R1 protein have not yet been reported. In this study, we aimed to perform sequence analysis, structure prediction and function elucidation using computational methods. The results revealed that (i) the highest proportions of α-helix and random coils were observed in several structural and functionally conserved sites of VN1R1; (ii) the comparative template-based VN1R1 model was predicted with >93% of the most favoured region in the Ramachandran Plot validation; (iii) with regard to the transmembrane topology, structural superimposition was additionally supported by the bovine VN1R1 computational model; and (iv) the molecular interactions of pheromones and VN1R1 output suggested the best-fit endogenous ligand 'p-Cresol', which is suitable for the binding of a VN1R1 model. For the first time, we performed comparative structure prediction of the bovine VN1R1 model to identify the putative isoforms of pheromones involved in signal transduction in animal reproduction and management.

Keywords: Chemical communication · Comparative modeling · Molecular docking · Pheromones · VNO · VN1R1

© Springer Nature Switzerland AG 2019
C. D. Buesching (Ed.): *Chemical Signals in Vertebrates 14*, pp. 221–237, 2019.
https://doi.org/10.1007/978-3-030-17616-7_16

1 Introduction

Pheromones are semiochemical molecules, which are secreted and emitted by an individual and received by another individual within the same species (Karlson and Lüscher 1959; Matsunami and Buck 1997; Wyatt 2010). Pheromones are evolved signals with signature mixtures to elicit a stereotyped behavioural and physiological innate response in conspecifics (Wyatt 2014, 2017). The mammalian olfactory system can recognize and discriminate the chemical signals of the external world (Buck and Axel 1991; Rodriguez et al. 2000). The chemosensory processing is mediated by two anatomically and functionally distinct olfactory sensory organs, the main olfactory epithelium (MOE) and the accessory olfactory bulb (vomeronasal system) (Dulac and Axel 1995). The pheromones are dissolved into the nasal mucus and then recognized by specific vomeronasal sensory neurons (VSNs), which are mainly located in the vomeronasal organ (VNO) (Dulac and Axel 1998; Rodriguez 2003). They project through the accessory olfactory bulb (AOB) to the amygdala, hypothalamus and areas of endocrine and behavioural responses (Matsunami and Buck 1997; Tirindelli et al. 2009). The involvement of VNO in signal processing is well understood in intraspecific communication (Dulac and Torello 2003; Tirindelli et al. 2009; Papes et al. 2010) and associated signal transduction functions (Dulac and Axel 1995; Keverne 1999).

There are two distinct superfamilies of pheromone receptors expressed in the VNO system that were classified as type-1 vomeronasal receptors (Dulac and Axel 1995) and type-2 vomeronasal receptors (Herrada and Dulac 1997; Matsunami and Buck 1997; Ryba and Tirindelli 1997) belonging to the superfamily of G-protein coupled receptors (GPCR). The vomeronasal receptors (VRs) are seven transmembrane proteins (7TM) but are highly deviated from the olfactory receptor (OR) superfamily (Buck and Axel 1991; Dulac and Axel 1995; Herrada and Dulac 1997; Matsunami and Buck 1997; Mombaerts 1999, 2004). The apical VSNs of the VNO are expressed pheromone receptors of type V1Rs and project to the anterior AOB. V1R expression is monoallelic and monogenic, thus achieving the expression of one receptor type per cell and so contributing to the narrow tuning of the vomeronasal neurons. Each V1R and VSN are co-expressed with the G protein subunits $G\alpha i2$, which are bound to V1R. V2R and the $G\alpha_o$ subunits of the G protein are co-expressed and bound in the basal VSNs of the VNO, which project to the posterior AOB. (Buck 2000; Zufall and Munger 2001; Tirindelli et al. 2009). The vomeronasal type-1 receptor 1 (VN1R1) includes one of the five individually derived functional classes of chemosensory receptor genes in the olfactory system of mammals (Yoder and Larsen 2014), which are partly responsible for pheromone detection and transduction of semiochemical cues in mammals (Young et al. 2010). The evolutionary diversity and expansion occurred independently from ancestral genes with larger pseudogenes of V1R genes in mammals (Rodriguez and Mombaerts 2002; Young et al. 2010, Kubo et al. 2016). In contrast, VN1R orthologous genes and sequence identity are highly conserved among cattle and small ruminants (Ohara et al. 2009). Also, the VN1R orthologs could be involved in the detection of analogous chemical cues as allomones (Wakabayashi et al. 2002; Ohara et al. 2009).

On the other hand, the source of chemical signals such as urine, faces, saliva and secretions from specialized scent glands have a potential role in chemical communication in animal reproduction and management (Albone 1984; Albone et al. 1986; Archunan 2009; Patra et al. 2012; Mozūraitis et al. 2017). Several putative pheromone compounds have been identified in cattle and buffalo species. In particular, 1-iodoundecane-1IDU was identified in female bovine urine and functioned as a stimulant in the bull response for coitus (Kumar et al. 2000; Archunan and Rameshkumar 2012). 4-Methylphenol (p-Cresol) (buffalo urine and saliva) functioned as attraction and mounting compounds in cattle (Rajanarayanan and Archunan 2011; Karthikeyan et al. 2014). In particular, 9-octadecenoic acid-9ODA is an inducing factor for the mounting response in buffalo bulls, and all the endogenous urinary sex pheromones enhance the production of sperm by exposing them to the nasal cavity (Archunan and Rajanarayanan 2010; Karthikeyan and Archunan 2013; Le Danvic et al. 2015).

Additionally, the family of appeasing pheromones was identified and patented in several animals (such as cattle, pigs, dogs, horses, avians and cats) by Patrick Pageat, IRSEA, France. In bovines, the synthetic analogue of the bovine appeasing pheromone (BAP) is a mixture of fatty acids that reproduces components of the natural pheromone (Pageat 1998). Recently, BAP was used to enhance Valdostana dairy cow performance in milk production especially during stressful situations (Osella et al. 2018). Understanding the biological role of receptors is required for elucidation of the structure, functions and binding interactions with various semiochemical cues. Since several endogenous pheromone compounds were identified and characterized via behavioural and physiological observations in cattle, it is noteworthy to build a comparative template-based VN1R1 model for the identification of molecular residue interactions of best-fit pheromones through computational analysis.

2 Methods

2.1 Computational Sequence Curation

All of the vomeronasal type-1 receptor 1 (VN1R1) protein sequences of cattle and other species were collected from the National Center for Biotechnology Information (NCBI) (https://www.ncbi.nlm.nih.gov/) and UniProt (http://www.uniprot.org/) databases. The low-quality and partial sequences were removed manually, and the intact VN1R1 gene corresponding proteins (~ 37) were collected from various ortholog species using the position-specific iterative basic local alignment search tool (PSI-BLAST) program (Altschul et al. 1997). The BLOSUM62 matrices were used as a scoring matrix with an E-value threshold of 0.005 for predicting sequence-scoring alignments between evolutionarily diverged protein sequences. The redundant protein dataset was filtered by $\geq 90\%$ identity limitation in the CD-HIT server (http://cd-hit.org/).

2.2 Evolutionary Tree Analysis

The selective VN1R1 (9) ortholog sequences were retrieved from the final collective dataset. The sequence was submitted to ClustalO (https://www.ebi.ac.uk/Tools/msa/clustalo/) and MultAlin (http://multalin.toulouse.inra.fr/) programs for multiple sequence alignment (MSA). The sequence alignment was visualized using ESPript V.3.

0 (http://espript.ibcp.fr/) (Robert and Gouet 2014). Further, the sequence alignment file (.aln) was transferred to MEGA V.6 software (Tamura et al. 2013). The phylogenetic tree was generated using the Maximum-Likelihood (ML) statistical method and the Jones–Taylor–Thornton (JTT) model with 1000 bootstrap replications (Jones et al. 1992). The representation and colouring of phylogeny were performed using MEGA tree explorer.

2.3 Sequence Analysis of VN1R1

The VN1R1 sequence of *Bos taurus* (NCBI ID: XP_005219876.1) was selected for all computational analyses. The protein sequence was retrieved from the gene entry (Gene ID: 790265) of the recently updated version. The physico-chemical properties and secondary structure were predicted using the ProtParam tool (https://web.expasy.org/protparam/) and self-optimized prediction method with alignment (SOPMA) web server (Geourjon and Deléage 1995), respectively. SOPMA is entirely based on the homolog method as suggested by the network protein sequence analysis (NPS) web-server. The conservation pattern of the VN1R1 sequence was predicted using the Consurf server (http://consurf.tau.ac.il/) (Landau et al. 2005). The functionally exposed and structurally buried residues were predicted according to the neural network algorithm. The evolutionary rate of VN1R1 conservation was computed using the empirical Bayesian method with a JTT evolutionary substitution model. Further, the analysis of the transmembrane region was predicted using various computational servers such as TMHMM, TMpred, Split and TOPCONS (data not shown). Specifically, the interactive protein feature visualization (PROTTER) server (http://wlab.ethz.ch/protter/start/) predicted the membrane topology using the Phobius tool and N-glycosylation motifs (Omasits et al. 2014). Additionally, the residue of N-glycosylation sites was confirmed using the NetNGlyc 1.0 (http://www.cbs.dtu.dk/services/NetNGlyc/) server.

2.4 Structure Annotation of VN1R1

In this present study, the bovine VN1R1 sequence was submitted to construct a structural model using a template-based tertiary structure prediction method (RaptorX) web portal (http://raptorx.uchicago.edu/) (Peng and Xu 2011; Källberg et al. 2012; Ma et al. 2013). It is a comparative structure-threading algorithm that predicts the protein three-dimensional (3D) structure as well as contact map, solvent accessibility, and disordered regions of the VN1R1 protein sequence. Then, the threaded VN1R1 model was validated using the Ramachandran Plot visualization from the Procheck program in the ProFunc server (https://www.ebi.ac.uk/thornton-srv/databases/profunc/) (Laskowski et al. 2005). The stereochemical quality of the threaded model and structure geometry was predicted based on residual torsion angles of Phi and Psi degrees in the plot. The analysis of structural stability and fluctuation simulation was carried out using the CABS-flex 2.0 (http://biocomp.chem.uw.edu.pl/CABSflex2) server for the VN1R1 model, where the simulation analysis helps to predict the residual mean square fluctuations (RMSF) profile and trajectory map (Kuriata et al. 2018). The validated VN1R1 threaded model was deposited in the Protein Model Database (PMDB) (https://bioinformatics.cineca.it/PMDB/). The unique model ID (PMDB ID: PM0081593) was provided for the VN1R1 threaded model, and the protein model was made available in the database.

2.5 Binding Site and Ligand Optimization

The receptor binding sites were predicted using the computed atlas of surface topography of proteins (CASTp) V.3.0 (http://sts.bioe.uic.edu/castp/) server (Tian et al. 2018), and the selected central binding site residues were tabulated with the area (A^2) and volume (A^3) of the pocket (Table 1). Besides, three endogenous semiochemicals and two standard compounds (ligands) were collected from earlier reports, and the ligand structural coordinates were retrieved from the PubChem database (https://pubchem.ncbi.nlm.nih.gov/). The selected standard ligands are Palmitic acid-PAL (CID_985) and Retinal-RET (CID_638015) (Murakami and Kouyama 2008 and Shimamura et al. 2008). The endogenous semiochemical ligands are 4-methylphenol (p-Cresol) (CID_2879), 1-iodoundecane-1IDU (CID_20276), and 9-octadecenoic acid-9ODA (CID_965), which were subjected to molecular docking analysis. The ligand details were included in the chemical formula, molecular weight (MW), and other physico-chemical properties as

Table 1. Major binding site cavity of bovine VN1R1 comparative model

Protein name	Area (A^2)	Volume (A^3)	Binding site residue with positions	Other binding site residues
VN1R1	108.7	34.9	Phe12, Val57, Lys61, Phe85, His88, Arg89, Thr92, Tyr167, Cys168, Ser169, Trp170, Phe246, Tyr247, Tyr250, Val272, Leu273, Ala275, Thr276, Phe278, Pro279	Leu26, Tyr30, His38, Asn39, Arg41, Ile43, Asp44, Ile46, Leu47, Leu50, Val51, Phe96, Thr98, Thr99, Phe102, Asn103, Phe105, Gln106, Lys109, Leu110, Met196, Ser200, Cys228, Glu229, Arg231, Ala232, Thr233, Arg234, Thr235, Val236, Leu237, Leu239, Val240, Ser282, Val285, Leu286, Lys289, Asp290, Thr291, Arg292

Table 2. Physico-chemical properties of selected pheromones and standard compounds

S. No	Compounds name	Short name	PubChem ID	Chemical formula	Molecular weight (g/mol)	H-bond donor	H-bond acceptor
1	4-methylphenol	p-Cresol	CID_2879	C_7H_8O	108.14	1	1
2	1-iodoundecane	1IDU	CID_20276	$C_{11}H_{23}I$	282.2	0	0
3	9-octadecenoic acid	9ODA	CID_965	$C_{18}H_{34}O_2$	282.468	1	2
4	Palmitic Acid	PAL	CID_985	$C_{16}H_{32}O_2$	256.43	1	2
5	Retinal	RET	CID_638015	$C_{20}H_{28}O$	284.443	0	1

shown in Table 2. The energy minimization of selected ligands and VN1R1 (receptor) was optimized using the Swiss-PDB viewer with the GROMOS96 Force Field algorithm.

2.6 Molecular Docking

The receptor–ligand interactions were predicted using DockingServer (https://www.dockingserver.com), which is a web-based interface module that consists of PM6 semi-empirical quantum-mechanical partial charge calculation to enhance the accuracy of the docking output (Bikadi and Hazai 2009). For the grid construction, we selected the central binding pocket fronting towards extracellular small molecule binding sites for grid construction as an earlier report (Harini and Sowdhamini 2015). The docking calculation was carried out on all the selected ligand and VN1R1 receptor. The grid parameter (nx = 27; ny = 27; nz = 27 and cx = −8.335; cy = −6.294; cz = 21.43) and 0.375 Å spacing were generated using Autogrid program (Morris et al. 1998; Morris et al. 2009). The docking simulation was performed using the Lamarckian genetic algorithm (LGA). All the docking analyses were derived from 100 different runs, maxi. 2500000 energy evaluations, maxi. 540000 number of generations with a population size of 150. Additionally, the receptor-ligand interactions were visualized using Discovery studio visualizer DSV 4.5 (Accelrys, San Diego, USA) free version, USCF Chimera (https://www.cgl.ucsf.edu/chimera/), and LIGPLOT V.4.5.3 program (Wallace et al. 1995). The evaluation of semi-empirical docking values was computed mainly with respect to free binding energy, H-bonds and other steric interactions.

3 Results and Discussion

An adequate number of 3D structures were designed and structurally characterized with the help of comparative template-based modeling (Mirkovic et al. 2007), threading and fold recognition prediction using an ab initio protocol. Until now, no computational study has been reported to analyse the bovine VN1R1 protein structure and its involvement in intra-species semiochemical communication. We predicted the rational computational model of bovine VN1R1 and elucidated the structural stability and endogenous pheromone binding function using *in silico* approaches. Thus, the present computational study of bovine VN1R1 consists of two different analyses such as sequence elucidation and structure annotation.

3.1 MSA and Phylogenetic Tree Analysis

VN1R1 ortholog sequence alignment was used to construct a phylogenetic tree of livestock animals. The traditional view of the unrooted phylogenetic tree and corresponding VN1R1 orthologous MSA analysis are shown in Fig. 1a, b. The 9 orthologous sequences were from *Bovidae* (cows, sheep, and goats), *Equidae* (horses and donkeys), and *Suidae* (pigs) for MSA and phylogenetic tree construction. Identical sites and conserved motifs were highly similar in all orthologous species except for pigs.

Specifically, the most of the VN1R1 sequence showed the signature of several conserved motifs such as -GN- and -SLxCLYxFTLLT-(Transmembrane (TM1),

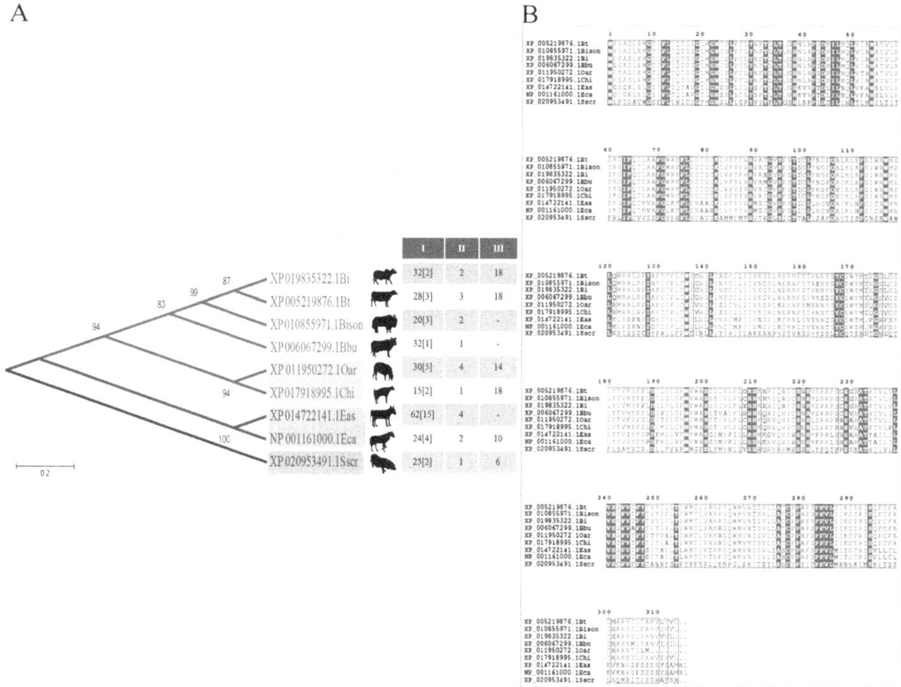

Fig. 1. The phylogenetic tree of selective VN1R1 protein sequence and MSA analysis. **a** The reconstruction of phylogenetic tree (straight view) shows evolutionary divergence of VN1R1 between livestock animals such as Bovidae family (*Bos indicus*-Bi; *Bos taurus*-Bt; *Bison bison bison*-Bison; *Bubalusbubalis*-Bbu; *Ovisaries*-Oar; *Capra hircus*-Chi); Equidae family (*Equus asinus*-Eas; *Equus caballus*-Eca); and Suidae (*Sus scrofa*-Sscr). The complete gene dataset was defined as I, II, III in a table. (I) Global members of vomeronasal receptor 1 with the intact VN1R1 genes in parenthesis. (II) Number of proteins corresponds to intact gene entries. (III) Chromosome number of VN1R1 gene entries. **b** The divergence tree was constructed from multiple sequence alignment (MSA) analysis. The identical and conserved sites were highlighted in the MSA using ESPript tool

-Ex**RAT**xT-(TM5 to TM6 intracellular loop), and -CxPAF**SPFVL**-(TM7). The highlighted conserved motifs are correlated with an earlier report (Dulac and Axel 1995). The VN1R1 sequence showed a similar profile pattern as the GPCR family 1 (Prosite ID: PS50262) and the VNO pheromone receptor family (Prosite ID: PF03402).

Further, the VN1R1 orthologous sequence from different species was analysed to understand the evolutionary diversity. The phylogenetic tree revealed that bovine VN1R1 has >90% identity within the *Bovidae* family, and the sister taxa were formed between *Bos taurus* (Bt) and *Bos indicus* (Bi). *Sus scrofa* (Sscr) highly deviated and displayed less than 30% sequence identity with the VN1R1_Bt protein. Hence, the tree suggests that numerous identical residues among the livestock species corroborate with previous studies (Ohara et al. 2009).

3.2 VN1R1 Sequence Elucidation

The physico-chemical properties of bovine VN1R1 consisted of 314 amino acids with a theoretical pI of 9.54. The VN1R1 bovine sequence contains about 26 positively charged residues (Arg and Lys) and 10 negatively charged residues (Asp and Glu). The aliphatic index was calculated as 107.71, and the instability index of the VN1R1 was computed as 39.01, which classifies the protein as stable. Leucine and Phenylalanine were the most numerous amino acids in the sequence: 12.7% and 8.9%, respectively. The secondary structural information of the VN1R1 sequence revealed that an alpha helix (Hh) was found at the highest level (37.58%), followed by the random coil (Cc) at 34.71%. The extended strand (Ee) and beta-turn were found to a lesser degree: 24.52% and 3.18%, respectively. More precisely, there were no 3_{10}, Pi helices or a Beta Bridge in the VN1R1 sequence analysis (Fig. 2 a).

Conservation properties of VN1R1 residues were predicted using an evolutionary substitution rate analysis among the homologous sequences, and rate differences were indicated based on a conservation scale (Glaser et al. 2003). We predicted 22 structurally buried and 21 functionally exposed residues in the VN1R1 sequence. The reported motif regions in the transmembrane were shown as more buried residues. The evolutionary conservation rates suggested that the predicted conserved residues might be involved in ligand binding functions (Berezin et al. 2004). The VN1R1 sequence had several higher conserved residues in the transmembrane segment, and the representation of conservation was displayed by grade scale (Fig. 2b).

Similarly, the protein topology prediction is an important analysis to understand the folds and structural motifs (Rajesh et al. 2016). We have identified and validated the transmembrane regions of VN1R1 using various computational analyses and illustrated them using the PROTTER server. The protein showed OUT topology (N-terminal loop in extracellular regions) and 7TM region with 3 external–internal loops. We have predicted the glycosylation sites (Asn23, Asn31, Asn54, and Asn158) and a disulphide bond between the Cys81-Cys168 residues (Fig. 2c). The glycosylation sites were verified using a NetNGlyc plot (data not shown).

3.3 Comparative Modeling and Simulation

The molecular computational model of VN1R1 was constructed based on the comparative template-based tertiary structure modeling, and the protein templates were human β2-adrenergic GPCR: 2RH1; mouse μ-opioid receptor: 5C1M; and other lower E-value receptors (bovine rhodopsin: 1GZM). The structure based sequence alignments were selected according to the P-value (1.06e-06) and sequence identity. The VN1R1 sequence identity ($\leq 30\%$) is below the "twilight zone" threshold (Rost 1999), whereas the structure threading method is most useful for recognizing the common folding pattern with template sequence comparison methods (Godzik 2003). We used the entropy-dependent scoring function for threading of the VN1R1 sequence, and the construction of a structural model was made by a multi-template threading procedure to model the target sequence (Källberg et al. 2012; Peng and Xu 2011).

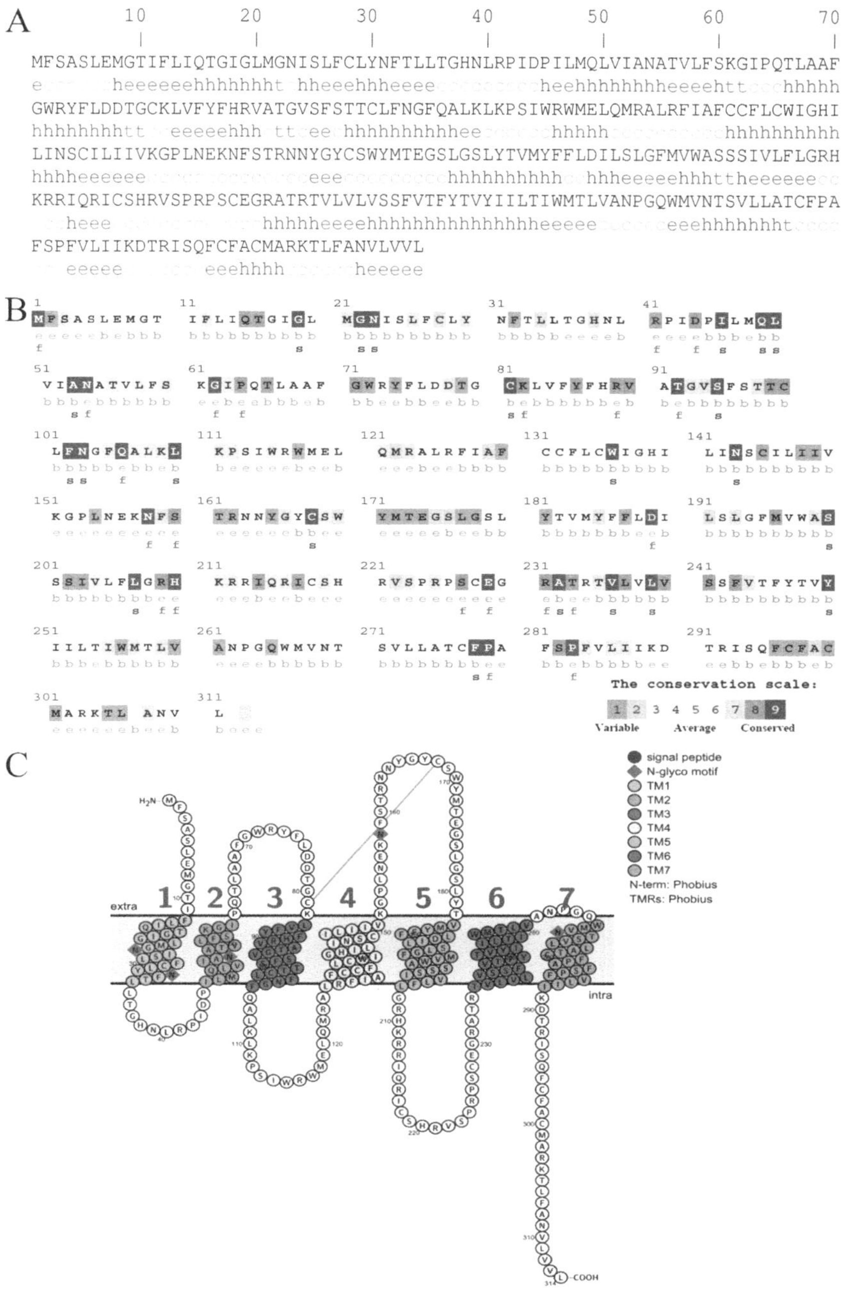

Fig. 2. The bovine VN1R1 sequence analysis. **a** The secondary structure analysis shows helices, coils and loops. **b** The sequence conservation was predicted based on conservation scale. **c** The transmembrane (TM) representation shows 7TM with cystine bridge. The sequence contains 3 extra- and intracellular loops between TM regions

In this study, the template-based threaded model was made for a bovine VN1R1 sequence using the above-mentioned templates (Fig. 3a), and the 3D backbone view of the pipes and plank structure is illustrated (Fig. 3b). In general, the best model would be expected to have >90% of the most favoured regions (Laskowski et al. 1993) and the conformation of the protein is validated using dihedral Phi/Psi angles (−180° to 180°). Then, the model was validated as 93.3% residues in the most favoured region, 4.9% of residues in additional allowed regions, and 1.1% of residues in generously

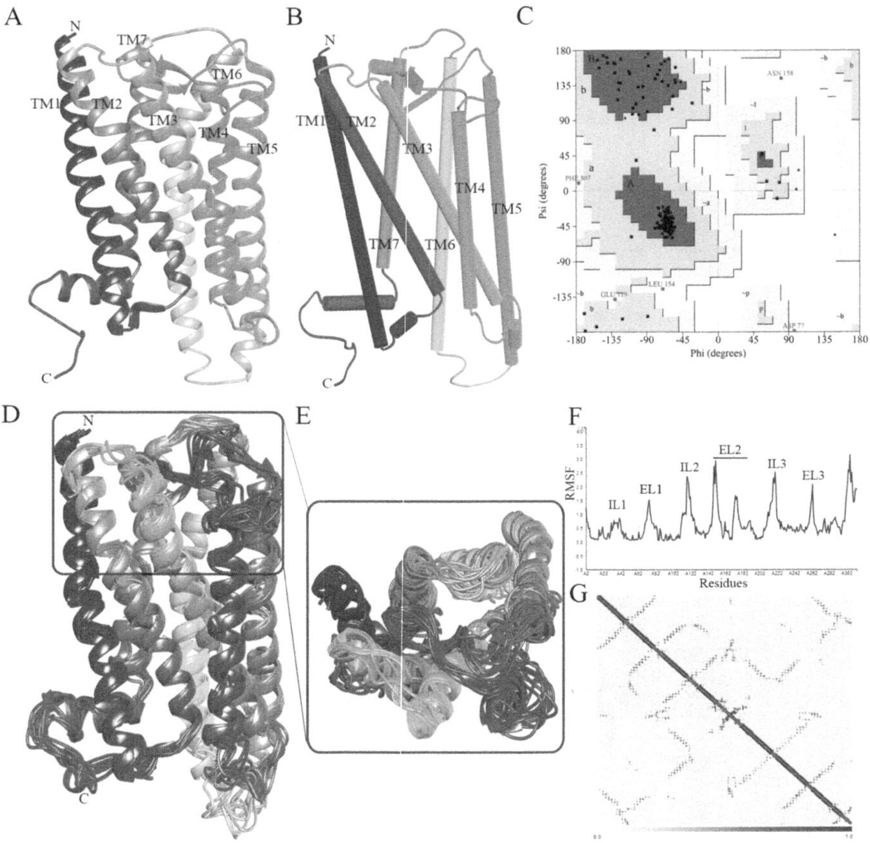

Fig. 3. Comparative template-based modeling, validation and simulation of **bovine VN1R1**. **a** The threaded VN1R1 model was constructed by multi-template modeling. The model was illustrated by helical-coil structure pattern and mentioned the TM1-7 regions. **b** The same structure illustrated using pipes and planks view. **c** The structure was validated by Ramachandran Plot, which showed the allowed and disallowed conformation of residues. **d** The structural superimposition of VN1R1 model with other trajectory conformations using molecular dynamic simulation. **e** The vertical view of superimposed VN1R1 model. **f** The root mean square fluctuation (RMSF) profile pattern was obtained from the structural simulation and the extra extracellular (EL) and intracellular regions were mentioned in the plot. **g** The residue–residue contacts were showed as trajectory map

allowed regions, but 2 amino acids (Asp77 and Asn158) occurred in disallowed regions (Fig. 3c). The TM regions of the VN1R1 computational model were predicted as high-quality resolution with validation reports of H-bond topology, main/side chain parameters, RMS distance planarity, and distorted geometry.

Subsequently, the residual fluctuation and stability of VN1R1 were observed using root mean square fluctuations (RMSF) profile and trajectory map analysis. Here, the VN1R1 model was simulated and superimposed on all other similar homologues of the proteins equivalent to 10 nanoseconds (ns) below 4 Å. The simulation results were affirmed through the best structural arrangements and superimposition of VN1R1 (Fig. 3d, e). We observed more residue fluctuation in all intra- and extracellular regions in the RMSF plot. Notably, the N- and C-terminal loops showed high residue flexibility (Fig. 3f). We predicted the most frequent residue–residue trajectory contact map throughout the simulation and 10 representatives of homologous proteins (Fig. 3g).

From the simulation output, we found that the TM regions of the protein remain in an active and stable conformation below 1.5 Å. The analysis allows the visualization and fluctuation trajectories for studying the biomolecular processes and stable structures for receptor-ligand interactions in molecular docking analysis (Kuriata et al. 2018). The binding sites of bovine VN1R1 showed plenty of hydrophobic and polar residues in the functionally conserved TM regions. The top 10 binding sites were observed based on the area and volume of the VN1R1 protein in which we have selected central binding pocket residues. The topological features, surface pockets, and interior cavities are essential to reveal the annotated functions of specific residues on the protein structure (Tian et al. 2018).

3.4 Receptor–Ligand Interactions

The extracellular ligand central binding pocket was selected for grid construction, and the cavity had an efficient binding area and volume for accommodating the suitable pheromone ligands. All of the semi-empirical values of receptor–ligand docking were tabulated including hydrophobic with other steric interactions (Table 3). The results reveal that the endogenous ligands (p-Cresol and 1IDU) have outstanding binding affinity and interactions in a similar binding site of the receptor. Interestingly, p-Cresol showed the highest frequency (90%) of binding and strong H-bond interaction (Thr92) below 3.01 Å distance with an acceptable total free binding energy (ΔG) of −4.36 kcal/mol towards the VN1R1 model. The lowest free binding energy (ΔG) is the highest binding affinity of ligands to the target protein. In particular, we observed that the Phe278 of the VN1R1 receptor has a cation–Pi interaction with p-Cresol. The other two pheromone compounds (1IDU and 9ODA) showed higher total free energies of −5.08 and −4.96 kcal/mol, respectively. All of the pheromones showed a better frequency of binding and strong hydrophobic interactions than standard ligands (PAL and RET). Specifically, PAL and RET showed an exterior parallel interaction of different ligand binding cavities, and the compound PAL has anion–pi contacts. The Pi interactions were identified by DSV using the molecular docking results, and we have obtained the binding affinity of ligands using MolDock rerank scores. The purpose of molecular modeling is to rank the best residue interactions with their ligand binding affinity with the protein (Golebiowski et al. 2012). Then, the estimation of the

Table 3. The receptor–ligand interactions. The table shows the binding energy and H-bond interactions between VN1R1, pheromone and standard compounds

Compound name	Total intermolecular energy (kcal/mol)	Total free energy (kcal/mol)	MolDock rerank score	Est. inhibition constant, Ki	Frequency (%)	H-bonds	Hydrophobic interaction	Pi interaction	Other interactions
P-Cresol	−4.66	−4.36	−48.181	652.70 μM	90	Thr92	Phe85, His88, Trp170, Tyr250, Val272, Ala275	Phe278 (Cation–Pi)	Ser169
1IDU	−7.44	−5.08	−35.857	189.19 μM	67	–	Val57, Phe85, His88, Thr92, Tyr167, Cys168, Val272, Ala275, Thr276, Phe278, Phe279	–	Lys61, Ser169, Trp170, Thr276
9ODA	−7.88	−4.96	−27.803	231.98 μM	38	–	Leu26, Leu47, Leu50, Val51, Phe102, Asn103, Val236, Leu239, Val240, Leu286	–	Thr99, Thr235, Ser282
PAL	−5.79	−2.15	−23.177	26.56 mM	30	–	Ile18, Ser242, Phe246, Leu274, Cys277, Ala280, Phe281	Phe284 (Anion–Pi)	Gln15, Thr245
RET	−5.08	−3.67	−49.278	2.04 mM	30	–	Ile11, Leu273, Leu274, Cys277	–	Gln15, Thr270, Ala280, Phe281

inhibition constant (Ki) suggested that the lower value indicates the most favourable compound (Table 3).

Further, we show, in Fig. 4, the molecular interactions of the pheromone, and standard compounds with the VN1R1 model were illustrated using Chimera and LigPlot visualizer. The surge of p-Cresol was observed in oestrus stages of mares and considered as a sex pheromone component (Buda et al. 2012). A recent report also suggested that p-Cresol has a favourable binding affinity with buffalo odorant binding protein (bunOBP) (Muthukumar et al. 2018) and the few oestrous pheromones were reviewed in the farm animals with reference to cow (Vyas et al. 2012; Kekan et al. 2017). Overall, the docking outputs showed best-fit residue interactions of pheromones towards VN1R1 compared to standard compounds. Thus, the understanding of receptor-pheromone interactions has enabled a better understanding of the functional significance of VN1R1 in bovine, and favourable binding with suitable pheromones highlights the importance of semiochemical communication.

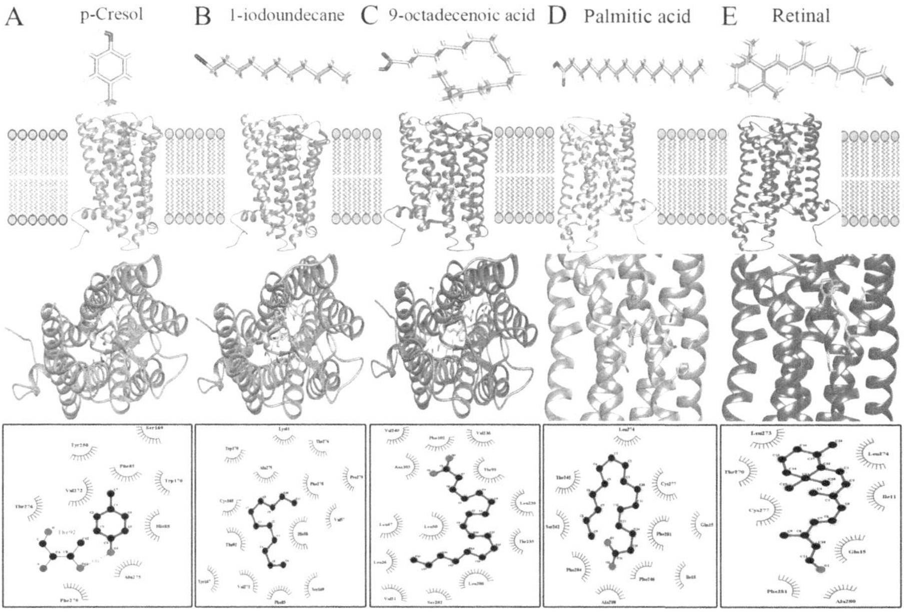

Fig. 4. The receptor–ligand interaction. **a–e** The selected pheromone (p-Cresol, 1-iodoundecane, 9-octadecenoic acid) and standard (Palmitic acid and Retinal) ligands were shown as well as their binding with receptor, respectively. The bovine VN1R1 model was illustrated between epithelial cytoplasmic membrane. All selective compounds were interacting with different binding sites. The vertical view shows the ligand binding on the central binding cavity. Also, we showed the 2-dimensional residue interaction of all ligands. Here, p-Cresol has a hydrogen bond (Thr92) and hydrophobic contacts. Hydrogen bonds are indicated by dashed lines between the receptor–ligand atoms involved, while the hydrophobic contacts are represented as an arc with spokes radiating towards the ligand atoms

4 Conclusion

The sex-specific pheromones in body fluids are serving a crucial role in mammalian behavioural response and mate recognition. Generally, the identification and confirmation of pheromone roles require animal experiments. In these experiments, the respective pheromones are exposed to bulls for measuring their behaviour as well as physiological indicators, which may induce discomfort and stress to the animals. In this study, we introduce a novel interesting attempt to validate potential pheromone compounds using virtual computational screening without the need for experimental testing in live animals. The computational methods are helpful for confirming or excluding the potential consistency of the pheromones by predicting their binding affinity with the receptor. Hence, the high-affinity pheromones of cattle could be used as potential semiochemical cues to promote healthier reproduction in bovines as well as livestock management. Therefore, this is a significant means to conserve animals, time and financial investments.

Acknowledgements. DR gratefully thanks the Research Institute in Semiochemistry and Applied Ethology (IRSEA), APT, France, for infrastructural facilities and support. We thank Dr. Christina D. Buesching and Dr. Carsten T. Mueller for providing valuable suggestions to improve the manuscript. We thank the Council of Research Education Board Members, IRSEA for valuable discussion and critical reading of this manuscript.

References

Albone ES (1984) Mammalian semiochemistry: the investigation of chemical signals between mammals. Wiley, New York

Albone ES, Blazquez NB, French J, Long SE, Perry GC (1986) Mammalian semiochemistry: issues and futures, with some examples from a study of chemical signalling in cattle. In: Duvall D, Müller-Schwarze D, Silverstein RM (eds) Chemical signals in vertebrates 4. Springer, Boston, MA, pp 27–36

Altschul SF, Madden TL, Schäffer AA, Zhang J, Zhang Z, Miller W, Lipman DJ (1997) Gapped BLAST and PSI-BLAST: a new generation of protein database search programs. Nucleic Acids Res 25:3389–3402

Archunan G (2009) Vertebrate pheromones and their biological importance. J Exp Zool India 12:227–239

Archunan G, Rajanarayanan S (2010) Composition for enhancing bull sex libido. Indian Patent No. 244991, 28 Dec 2010

Archunan G, Rameshkumar K (2012) 1-Iodoundecae an estrus indicating urinary chemosignal in bovine (*Bos taurus*). J Vet Sci Tech 3:121–123

Berezin C, Glaser F, Rosenberg J, Paz I, Pupko T, Fariselli P, Casadio R, Ben-Tal N (2004) ConSeq: the identification of functionally and structurally important residues in protein sequences. Bioinformatics 20:1322–1324

Bikadi Z, Hazai E (2009) Application of the PM6 semi-empirical method to modeling proteins enhances docking accuracy of AutoDock. J Cheminform 1:1–16

Buck L, Axel R (1991) A novel multigene family may encode odorant receptors: a molecular basis for odor recognition. Cell 65:175–187

Buck LB (2000) The molecular architecture of review odor and pheromone sensing in mammals. Cell 100:611–618

BūdaV Mozūraitis R, Kutra J, Borg-Karlson AK (2012) p-Cresol: a sex pheromone component identified from the estrous urine of mares. J Chem Ecol 38:811–813

Dulac C, Axel R (1995) A novel family of genes encoding putative pheromone receptors in mammals. Cell 83:195–206

Dulac C, Axel R (1998) Expression of Candidate pheromone receptor genes in vomeronasal Neurons. Chem Senses 23:467–475

Dulac C, Torello AT (2003) Molecular detection of pheromone signals in mammals: from genes to behaviour. Nat Rev Neurosci 4:551–562

Geourjon C, Deléage G (1995) SOPMA: significant improvements in protein secondary structure prediction by consensus prediction from multiple alignments. Bioinformatics 11:681–684

Glaser F, Pupko T, Paz I, Bell RE, Bechor-Schental D, Martz E, Ben-Tal N (2003) ConSurf: identification of functional regions in proteins by surface-mapping of phylogenetic information. Bioinformatics 19:163–164

Godzik A (2003) Fold recognition methods. Methods Biochem Anal 44:525–546

Golebiowski J, Topin J, Charlier L (2012) Interaction between odorants and proteins involved in the perception of smell: the case of odorant-binding proteins probed by molecular modelling and biophysical data. Flavour Fragr J 27:445–453

Harini K, Sowdhamini R (2015) Computational approaches for decoding select odorant-olfactory receptor interactions using mini-virtual screening. PLoS ONE 10:1–30

Herrada G, Dulac C (1997) A novel family of putative pheromone receptors in mammals with a topographically organized and sexually dimorphic distribution. Cell 90:763–773

Jones DT, Taylor WR, Thornton JM (1992) The rapid generation of mutation data matricies from protein sequences. Comput Appl Biosci 8:275–282

Källberg M, Wang H, Wang S (2012) Template-based protein structure modeling using the RaptorX web server. Nat Protoc 7:1511–1522

Karlson P, Lüscher M (1959) "Pheromones": a new term for a class of biologically active substances. Nature 183:55–56

Karthikeyan K, Archunan G (2013) Gas chromatographic mass spectrometric analysis of estrus-specific volatile compounds in buffalo vaginal mucus after initial sexual foreplay. J Buffalo Sci 2:1–7

Karthikeyan K, Manivannan P, Rajesh D, Muthukumar S, Muralidharan G, Akbarsha MA, Archunan G (2014) Identification of p-cresol as an estrus-specific volatile in buffalo saliva: comparative docking analysis of buffalo OBP and β-lactoglobulin with p-cresol. Zoolog Sci 31:31–36

Kekan PM, Ingole SD, Sirsat SD, Bharucha SV, Kharde SD, Nagvekar AS (2017) The role of pheromones in farm animals-a review. Agri Rev 38:83–93

Keverne EB (1999) The vomeronasal organ. Handbook of olfaction gustation, vol 286, pp 967–980

Kubo H, Otsuka M, Kadokawa H (2016) Sexual polymorphisms of vomeronasal 1 receptor family gene expression in bulls, steers, and estrous and early luteal-phase heifers. J Vet Med Sci 78:271–279

Kumar KR, Archunan G, Jeyaraman R, Narasimhan S (2000) Chemical characterization of bovine urine with special reference to oestrus. Vet Res Commun 24:445–454

Kuriata A, Gierut AM, Oleniecki T, Ciemny MP, Kolinski A, Kurcinski M, Kmiecik S (2018) CABS-flex 2.0: a web server for fast simulations of flexibility of protein structures. Nucleic Acids Res 46:338–343

Landau M, Mayrose I, Rosenberg Y, Glaser F, Martz E, Pupko T, Ben-Tal N (2005) ConSurf 2005: the projection of evolutionary conservation scores of residues on protein structures. Nucleic Acids Res 33:299–302

Laskowski RA, MacArthur MW, Moss DS (1993) PROCHECK: a program to check the stereochemical quality of protein structures. J Appl Crystallogr 26:283–291

Laskowski RA, Watson JD, Thornton JM (2005) ProFunc: a server for predicting protein function from 3D structure. Nucleic Acids Res 33:89–93

Le Danvic C, Gérard O, Sellem E, Ponsart C, Chemineau P, Humblot P, Nagnan-Le Meillour P (2015) Enhancing bull sexual behavior using estrus-specific molecules identified in cow urine. Theriogenology 83:1381–1388

Ma J, Wang S, Zhao F, Xu J (2013) Protein threading using context-specific alignment potential. Bioinformatics 29:257–265

Matsunami H, Buck LB (1997) A multigene family encoding a diverse array of putative pheromone receptors in mammals. Cell 90:775–784

Mirkovic N, Li Z, Parnassa A, Murray D (2007) Strategies for high-throughput comparative modeling: applications to leverage analysis in structural genomics and protein family organization. Proteins 66:766–777

Mombaerts P (2004) Genes and ligands for odorant, vomeronasal and taste receptors. Nat Rev Neurosci 5:263–278

Mombearts P (1999) Seven-transmembrane proteins as odorants and chemosensory receptors. Science 286:707–711

Morris GM, Goodsell DS, Halliday RS, Huey R, Hart WE, Belew RK, Olson AJ (1998) Automated docking using a Lamarckian genetic algorithm and an empirical binding free energy function. J Comput Chem 19:1639–1662

Morris GM, Goodsell DS, Pique ME, Lindstrom W, Huey R, Forli S, Hart WE, Halliday S Belew R, Olson AJ (2009) AutoDock version 4.2. User guide 1–49. http://autodock.scripps.edu/faqs-help/manual/autodock-4-2-user-guide. Accessed 18 Dec 2018

Mozūraitis R, Kutra J, Borg-Karlson AK, Būda V (2017) Dynamics of putative sex pheromone components during heat periods in estrus-induced cows. J Dairy Sci 100:7686–7695

Murakami M, Kouyama T (2008) Crystal structure of squid rhodopsin. Nature 453:363–367

Muthukumar S, Rajesh D, Selvam RM, Saibaba G, Suvaithenamudhan S, Akbarsha MA, Padmanabhan P, Gulyas B, Archunan G (2018) Buffalo nasal odorant-binding protein (bunOBP) and its structural evaluation with putative pheromones. Sci Rep 8:9323

Ohara H, Nikaido M, Date-Ito A, Mogi K, Okamura H, Okada N, Takeuchi Y, Mori Y, Hagino-Yamagishi K (2009) Conserved repertoire of orthologous vomeronasal type 1 receptor genes in ruminant species. BMC Evol Biol 9:233

Omasits U, Ahrens CH, Müller S, Wollscheid B (2014) Protter: interactive protein feature visualization and integration with experimental proteomic data. Bioinformatics 30:884–886

Osella MC, Cozzi A, Spegis C, Turille G, Barmaz A, Lecuelle CL, Teruel E, Bienboire-Frosini C, Chabaud C, Bougrat L, Pageat P (2018) The effects of a synthetic analogue of the Bovine Appeasing Pheromone on milk yield and composition in Valdostana dairy cows during the move from winter housing to confined lowland pastures. J Dairy Res 85:174–177

Pageat P (1998) Appeasing pheromones to decrease stress, anxiety and aggressiveness. European patent EP 0 948 963 A1; US Patent 6,054,481; US Patent 6,077,867; US Patent US 6,169,113 B1; Japan Patent 2000-528279; Japan Patent 980298

Papes F, Logan DW, Stowers L (2010) The vomeronasal organ mediates interspecies defensive behaviors through detection of protein pheromone homologs. Cell 141:692–703

Patra MK, Barman P, Kumar H (2012) Potential application of pheromones in reproduction of farm animals–a review. Agri Rev 33:82–86

Peng J, Xu J (2011) A multiple-template approach to protein threading. Proteins 79:1930–1939

Rajanarayanan S, Archunan G (2011) Identification of urinary sex pheromones in female buffaloes and their influence on bull reproductive behaviour. Res Vet Sci 91:301–305
Rajesh D, Muthukumar S, Saibaba G, Siva D, Akbarsha MA, Gulyas B, Padmanabhan P, Archunan G (2016) Structural elucidation of estrus urinary lipocalin protein (EULP) and evaluating binding affinity with pheromones using molecular docking and fluorescence study. Sci Rep 6:35900
Robert X, Gouet P (2014) Deciphering key features in protein structures with the new ENDscript server. Nucleic Acids Res 42:320–324
Rodriguez I, Greer CA, Mok MY, Mombaerts P (2000) A putative pheromone receptor gene expressed in human olfactory mucosa. Nat Genet 26:18–19
Rodriguez I, Mombaerts P (2002) Novel human vomeronasal receptor-like genes reveal species-specific families. Curr Biol 12:409–411
Rodriguez I (2003) Nosing into pheromone detectors. Nat Neurosci 6:438–440
Rost B (1999) Twilight zone of protein sequence alignments. Protein Eng 12:85–94
Ryba NJP, Tirindelli R (1997) A new multigene family of putative pheromone receptors. Neuron 19:371–379
Shimamura T, Hiraki K, Takahashi N, Hori T, Ago H, Masuda K, Takio K, Ishiguro M, Miyano M (2008) Crystal structure of squid rhodopsin with intracellularly extended cytoplasmic region. J Biol Chem 283:17753–17756
Tamura K, Stecher G, Peterson D, Filipski A, Kumar S (2013) MEGA6: molecular evolutionary genetics analysis version 6.0. Mol Biol Evol 30:2725–2729
Tian W, Chen C, Lei X, Zhao J, Liang J (2018) CASTp 3.0: computed atlas of surface topography of proteins. Nucleic Acids Res 1–5
Tirindelli R, Dibattista M, Pifferi S, Menini A (2009) From pheromones to behavior. Physiol Rev 89:921–956
Vyas S, Briant C, Chemineau P, Danvic Le, Nagnan-Le Meillour P (2012) Oestrus pheromones in farm mammals, with special reference to cow. Indian J Anim Sci 82:256–267
Wakabayashi Y, Mori Y, Ichikawa M, Yazaki K, Hagino-Yamagishi K (2002) A putative pheromone receptor gene is expressed in two distinct olfactory organs in goats. Chem Senses 27:207–213
Wallace AC, Laskowski RA, Thornton JM (1995) LIGPLOT: a program to generate schematic diagrams of protein-ligand interactions. Protein Eng 8:127–134
Wyatt TD (2010) Pheromones and signature mixtures: defining species-wide signals and variable cues for identity in both invertebrates and vertebrates. J Comp Physiol A 196:685–700
Wyatt TD (2014) Proteins and peptides as pheromone signals and chemical signatures. Anim Behav 97:273–280
Wyatt TD (2017) Pheromones. Curr Biol 27:739–743
Yoder AD, Larsen PA (2014) The molecular evolutionary dynamics of the vomeronasal receptor (class 1) genes in primates: a gene family on the verge of a functional breakdown. Front Neuroanat 8:1–9
Young JM, Massa HF, Hsu L, Trask BJ (2010) Extreme variability among mammalian V1R gene families. Genome Res 20:10–18
Zufall F, Munger SD (2001) From odor and pheromone transduction to the organization of the sense of smell. Trends Neurosci 24:191–193

Detecting the Smell of Disease and Injury: Scoping Evolutionary and Ecological Implications

Chris Newman and Christina D. Buesching(✉)

Wildlife Conservation Research Unit, Department of Zoology,
University of Oxford, Recanati-Kaplan Centre,
Tubney, Abingdon, Oxfordshire, UK
christina.buesching@zoo.ox.ac.uk

Abstract. Sickness, along with most other health conditions, smell; and in order to avoid infection and contamination, humans and other animals alike have evolved to find these odours repugnant. Indeed, so evolved is the selective advantage offered by detecting disease, that even changes to metabolites triggered by immune and stress responses are detectable. Here, we first review mechanisms of how different health conditions can modify host odour profiles, and how these changes can affect social interactions in general. In the second part, we focus on the two contexts that render individuals most vulnerable to evolutionary selection pressures: reproduction and predation. Linked to sexual selection and inter-specific eavesdropping, we discuss the ecological and game-theoretic consequences of olfactory detectability of health conditions, where we use examples from the published literature to illustrate what host behaviours may have evolved to avoid and/or conceal their disadvantageous odour changes.

1 Introduction

Disease, parasites, cellular mutations/cancer, injury and poisoning typically not only acutely affect an animal's (in the following termed 'host') somatic condition, energy requirements and general well-being, but—dependent on the nature of the pathogen/-toxin involved—can also have substantial impacts on (present and future) reproductive output and even threaten the immediate or longer term survival of the afflicted individual. Game theory (Weibull 1997) thus predicts a strong selection pressure on detecting such health issues in self and others: Recognising and identifying the nature of one's own health concern can aid in healing (e.g. licking wounds: Bodner 1991), help improve somatic health (e.g. targeted self-grooming to eliminate ectoparasites: Boccia 1983) and reproductive fitness (e.g. licking/cleaning infected genitals: Nunn 2003) as well as enhancing survival chances (e.g. seeking out antidotes to toxins: Pfister 1999). Detecting disease and parasitic infection in conspecifics can help avoid getting infected (Curtis 2014), recognising injuries or fear, can aid in identifying hazards (Lee and Noh 2016) and—particularly in mothers with dependent offspring—can help improve her fitness by inducing her to clean to any infected wounds among her offspring, before they become debilitating/deadly(Hart and Powell 1990); similarly

© Springer Nature Switzerland AG 2019
C. D. Buesching (Ed.): *Chemical Signals in Vertebrates 14*, pp. 238–250, 2019.
https://doi.org/10.1007/978-3-030-17616-7_17

recognising poisoning can help conspecifics learn to discriminate toxic plants (Galef 1987).

Surprisingly, while many behavioural ecology studies have focused on genetic disease and parasite resistance through MHC variability (e.g. reviewed in Knapp et al. 2006), and the study of olfactory recognition of MHC compatibility in mate choice is a fast-growing field of research, far fewer studies have investigated an animal's ability to recognise health problems directly. However, because health issues are either caused by living organisms (parasites, bacterial/fungal infections) and/or alter the host's metabolism (viral infections, poison) or general body odour (wounds), almost all are associated with changes in host odour profiles (e.g. reviewed in Penn and Potts 1998). This phenomenon has long been used in human medicine, where an ancient Sanskrit text by Sushruta Samhita (c. Sixth Century BCE) states: '*By the sense of smell we can recognise the peculiar perspiration of many diseases, which has an important bearing on their identification*', and in recent years, the use of dog ability to sniff out various cancers (e.g. lung cancer: Amundsen et al. 2014) and diseases (e.g. diabetes: Hardin et al. 2015) has been taken advantage of diagnostically in human medicine. Nevertheless, the ecological and evolutionary implications of this sensitivity to the scent of sickness are largely ignored. Being able to recognise a sick individual can have profound survival advantages in the natural world, both in terms of avoiding infectious conspecifics and being choosey with whom to mate, as well as taking advantage of 'dis-ease' among rivals, and thus maximising one's own dominance advantages, whereas predators may use this ability to recognise sick prey that may be easier to hunt down.

Thus, in the first part of this chapter, we review how different health issues can modify host odour profiles, and how these changes can affect social interactions in general. In the second part, we discuss the ecological and game-theoretic consequences of olfactory detectability of health issues. We focus on two contexts that render individuals most vulnerable to evolutionary selection pressures: reproduction and predation, and use examples from the published literature to illustrate what host behaviours have evolved to avoid and/or conceal their disadvantageous odour changes.

2 How Health Issues Can Change Host Odour

Many poisons smell, and most parasites, fungi, and bacteria emit chemical compounds themselves that can be detected as (non-)volatile scent providing primary olfactory cues of the respective affliction. In principle, the vertebrate body can also generate five principal sources of odour: (1) Metabolic and physiological by-/waste products, excreted in urine and faeces or as respiratory gases from lungs and skin, such as various respiratory and dietary components, but also hormones and their metabolites; (2) Vomit; (3) Secretions/excretions with a primary function other than olfactory communication, such as tears, saliva, sweat, skin exudates, mucosal secretions; (4) Secretions from specialised scent-glands; and (5) the organic building blocks of the body itself, such as tissues, digestive fluids and blood/interstitial fluid. Detecting the scent of some of these provides in itself a primary cue to injury/sickness (i.e. organic building blocks, particularly internal tissues and fluids/blood; vomit), whereas for others (i.e. exudates, excretions, egestions and secretions) only a change in their natural odour profiles indicates an affliction. Such changes can be caused either by changes in the host's own physiology (e.g. immune response, endocrinological changes) or metabolism (undigested food, toxins exuded through skin) or by pathogens (microbiota such as bacteria and fungi or larger parasites) metabolising primary glandular secretions or excreta resulting in an intricate network of olfactory cues (Fig. 17.1).

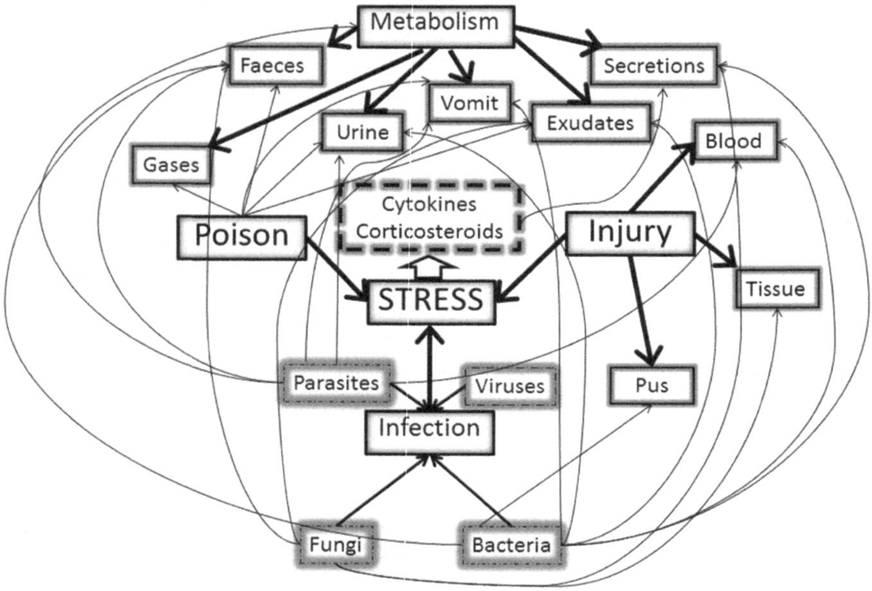

Fig. 17.1. Interaction of health threats and animal odour profiles

3 How Health-Related Odour Changes Modify Social Behaviour of Conspecifics

3.1 General Odour of Ill-Health

In vertebrates, all health problems result in the release of pro-inflammatory cytokines (principally interleukin-1beta (IL-1ß) and tumour necrosis factor alpha (TNF-α), which have been shown to affect the central nervous system (CNS) and modify the physiology of the sick individual to yield aversive cytokine odour cues (Kent et al. 1992; Aubert 1999; Dantzer 2001): For instance, lab rats avoid sick individuals that emit airborne olfactory cues when cytokines are released during an inflammatory response, where the trigger of this inflammation (i.e. bacterial/fungal infection, parasites, simple injury/contusion or even exposure to psychological stress challenges) has been proven to be largely irrelevant (reviewed in Arakawa et al. 2011). Even humans respond in this way: after just four hours the body odour (BO) of trial participants injected with an endotoxin (lipopolysaccharide; LPS) to trigger an innate immune response becomes aversive compared to the BO of participants treated with a placebo (Olsson et al. 2014). This chemosensory aversion appears to be linked directly to the level of immune activation, where changes in the chemical compounds comprising BO (such as E-3-methyl-2-hexenoic acid, 3-methyl-3-sulfanylhexan-1-ol, and sulfanylalkanols) affect the fermentation products of the microbiota living on the skin (Ackerl et al. 2002; Olsson et al. 2014).

3.2 Parasites

Although some parasites evoke a continuous systemic inflammatory response, most induce neuroendocrine alterations in the host, suppressing the pro-inflammatory cytokine response after the initial stage of infection (Fiore and Aloe 2001; Klein 2003). Nevertheless, some parasites (e.g. ticks: Sonenshine 2004) have been shown to produce arrestment (= assembly) pheromones and attraction–aggregation–attachment (AAA) pheromones to signal to other parasites that a walking meal has arrived. It is thus possible that vertebrate hosts have evolved the ability to eavesdrop on their invertebrate ectoparasites using olfactory cues in the same way as these same parasites use chemosensory cues to find and choose their vertebrate hosts (Osterkamp et al. 1999). Indeed, mandrills have been shown to use olfaction to avoid social interactions with parasitized conspecifics (e.g. Poirotte et al. 2017), and female mice actively reject urine from males infected with the intestinal nematode, *Heligmosomoides polygyrus* (Ehman and Scott 2001) as well as urine from males infested with the mouse louse *Polyplax serrata* (Kavaliers et al. 2003). This olfactory recognition and subsequent avoidance of infected individuals involves oxytocin and oestrogen receptors, where mice with deletions of the oxytocin gene and oestrogen receptor genes ER-α and/or ER-β are aversive to odours of infected conspecifics, with an impaired ability to recognise and remember that specific individual's scent profile (Kavaliers et al. 2005).

3.3 Excreta, Egestions and Secretions

Many specialised scent secretions comprise a mixture of proteins and carboxylic acids (Brown and Macdonald 1985), which are then metabolised by site-specific microbiota ('Fermentation hypothesis': Albone and Perry 1976; Theis et al. 2013; Ezenwa and Williams 2014). Any change in either the primary gland products or the microbiome community structure will thus result in modification of the host odour profile. Many dietary compounds are exuded through the skin (e.g. garlic ~ allyl methyl sulphide: Fialová et al. 2016), or can be detected in specialised scent secretions (Walkden-Brown et al. 1994). Some diseases have even been shown to change skin pH, with likely knock-on effects on skin microbiome composition (SanMiguel and Grice 2015). Mammals are able to smell changes in volatile urinary and faecal acids (Zeng et al. 1996) as well as specific gut products, such as guanylin, with great sensitivity (Leinders-Zufall et al. 2007), thus finding it easy to detect faecal scent-marks (reviewed in Brown and Macdonald 1985; Buesching and Jordan 2019, this volume). Infection can induce changes in the intestinal microbiome structure—the so-called '*microflora hypothesis*' (Schellinck et al. 1991), and/or either inhibit gastric emptying (preventing faecal scent-marking altogether) or stimulate colonic propulsive motility causing diarrhoea (Brown and Schellinck 1995) and the defecation of less/differently digested excrement (Conte et al. 2006; Hurst and Beynon 2004); changes that can affect the information encoded in faecal scent marks (reviewed in Buesching and Jordan 2019, this volume). In terms of detecting these olfactory cues, stress—hormones (e.g. from physiological stress following sickness or injury), and related metabolites excreted in faeces, elicit interest and increased vigilance, especially in prey species (Fourie 2012), while high levels of faecal/urinary sex-steroids (which, aside from being an indicator of oestrus in healthy females, are also elevated when suffering from certain types of ovarian cancer: Zhao et al. 2013) typically elicit characteristic flehmen responses in species with functional vomero-nasal organs (Dehnhard 2011).

Collectively, the evolved interpretation of such chemosensory cues, therefore, allows potential social partners to realise that they should 'stay at a healthy distance' and/or avoid social interactions with infected conspecifics. This selection pressure of avoiding interactions with sickly individuals is particularly intense in the context of reproduction, where health issues can affect immediate and also future reproductive success, as well as inclusive fitness.

4 Sexual Selection: The Fragrance of Rejection

In 1982, Hamilton and Zuk proposed their landmark hypothesis that animals should inspect a potential mate's urine and faecal odour to select for parasite-free/-resistant status. Particularly for females, selecting for pathogen resistant genes will be advantageous to her offspring not only through increasing genetic fitness, but because a healthy partner may also help provision progeny through neonatal development (monogamous species: Kokko et al. 2003; birds: Tomás et al. 2007). Building on this, substantial work has shown that mammals base part of their selection for mating partners on major histocompatibility complex (MHC) cell surface proteins that in part determine individual immuno-competency; an olfactory signal readily detect from urine (e.g. Ehman and Scott 2001). There is a wealth of examples for this type of scent-based mate choice—although mostly from controlled lab experiments: for example, healthy male mice have been shown to avoid, and refuse to copulate with, females infected with the indirectly transmitted parasite, *Trichinella spiralis* (Edwards and Barnard 1987), where Moore (2002) showed that females infected with *T. crassiceps*, *Echinostoma revolutum*, and *E. caproni* exhibited decreased reproductive output.

4.1 Sexually Transmitted Diseases (STDs)

One surprisingly under-investigated arena where the scent of disease might be expected to influence mate choice directly is that of sexually (i.e. during coitus) transmittable diseases (STDs; Kokko et al. 2002) . Aside from potential impacts on sexual performance that preclude mating attempts, STDs typically have subsequent cryptic implications, such as vertical transmission to offspring either causing miscarriage, still-birth or reduced lifetime fitness, or causing permanent damage to the reproductive organs of sexual partners through horizontal transmission, extinguishing their future reproductive success, thus acting as major driver of population dynamics (Kokko et al. 2002; Fig. 17.2).

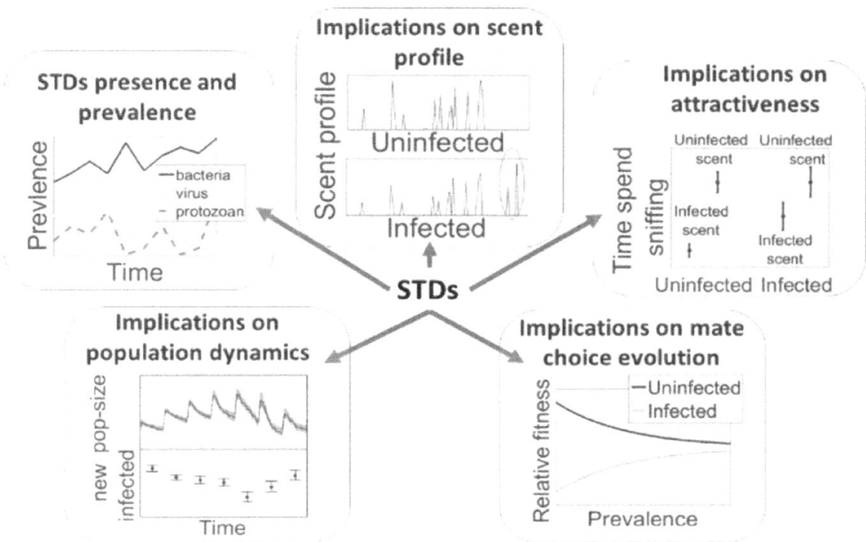

Fig. 17.2. Potential effects of sexually transmittable diseases (STDs) on animal ecology

In humans, physiological conditions such as cervical or vaginal cancer or recto-vaginal fistulae cause distinctive smells, indicative of a compromised female reproductive system (Villasanta 1972). Additionally, various STDs are associated with vaginal discharge with a distinctive 'fishy' odour and associated discomfort. Pathogenic agents causing odorous infection may be protozoan (e.g. Trichomoniasis), bacterial (e.g. Chlamydia, syphilis/*Treponema pallidum*, gonorrhoea/*Neisseria gonorrhoeae*—often linked to bloody discharge/vaginosis), or fungal (Candida/yeast), whereas viral STDs (e.g. genital herpes, HIV, papillomavirus) tend to cause less immediate odour. Primary STDs, however, can lead to serious secondary conditions, again smelling nefarious, such as mucopurulent cervicitis and pelvic inflammatory disease. In males, gonorrhoea, nongonococcal urethritis, syphilis and Chlamydia cause smelly pus to drip from the penis, and can cause associated testicular swelling. Even highly odourous secondary infections arising from poor hygiene can cause odour, such as the build-up of smegma under the foreskin from accumulating un-cleaned oils and skin cells around the shaft of the penis (Meier et al. 2006). Targeted behavioural studies investigating behavioural aversion of STD-generated genital odour cues are rare, but dogs have been shown to be able to distinguish a variety of venereal diseases in human patients, such as cervical/ovarian cancer from exhaled breath (Guerrero-Flores et al. 2017), or prostate cancer from urine (Cornu et al. 2011). Even the smell of axillary sweat from men infected with gonorrhoea has been proven to be distinguishable and aversively 'putrid' to women sniffers (Moshkin et al. 2012), indicating that STDs are likely recognisable by altered odour profiles also in—and by—non-human animals.

Game-theory predicts that if STD-risk matings can be avoided, then the rate of spread of STDs within a population would be curbed. In addition, any reduction in matings involving infected females would limit the vertical spread of infection to offspring. However, if reproductive opportunities are scarce, or survival until the next mating season is unlikely, the terminal investment theory predicts that individuals (esp. males) might risk infection to achieve some reproductive output, thus mating indiscriminately—a tactic modified by the pathogenicity of the STD, the risk of infection, recipient immune status, and the extent to which the disease impacts fecundity and subsequent offspring fitness.

In mice, the odour of parasitized male mice triggers a neuroendocrine analgesic mechanism in females, elevating their endorphin levels, enkephalins and non-opioid serotogenic levels to enhance pain tolerance (Kavaliers and Colwell 1993). This may simply indicate a general stress response to being faced with an infected mating partner (Kavaliers and Colwell 1995a, b), but could also be an odour-mediated preparatory defence mechanism, because opiods bind readily to receptors in the intestine, which could alleviate some of the otherwise debilitating discomfort arising from coccidial diarrhoea (Stefano et al. 1996). Furthermore, opiods inhibit libido, and thus reduce lordosis, sexual receptivity and oestrous in rodents (Hammer et al. 1994). Counter-intuitively though, Stevenson et al. (2011) report that male arousal causes a distinct drop in sensitivity to sex-related disgust cues.

5 Predation: Olfactory 'Come Catch-Me' Cues

Many predators (from mammals to snakes to sharks) use olfaction to track down prey (indeed this is the very basis for how dogs became 'man's best friend' in the hunt, scenting quarry and following blood trails: Budgett and Edwards 1933). Nevertheless, while it is well established that wolves follow blood trails (Gable et al. 2016), and sharks can detect—and are attracted by—blood in water (Tester 1963), evidence remains limited that predators seek out prey in poorer physical or behavioural condition (Gosling and Petrie 1981). Game theory would predict a strong selection pressure, but studies examining if predators scent prey quality are conspicuously under-represented.

Across taxa, pathogens and injury (indeed even just fear of injury/attack) cause activation of the hypothalamus–pituitary–adrenal (HPA) axis and the sympathetic nervous system (Bernard et al. 1996; Morales-Montor et al. 2001; Corrêa-de-Santana et al. 2006), cueing a stress response with the release of corticosteroids and epinephrine —the smell of fear. These alarm odours ('*Schreckstoffe*') typically elicit a fright (freeze) or flight response in conspecifics (e.g. rodents; reviewed in Kim and Jung 2018). Likewise, Nile tilapia (*Oreochromis niloticus*) display an anti-predator response to chemical cues present in the blood (likely epidermal 'club' cells) of fearful conspecifics (Barreto et al. 2013). Predators can use these alarm odours to localise prey and to assess their defensive/escape capabilities (Cocke and Del Thiessen 1986): For example, least weasels (*Mustela nivalis*) can assess the quality of deer mice (*Peromyscus bairoli*: Cushing 1984), and garter snakes (*Thamnophis sirtalis*) can scent out which individual prey would be energetically easiest to hunt (Burghardt 1970). Grouse killed by predators have been found to suffer from higher caecal nematode (*Trichostrongylus tenuis*) counts than those shot during hunts, where the proportion of grouse with high levels of parasite infection increased with the intensity of predator control as measured indirectly through gamekeeper effort, suggesting that predators selectively predated on heavily infected grouse (Hudson et al. 1992). Fascinatingly, when grouse nematode burdens were reduced experimentally through the use of oral anthelmintic, it proved much harder for specially trained dogs to locate the nests of treated birds, implying that parasite-free grouse smelt less (Hudson et al. 1992).

Interestingly, rodents appear well aware of their predators eavesdropping on their scent marks, with domestic mice reducing scent investigation (but not scent-marking) behaviour in the presence of cat urine (Hughes et al. 2009), potentially indicating an awareness that the scent-marking animal has by definition left the area, but that the investigating animal is standing by the 'come eat me' cue. Similarly, two competing vole species, *Microtus rossiameridionalis* and *M. agresti*, trade-off conspicuous resource defence scent-marks under weasel predation (Hughes et al. 2010), where, in the marine environment, Lönnstedt et al. (2012) found that fish conditioned with visual and/or chemical predator cues survived eight times better than those with no experience of benthic predator cues. This is particularly relevant, if sickness-related odour cues might advertise to predators that the marking individual is suffering from a debilitating condition making it particularly easy to catch. For instance, foxes (*Vulpes vulpes*), ferrets (*Mustela putorius furo*) and domestic cats (*Felis domestica*) actively seek out blood odour from stressed Mongolian gerbils (*Meriones unguiculatus*; Cocke and Del Thiessen 1986), while other gerbils become anxious upon smelling the 'blood odour' collected from individuals subjected to a stressor—indicating that stressed or injured rodents are avoided by conspecifics. In this context, Pfeiffer (1962) and Müller-Schwarze (1980) demonstrated that such stress-cues can even cause mobbing by conspecifics; possibly explaining evolved blood-injury phobia in people Marks (1988).

6 Conclusion: The Ultimate Red Queen Game

As we have illustrated above, most health conditions pose risks in terms of infection as well as predation, not only to the afflicted individual, but often also to conspecifics, particularly their social partners. Many infectious diseases and parasites are easily transmitted during any social interaction, and the scent of injured or sickly individuals can actively attract predators, also jeopardising healthy bystanders. In an evolutionary arms race, both predators and healthy conspecifics are therefore under selective pressure to recognise afflicted individuals, concomitantly driving pressures for individuals to hide their sickly scents. It is therefore not surprising that long-term datasets increasingly reveal evidence that vertebrates indeed use scent cues to inform themselves on the disease, injury and immune status of conspecifics in the context of social contact networks and mate choice.

In contrast, although the abilities of prey eavesdropping on their predators have received much attention in recent years, one area currently understudied in ecological research, is the arms-race between eavesdropping predators and their prey, and how conspicuous prey scent marks might be deployed and spied upon. Nevertheless, this represents a vital component of understanding not only how predators select which individual prey to hunt (or not), but might also help explain evolutionary pressures on many aspects of use and function of semiochemical cues: For instance, inevitably, there are also caveats to choosing weak, injured or diseased prey if that prey carries a communicable parasite, and under these circumstances, predators may face selective pressure to recognise and avoid hunting that particular prey to avoid infection. For example, *Toxoplasma gondii* infection in rats is hypothetically detectable by predators insofar as uninfected female rats prefer the smell of infected males (likely through *T. gondii* heightening testosterone odour in urine: Vyas 2013). Nevertheless, although predators might find infected prey less attractive, in the ultimate evolutionary Red Queen game (Van Valen 1977) *Toxoplasma* also blocks the innate aversion of rodents to cat urine, instead producing an attraction to pheromones that increase the likelihood of being predated (Vyas et al. 2007), highlighting the intricacies of evolutionary arms races that thus far are often overlooked in the field of semiochemical research.

As Dietland Müller-Schwarze (2016) illustrated in his recent outlook on the future of semiochemical research, the advances in neuroscience, proteomics and physiology coupled with ever more sensitive and affordable chemical analysis tools, open new avenues to explore ever more fundamental questions in the context of evolutionary ecology in parallel to 'just' listening in on olfactory communication.

Acknowledgements. C.N holds a H.N. Southern Memorial Fellowship in Ecology (at Lady Margaret Hall), and C.D.B. held a Research Fellowship from the Poleberry foundation.

References

Ackerl K, Atzmueller M, Grammer K (2002) The scent of fear. *Neuroendocrinol Let* 23(2):79–84
Albone ES, Perry GC (1976) Anal sac secretion of the red fox, *Vulpes vulpes*; volatile fatty acids and diamines: implications for a fermentation hypothesis of chemical recognition. J Chem Ecol 2(1):101–111
Amundsen T, Sundstrøm S, Buvik T, Gederaas OA, Haaverstad R (2014) Can dogs smell lung cancer? First study using exhaled breath and urine screening in unselected patients with suspected lung cancer. Acta Oncol 53(3):307–315
Arakawa H, Cruz S, Deak T (2011) From models to mechanisms: odorant communication as a key determinant of social behavior in rodents during illness-associated states. Neurosc Biobehav Rev 35(9):1916–1928
Aubert A (1999) Sickness and behaviour in animals: a motivational perspective. Neurosc Biobehav Rev 23(7):1029–1036
Barreto RE, Miyai CA, Sanches FHC, Giaquinto PC, Delicio HC, Volpato GL (2013) Blood cues induce antipredator behavior in Nile tilapia conspecifics. PLoS ONE 8(1):e54642
Bernard JF, Bester H, Besson JM (1996) Involvement of the spino-parabrachio-amygdaloid and-hypothalamic pathways in the autonomic and affective emotional aspects of pain. In: Progress in brain research, vol 107. Elsevier, pp 243–255
Boccia ML (1983) A functional analysis of social grooming patterns through direct comparison with self-grooming in rhesus monkeys. Int J Primatol 4(4):399–418
Bodner L (1991) Effect of parotid submandibular and sublingual saliva on wound healing in rats. Comp Biochem Physiol A 100(4):887–890
Brown RE, Macdonald DW (eds) (1985) Social ddours in mammals. Clarendon Press, Oxford
Brown RE, Schellinck HM (1995) Effects of selective depletion of gut bacteria on the odours of individuality in rats. In: Apfelbach R, Muller-Schwarze D, Reutter K, Weiler E (eds) Chemical signals in vertebrates 7. Pergamon Press, New York, pp 267–271
Budgett HM, Edwards L (1933) Hunting by scent. Charles Scribner's Sons

Buesching CD, Jordan N (2019) Chapter 8: the social function of latrines: a hypothesis-driven research approach. In: Buesching CD, Mueller CT (eds) Chemical signals in vertebrates 14. Springer, Cham

Burghardt GM (1970) Intraspecific geographical variation in chemical food cue preferences of newborn garter snakes (*Thamnophis sirtalis*). Behaviour 36(3)

Cocke R, Del Thiessen D (1986) Chemocommunication among prey and predator species. Anim Learn Behav 14(1):90–92

Conte MP, Schippa S, Zamboni I, Penta M, Chiarini F, Seganti L, Osborn J, Falconieri P, Borrelli O, Cucchiara S (2006) Gut-associated bacterial microbiota in paediatric patients with inflammatory bowel disease. Gut 55(12):1760–1767

Cornu JN, Cancel-Tassin G, Ondet V, Girardet C, Cussenot O (2011) Olfactory detection of prostate cancer by dogs sniffing urine: a step forward in early diagnosis. Europ Urol 59 (2):197–201

Corrêa-de-Santana E, Paez-Pereda M, Theodoropoulou M, Nihei OK, Gruebler Y, Bozza M, Arzt E, Villa-Verde DMS, Renner U, Stalla J, Stalla GK (2006) Hypothalamus–pituitary–adrenal axis during *Trypanosoma cruzi* acute infection in mice. J Neuroimmunol 173(1–2):12–22

Curtis VA (2014) Infection-avoidance behaviour in humans and other animals. Tr Immunol 35 (10):457–464

Cushing BS (1984) A selective preference by least weasels for oestrous versus dioestrous urine of prairie deer mice. Anim Behav 32(4):1263–1265

Dantzer R (2001) Cytokine-induced sickness behavior: mechanisms and implications. Ann NY Acad Sci 933(1):222–234

Dehnhard M (2011) Mammal semiochemicals: understanding pheromones and signature mixtures for better zoo-animal husbandry and conservation. Int Zoo Yearb 45(1):55–79

Edwards JC, Barnard CJ (1987) The effects of Trichinella infection on intersexual interactions between mice. Anim Behav 35(2):533–540

Ehman KD, Scott ME (2001) Urinary odour preferences of MHC congenic female mice, *Mus domesticus*: implications for kin recognition and detection of parasitized males. Anim Behav 62(4):781–789

Ezenwa VO, Williams AE (2014) Microbes and animal olfactory communication: where do we go from here? Bioessays 36(9):847–854

Fialová J, Roberts SC, Havlíček J (2016) Consumption of garlic positively affects hedonic perception of axillary body odour. Appetite 97:8–15

Fiore M, Aloe L (2001) Neuroinflammatory implication of Schistosoma mansoni infection in the mouse. Arc Physiol Biochem 109(4):361–364

Fourie CE (2012) Vigilance behaviour and its endocrine correlates in Plains zebra (*Equus burchelli*) living in a predator-free landscape (Doct diss), Univ Pretoria

Gable TD, Windels SK, Bruggink JG, Homkes AT (2016) Where and how wolves (*Canis lupus*) kill beavers (*Castor canadensis*). PLoS ONE 11(12):e0165537

Galef BG (1987) Social influences on the identification of toxic foods by Norway rats. Anim Learn Behav 15(3):327–332

Gosling LM, Petrie M (1981) The economics of social organization. In: Townsend CR, Calow P (eds) Physiological ecology. Blackwell Scientific Publications, Oxford, pp 315–345

Guerrero-Flores H, Apresa-García T, Garay-Villar Ó, Sánchez-Pérez A, Flores-Villegas D, Bandera-Calderón A, García-Palacios R, Rojas-Sánchez T, Romero-Morelos P, Sánchez-Albor V, Mata O (2017) A non-invasive tool for detecting cervical cancer odor by trained scent dogs. BMC Cancer 17(1):79

Hamilton WD, Zuk M (1982) Heritable true fitness and bright birds: a role for parasites? Science 218(4570):384–387

Hammer RP Jr, Zhou L, Cheung S (1994) Gonadal steroid hormones and hypothalamic opioid circuitry. Horm Behav 28(4):431–437
Hardin DS, Anderson W, Cattet J (2015) Dogs can be successfully trained to alert to hypoglycemia samples from patients with type 1 diabetes. Diab Ther 6(4):509–517
Hart BL, Powell KL (1990) Antibacterial properties of saliva: role in maternal periparturient grooming and in licking wounds. Physiol Behav 48(3):383–386
Hudson PJ, Dobson AP, Newborn D (1992) Do parasites make prey vulnerable to predation? Red grouse and parasites. J Anim Ecol:681–692
Hughes NK, Kelley JL, Banks PB (2009) Receiving behaviour is sensitive to risks from eavesdropping predators. Oecolog 160(3):609–617
Hughes NK, Korpimäki E, Banks PB (2010) The predation risks of interspecific eavesdropping: weasel–vole interactions. Oikos 119(7):1210–1216
Hurst JL, Beynon RJ (2004) Scent wars: the chemobiology of competitive signalling in mice. Bioessays 26(12):1288–1298
Kavaliers M, Colwell DD (1993) Aversive responses of female mice to the odors of parasitized males: neuromodulatory mechanisms and implications for mate choice. Ethology 95(3):202–212
Kavaliers M, Colwell DD (1995a) Discrimination by female mice between the odours of parasitized and non-parasitized males. Proc R Soc Lond B 261(1360):31–35
Kavaliers M, Colwell DD (1995b) Odours of parasitized males induce aversive responses in female mice. Anim Behav 50(5):1161–1169
Kavaliers M, Fudge MA, Colwell DD, Choleris E (2003) Aversive and avoidance responses of female mice to the odors of males infected with an ectoparasite and the effects of prior familiarity. Behav Ecol Sociobiol 54(5):423–430
Kavaliers M, Choleris E, Pfaff DW (2005) Genes, odours and the recognition of parasitized individuals by rodents. Tr Parasitol 21(9):423–429
Kent S, Bluthe RM, Dantzer R, Hardwick AJ, Kelley KW, Rothwell NJ, Vannice JL (1992) Different receptor mechanisms mediate the pyrogenic and behavioral effects of interleukin 1. Proc Nat Acad Sci 89(19):9117–9120
Kent A, Ehlers B, Mendum T, Newman C, Macdonald DW, Chambers M, Buesching CD (2018) Genital tract screening finds widespread infection with Mustelid Gammaherpesvirus 1 in the European Badger (*Meles meles*). J Wildl Dis 54(1):133–137
Kim JJ, Jung MW (2018) Fear paradigms: the times they are a-changin'. Curr Op Behav Sci 24:38–43
Klein SL (2003) Parasite manipulation of the proximate mechanisms that mediate social behavior in vertebrates. Physiol Behav 79(3):441–449
Knapp LA, Robson J, Waterhouse JS (2006) Olfactory signals and the MHC: a review and a case study in *Lemur catta*. Am J Primatol 68(6):568–584
Kokko H, Ranta E, Ruxton G, Lundberg P (2002) Sexually transmitted disease and the evolution of mating systems. Evol 56(6):1091–1100
Kokko H, Brooks R, Jennions MD, Morley J (2003) The evolution of mate choice and mating biases. Proc Roy Soc Lond B 270(1515):653–664
Lee H, Noh J (2016) Pair exposure with conspecific during fear conditioning induces the link between freezing and passive avoidance behaviors in rats. Neurosci Res 108:40–45
Leinders-Zufall T, Cockerham RE, Michalakis S, Biel M, Garbers DL, Reed RR, Zufall F, Munger SD (2007) Contribution of the receptor guanylyl cyclase GC-D to chemosensory function in the olfactory epithelium. Proc Nat Acad Sci 104(36):14507–14512
Lönnstedt OM, McCormick MI, Meekan MG, Ferrari MC, Chivers DP (2012) Learn and live: predator experience and feeding history determines prey behaviour and survival. Proc Roy Soc Lond B: rspb20112516

Marks I (1988) Blood-injury phobia: a review. Am J Psych 145(10):1207
Meier AS, Bukusi EA, Cohen CR, Holmes KK (2006) Independent association of hygiene, socioeconomic status, and circumcision with reduced risk of HIV infection among Kenyan men. JAIDS 43(1):117–118
Morales-Montor J, Mohamed F, Ghaleb AM, Baig S, Hallal-Calleros C, Damian RT (2001) In vitro effects of hypothalamic–pituitary–adrenal axis (HPA) hormones on Schistosoma mansoni. J Parasitol 87(5):1132–1139
Moshkin M, Litvinova N, Litvinova EA, Bedareva A, Lutsyuk A, Gerlinskaya L (2012) Scent recognition of infected status in humans. J Sex Med 9(12):3211–3218
Moore J (2002) Parasites and the behavior of animals. Oxford University Press, New York
Müller-Schwarze D (1980) Chemical signals in alarm behavior of deer. In Chemical signals. Springer, Boston, MA, pp 39–51
Müller-Schwarze D (2016) Chemical signals in vertebrates 13: where we stand and what might be next. In: Chemical signals, vol 13. Springer, Boston, MA, pp 11–16
Nunn CL (2003) Behavioural defences against sexually transmitted diseases in primates. Anim Behav 66(1):37–48
Penn D, Potts WK (1998) Chemical signals and parasite-mediated sexual selection. Tr Ecol Evol 13(10):391–396
Olsson MJ, Lundström JN, Kimball BA, Gordon AR, Karshikoff B, Hosseini N, Sorjonen K, Olgart Höglund C, Solares C, Soop A, Axelsson J (2014) The scent of disease: human body odor contains an early chemosensory cue of sickness. Psychol Sci 25(3):817–823
Osterkamp J, Wahl U, Schmalfuss G, Haas W (1999) Host-odour recognition in two tick species is coded in a blend of vertebrate volatiles. J Comp Physiol 185(1):59–67
Pfeiffer W (1962) The fright reaction of fish. Biol Rev 37(4):495–511
Pfister JA (1999) Behavioral strategies for coping with poisonous plants. In: Launchbaugh K, Sanders KD, Mosley JC (eds) Grazing behavior of livestock and wildlife. Idaho Forest, Wildlife, and Range Experimental Station Bulletin, vol 70, pp 45–59
Poirotte C, Massol F, Herbert A, Willaume E, Bomo PM, Kappeler PM, Charpentier MJ (2017) Mandrills use olfaction to socially avoid parasitized conspecifics. Sci Adv 3(4):e1601721
SanMiguel A, Grice EA (2015) Interactions between host factors and the skin microbiome. Cell Mol Life Sci 72(8):1499–1515
Schellinck HM, Brown RE, Slotnick BM (1991) Training rats to discriminate between the odors of individual conspecifics. Anim Learn Behav 19(3):223–233
Sonenshine DE (2004) Pheromones and other semiochemicals of ticks and their use in tick control. Parasitology 129(S1):S405–S425
Stefano GB, Scharrer B, Smith EM, Hughes Jr TK, Magazine HI, Bilfinger TV, Hartman AR, Fricchione GL, Liu Y, Makman MH (1996) Opioid and opiate immunoregulatory processes. Crit Rev Immunol 16(2)
Stevenson RJ, Case TI, Oaten MJ (2011) Effect of self-reported sexual arousal on responses to sex-related and non-sex-related disgust cues. Arch Sex Behav 40(1):79–85
Tester AL (1963) The role of olfaction in shark predation. Pacif Sci 17:148–170
Theis KR, Venkataraman A, Dycus JA, Koonter KD, Schmitt-Matzen EN, Wagner AP, Holekamp KE, Schmidt TM (2013) Symbiotic bacteria appear to mediate hyena social odors. Proc Nat Acad Sci 110(49):19832–19837
Tomás G, Merino S, Moreno J, Morales J, Martinez-De La Puente J (2007) Impact of blood parasites on immunoglobulin level and parental effort: a medication field experiment on a wild passerine. Funct Ecol 21(1):125–133
Van Valen L (1977) The red queen. Am Nat 111(980):809–810
Villasanta U (1972) Complications of radiotherapy for carcinoma of the uterine cervix. Am J Obstetr Gynecol 114(6):717–726

Vyas A (2013) Parasite-augmented mate choice and reduction in innate fear in rats infected by *Toxoplasma gondii*. J Exp Biol 216(1):120–126

Vyas A, Kim SK, Giacomini N, Boothroyd JC, Sapolsky RM (2007) Behavioral changes induced by Toxoplasma infection of rodents are highly specific to aversion of cat odors. Proc Nat Acad Sci 104(15):6442–6447

Walkden-Brown SW, Restall BJ, Norton BW, Scaramuzzi RJ, Martin GB (1994) Effect of nutrition on seasonal patterns of LH, FSH and testosterone concentration, testicular mass, sebaceous gland volume and odour in Australian cashmere goats. Reproduction 102(2):351–360

Weibull JW (1997) Evolutionary game theory. MIT Press

Zhao D, Zhang F, Zhang W, He J, Zhao Y, Sun J (2013) Prognostic role of hormone receptors in ovarian cancer: a systematic review and meta-analysis. Int J Gynecol Canc 23(1):25–33

Zeng XN, Leyden JJ, Spielman AI, Preti G (1996) Analysis of characteristic human female axillary odors: qualitative comparison to males. J Chem Ecol 22(2):237–257

Index

A
Abdominal glands, 213
Accessory Olfactory Bulb (AOB), 221, 222
Acetyl pyrroline, 33
Acinonyx jubatus, 28, 185
Activity, 8, 9, 14, 17, 19, 22, 23, 40, 60, 63, 74, 77, 96, 97, 99, 100, 118, 120, 123, 129, 131–133, 140, 142, 145, 167, 168, 172, 196, 197, 208, 221
Adolescence, 39
Advertisement, 54, 94, 95, 97, 132
Affective feelings, 143, 144
Africa, 146, 182, 188, 191, 192
African wild dog, 185
Age, 8, 38, 40–45, 47–54, 60, 75, 87, 97, 98, 130, 142, 144, 153–156, 159, 166
Alarm, 61, 140, 195, 244
Alarm odour, 244
Allyl methyl sulphide, 241
α-helix, 221
α-tocopherol, 82
Alzheimer's disease, 68
Amino acid, 68, 168, 177, 207, 210
Amniotic fluid, 83, 84
Amphibia, 209
Amplexus, 205, 208, 213–216
Amygdala, 222
Anal gland, 4, 39, 54, 97, 167, 170, 174, 198
Androgen, 141, 145
Androstene, 43, 144
Animal communication, 14, 67
Anogenital area, 6, 85
Anosmic, 107, 153
Anti-inflammatory, 241
Anti-microbial, 96
Anti-predator cue, 167
Apocrine, 6, 120–122, 142, 145, 147
Apodemus, 168, 175
Appeasing pheromone, 223
Areola, 118, 120–125, 129, 131, 134, 147
Areolar glands, 118, 122, 124, 127, 128, 140
Armpit, 84, 142
Arms race, 166, 245
Arrestment pheromone, 241
Artificial fragrances, 107, 113–115
Association, 83, 96, 154
Attractive, 5, 6, 67, 73, 83–85, 108, 113, 114, 118, 121, 125, 140, 141, 144, 146, 147, 246
Australia, 14, 17, 72, 182, 191
Aversion, 241, 243
Aversive, 140, 177, 240, 241
Avian Influenza Virus, 68
Avoidance, 87, 95, 98, 119, 143, 165–168, 170–172, 174–177, 241
Axillary, 83, 85, 113, 121, 131, 145, 146, 243

B
Bacteria, 60, 61, 89, 122, 142, 239, 240, 243
Badger, 38–42, 44, 54, 97, 100, 191
Banded mongoose, 185
Bar code, 4
Bat-eared fox, 185
Baylisascaris procyonis, 98
Behavior, 7, 8, 14–20, 22, 23, 27, 29, 38–41, 43–45, 47–49, 52, 53, 54, 60–63, 65–69, 71–73, 74, 76–78, 81–83, 88, 90, 94–99, 107, 118, 125, 128–134,

© Springer Nature Switzerland AG 2019
C. D. Buesching (Ed.): *Chemical Signals in Vertebrates 14*, pp. 251–260, 2019.
https://doi.org/10.1007/978-3-030-17616-7

133, 159, 165–168, 170–172, 174, 175, 190, 196, 198, 205, 207–209, 212–216, 222, 223, 234, 240, 243–245
Behavioral test, 171, 172, 205
Behavioral ecology, 239
Behavioral observation, 41, 43, 44, 52
Behavioral response, 127, 234
Big cats, 28, 36
Bioassay, 133
Biomarkers, 60, 61, 67
Biosensor, 67
Bird, 16, 242, 245
Black-backed jackal, 185
Blending, 107, 109, 111, 112
Blood, 84, 85, 176, 209, 239, 244, 245
Body odour, 61, 67, 81, 83–85, 87, 107–115, 120, 239, 240
Body rubbing, 166
Body size, 39, 42
Body temperature, 60, 63, 67
Book keeping, 196
Border, 96, 100
Border latrine, 96, 100
Bos taurus, 224, 227
Botswana, 182, 183, 186, 188, 189, 192, 195, 198
Bovidae, 226, 227
Bovine, 125, 221, 223–231, 233
Bovine Appeasing Pheromone (BAP), 223
Breast, 118–126, 122–124, 128–133, 147
Breast crawl, 120
Breastfeeding, 84, 118, 119, 123, 128, 130–133, 140
Breeding, 6, 15, 20–23, 36, 71–77, 95–97, 169, 206, 208
Breeding site, 95, 96
Brown bear, 189
Brown hyaena, 185, 192
Bulletin board, 5, 13, 195
Byproduct, 82
By-standers, 9

C
Caecotrophe, 88
Camera, 14, 15, 17, 19–22, 42, 72, 74–77, 172, 182–184, 188, 192, 194, 196–198
Camera trap, 71, 74–76, 184, 192, 196, 198
Cancer, 61, 238, 239, 242, 243
Candida, 243
Canid, 166
Canis familiaris, 168, 171, 188
Canis mesomelas, 185
Cape fox, 185, 196
Caracal, 185, 194, 197
Caracal caracal, 185
Carboxylic acid, 33, 141, 145, 241
Carnivore, 168, 182–184, 188, 190, 195, 196, 198
Carnivore guild, 182
Carrier gas, 29, 62
Cat, 27, 166, 167, 171, 172, 174, 175, 177, 188, 189, 191, 192
Catabolism, 68
Cat allergen, 175
Cat urine, 165, 168, 172, 174, 176, 177, 245, 246
Cauxin, 168
Cd-1 mice, 167
Cheek rubbing, 184
Cheetah, 28, 35, 36, 185, 191, 192
Chemical bulletin, 3, 5
Chemical communication, 27, 28, 35, 139–143, 145–147, 166, 175, 182, 183, 190, 196, 198, 205, 207, 208, 215, 221, 222
Chemical composition, 97, 98, 100, 144, 145, 147
Chemical profile, 34
Chemical signals, 3, 6, 27, 71, 128, 141, 167, 205, 222
Civettictis civetta, 185
Chemosensation, 120
Chemosensory cues, 241, 242
Chlamydia, 243
Cholestadienol, 82
Chromatogram, 32
Civet, 185, 189, 191
Cloacal gaping, 205, 208
Cloacal scent gland, 14
Cloacal secretion, 73, 78
Cognition, 142, 144
Cognitive processing, 3, 4
Colostrum, 118, 119, 122, 125, 133
Comparative modeling, 228
Competitors, 7, 13, 14, 23
Composite latrines, 94
Composition, 28, 71, 82, 89, 174, 241
Compound(s), 14, 27–35, 60–63, 66–68, 71, 81–85, 108–111, 119, 122, 126, 128, 131, 139, 141, 142, 144–147, 167, 168, 170, 174–177, 197, 223, 225, 231–234, 239, 241

Computational sequence curation, 223
Condition, 3, 4, 8, 14, 28, 53, 61, 62, 64–66, 68, 71, 72, 84, 89, 90, 97, 99, 114, 120, 126, 128, 129, 132, 141, 153, 166, 169, 172, 174, 184, 238, 243–245
Conspecific, 6–8, 39, 47, 54, 71, 73, 74, 76, 77, 85, 86, 98, 190, 195, 212
Conspecific odour, 72, 77
Contamination, 41, 108, 125, 172, 238
Content, 42, 53, 81, 85, 90, 95, 97
Contribution, 83, 94–100, 125
Core area, 14
Cortico-steroid, 98
Cosmetic(s), 113, 114, 140, 142
Countermark (ing), 5, 38, 43–45, 47, 48, 50, 52, 53, 96, 182–186, 188, 190, 192–198
Courtship, 86, 95, 205–216
Courtship pheromone, 208
Crocuta crocuta, 185
Cross-cultural, 139, 143–147
Cub, 40, 41, 43, 45, 47, 48, 50, 51, 53
Cue(s), 3, 4, 13, 22, 23, 27, 38, 39, 54, 60, 67–69, 71–73, 76–78, 81–83, 85–90, 107–109, 112, 120–122, 130, 132, 133, 146, 165–168, 171, 175–177, 208, 222, 223, 234, 239–241, 243–245
Culture, 108, 114, 143–147, 154
Cynictis penicillata, 185
Cynops, 210–212
Cynops ensicauda, 212
Cynops pyrrhogaster, 207, 209, 212, 213
Cytokine, 240, 241

D
Daily patterns, 16
Danger, 98, 140, 143, 167, 168
Darjeeling hills, 27, 28
Dear enemy, 54
Dear Enemy Phenomenon (DEP), 54
Decapeptide(s), 205, 207, 210–212
Decision, 9, 81, 120
Deer mouse, 167
Defecation, 13, 20, 21, 38–40, 94, 95, 166, 184, 242
Defence, 14, 22, 23, 54, 95, 96, 196, 244
Den, 41, 96, 195
Deposition, 22, 94, 97, 214
Deprivation, 81, 82, 85, 86
Diarrhoea, 242, 244
Diet, 81–86, 88–90, 108, 109, 142, 166, 168, 169, 174, 175
Dimethyl disulphide, 31, 33, 35
Dimethyl thietane, 167, 175
Dimethyl trisulfide, 27, 31, 33
Dinosaurs, 94
Discrimination, 4–6, 8, 53, 54, 88, 96, 98, 109, 112, 195
Disease, 61, 67, 68, 109, 142, 154, 170, 198, 238, 239, 241–245
Disturbance, 21, 100
Djungarian hamsters, 6
Docosenoic acid, 34, 35
Dog, 165, 166, 168, 171, 174, 176, 189, 191, 239
Dominance, 40, 96, 97, 111, 113, 114, 239
Donor, 17, 38, 40–44, 47–54, 74, 114, 225
Dorsal gland, 206, 207, 212, 213, 215
Dwarf mongoose, 183, 185

E
Eavesdropping, 5, 99, 182, 190, 195, 238, 245
Eccrine, 6, 121, 122
Echidna, 13–23, 72, 74–78
Echinostoma caproni, 242
Echinostoma revolutum, 242
Ecological implications, 239
Ectoparasite, 238, 241
Egestion, 240, 241
Eicosatrienoic acid, 34, 35
Eicosenoic acid, 34, 35
Emotional state, 107, 109
Emotion and Odor Scales, 143
Encoded information, 98
Encounter, 4, 5, 7, 8, 77, 194
Endocrinology, 38, 97, 240
Enzyme Immunoassay (EIA), 38, 41–44
Epinephrine, 244
Episodic memory, 4, 8
Equidae, 226, 227
ER-α, 241
ER-β, 241
Ethylphenol, 60, 62, 64, 66–68
Evolution, 38, 54, 81, 82, 107, 118, 132, 139, 156, 166, 198, 205, 207, 209–215
Evolutionary implications, 239
Evolutionary tree analysis, 223
Excretion, 68, 83, 121
Excretion filtration pathways, 39
Exhaled breath, 61, 243
Experiment, 38, 40, 41, 43, 44, 82, 86, 125, 126, 140, 166, 168, 171, 172, 175–177, 207–209, 212, 234, 242
Exudate, 61, 239, 240
Eye temperature, 62, 63, 67

F

Faecal, 38, 39, 84, 95–97, 99, 171, 175, 241, 242
Faeces, 4, 6, 13, 23, 38, 39, 54, 81, 82, 85, 86, 94, 97, 165, 166, 168, 170, 171, 174, 176, 183, 191, 192, 197, 239, 242
Familiarity, 8, 38, 40–43, 45–50, 52–54, 86–88, 96, 127, 128, 154, 155, 159, 195
Family, 27, 87, 94, 111, 113, 146, 171, 175, 205–207, 209, 210, 213, 214, 222, 223, 227
Farming, 61, 132
Fatty acids, 27, 32, 34–36, 167, 223
Fear, 140, 176, 238, 244
Fel d 4, 175
Felids, 27, 35, 36
Felis catus, 167
Felis domestica, 245
Femininity, 113
Femoral gland, 82
Fermentation, 241
Fermentation hypothesis, 241
Ferret, 36, 165–167, 170, 171, 173–175, 177
Fiber, 62
Field study, 71, 72
Fish, 81, 85, 88, 89, 245
Fitness, 3, 4, 7–9, 38, 71, 82, 97, 118, 130, 131, 238, 242, 244
Fitness trade-off, 7–9
Flehmen, 53, 97, 242
Food, 81–83, 85, 86, 88, 89, 95–97, 100, 140, 153, 154, 165, 166, 169, 176, 240
Food availability, 14, 22, 23, 89
Foot pads, 4
Fox, 165, 166, 168, 171, 174, 176, 182, 189, 191
Fragrance, 107–109, 111–114, 242
Fragrance preference, 109
Fungi, 239, 240
Fur, 6, 41, 165–167, 170, 174, 175, 177

G

Galerella sanguinea, 185
Game theory, 238, 244
Garlic, 83–85, 241
Gas chromatography, 27, 29, 73
Gastrointestinal, 89
GC-MS, 53, 60, 62
Genetic compatibility, 107, 108
Genetta genetta, 188
Genetta maculata, 185
Genotype, 3, 4, 89, 90, 112, 145
Geographic mapping, 13–15
Gerbil, 245
Ghana, 119
Gland(s), 4, 6, 14, 39, 81–84, 86, 97, 118, 120–122, 127, 134, 141, 142, 145, 147, 167, 192, 205–209, 212, 213, 215, 222, 239, 241
Glandular, 94, 97, 99, 121, 131, 132, 192, 240
Gonorrhoea, 243
G-Protein Coupled Receptor (GPCR), 221, 222, 227, 228
Group, 31, 32, 38, 39, 48, 53, 63, 96, 97, 100, 129, 145, 154–156, 158, 159, 169, 172, 177, 190, 207, 209
Group membership, 97
Group-size, 96
Grouse, 244, 245
Guanylin, 241

H

Habitat type, 15, 16, 18, 73
Habitat use, 13
Hand, 16, 88, 141, 147, 177, 194, 208, 214, 222
Headspace, 28
Headspace solid phase microextraction, 27
Headspace volatile extraction, 62
Health, 38, 60, 61, 67, 84, 85, 97, 108, 111, 115, 119, 140, 142, 153, 159, 238–240, 242, 245
Heligmosomoides polygyrus, 241
Heptanol, 30, 35
Heterospecifics, 3, 4
Hexadecenoic acid, 34, 35
Hexanol, 30, 35
Hibernation, 14, 15, 18, 20, 22, 23, 73, 171
High protein diet, 86
Himalayan snow leopard, 27
Hinterland latrine, 100
Home range, 16, 18, 19, 22, 98–100
Honest signal, 7
Honey badger, 185, 191
Hormones, 38, 40, 43, 44, 67, 73, 239, 242
Host, 67, 68, 198, 238–241
Human(s), 27, 33, 35, 61, 68, 83, 84, 87, 88, 98, 100, 107–109, 111, 113–115, 118–122, 124, 125, 127, 128, 130–134, 139–147, 153, 165, 166, 172, 176, 228, 238, 240, 243
Human Leukocyte Antigen (HLA), 111, 112
Human medicine, 239
Hygiene, 140, 146, 243

Hypothalamus, 141, 222, 244

I
Ichthyosaura alpestris, 212
Identification, 15, 30, 32, 88, 96, 154, 156, 159, 166, 223, 234, 239
Identity, 3–6, 8, 20, 23, 29, 40, 44, 60, 73, 78, 87, 97, 98, 154, 222, 223, 227, 228
Immune-competency, 242
Immune response, 83, 240
Immune system, 61, 67, 89
Immune System Stimulation (ISS), 60–64, 66–68
Immunogenetics, 112
Index signal, 82
India, 29, 118
Individual advertisement, 38, 39, 54, 97
Individual pattern, 94, 96, 99
Individual recognition, 54, 82, 86–88, 90, 166, 195
Individual scent, 88
Indole, 31, 35, 62, 66–68
Infant Breastfeeding Assessment Tool (IBFAT), 129, 130
Infection, 61, 67, 68, 98, 99, 238, 240, 241, 243–246
Influenza virus, 67
Information, 3–9, 13, 14, 22, 23, 27, 38–40, 43, 45, 47, 53, 54, 60, 63, 71, 73, 77, 78, 82, 89, 95–97, 100, 107–109, 114, 115, 118, 140, 154, 168, 182, 183, 195, 216, 223, 228, 242
Information centre, 94, 95, 97
Information exchange, 97
Ingredients, 112, 115, 197
Injury, 238–240, 242, 244, 245
Innate aversion, 246
Innate response, 222
Insectivore, 195
In silico, 226
Integument, 4
Interspecific, 182, 183, 185, 190, 194–198, 212
Intra-guild, 99, 183, 196
Intruder, 96

K
Kairomone, 167, 175, 177
Kin, 35, 81, 86–88, 90
Kinship, 88, 96, 97

L
Lagomorphs, 94
Landmarks, 94, 95, 98, 242
Large-spotted genet, 185, 192, 194
Latency, 75, 123, 174
Latrine, 15–22, 40, 94–100
Latrine pits, 94
Latrine use, 13–18, 20, 22, 23, 94, 95, 98–100
Learning, 3
Lemur catta, 72, 77, 78
Leopard, 36, 185, 189, 192, 194, 196
L-Felinine, 168
Ligand, 221, 225, 226, 228, 231–233
Ligand optimization, 225
Lion, 28, 36, 191, 194, 196
Lipocalin, 87, 175
Lipopolysaccharide (LPS), 60, 61, 63–68, 240
Lissotriton helveticus, 212
Lizard, 82–84
Location, 8, 13, 15–19, 41, 74, 77, 96, 189
Longevity, 77, 94–96
Long-term signal, 14
Lordosis, 244
Louse, 241
Lycaon pictus, 185

M
Macrosmatic, 39, 165, 166
Main Olfactory Epithelium (MOE), 222
Major Histocompatibility Complex (MHC), 54, 87–89, 111, 112, 140, 239, 242
Major Urinary Proteins (MUP(s)), 4, 5, 87, 167, 168, 175
Mamilla, 120
Mammal, 55, 94, 165, 168, 182, 183, 190, 195–198
Mammary odour, 118, 121, 125, 132, 133
Marker, 97, 142, 195, 197
Marking, 14, 15, 20, 22, 38, 39, 42–44, 54, 55, 73, 86, 95–98, 100, 174, 182, 183, 185, 188, 190–192, 194–197, 241, 245
Marking Fluid (MF), 27, 28, 33, 35, 36
Marsupials, 94
Masculinity, 113
Mate choice, 5, 81, 83, 85, 87, 108, 109, 111–113, 144, 145, 221, 239, 242, 245
Masking, 107, 109, 111, 112
Mass spectrometry, 27, 29, 73
Maternal behaviour, 129
Mate(s), 5, 7, 8, 13, 14, 23, 38, 39, 54, 71, 72, 76, 77, 82, 86–89, 95, 97, 98, 109, 113, 140, 146, 195, 206, 234, 239, 242
Mating behaviour, 71, 72, 77
Meadow voles, 6, 7, 85

Mediator, 54
Meles meles, 38, 39, 77, 94, 100, 188
Mellivora capensis, 185
Memory, 4, 8, 87, 194
Menstrual cycle, 108, 114, 140, 142
Meriones unguiculatus, 245
Mesocricetus auratus, 6
Metabolic by-products, 38, 68
Metabolic rate, 39
Metabolism crates, 61
Metabolite(s), 27, 33, 38–40, 42, 44, 53, 54, 67, 68, 81, 84, 90, 97, 109, 238, 239, 242
Metabolome, 61, 68
Methyl butanethiol, 167
Methyl hepten, 27, 30, 33, 35
Methyl phenol, 60, 62, 64, 66–68, 223, 225
Microbiome, 141, 241
Microbiota, 118, 119, 240, 241
Microflora, 84, 89, 90
Microflora hypothesis, 241
Microsmatic, 107
Microtus agresti, 245
Microtus pennsylvanicus, 6, 71, 85
Microtus rossiameridionalis, 245
Midden, 94, 191, 192
Milk, 83, 84, 118, 121, 122, 125–128, 130–133, 223
Molecular docking, 225, 226, 231
Molecular interactions, 221, 233
Monotreme, 13, 72
Montgomerian secretion, 122
Montgomery glands, 121
Morphology, 221
Mouse/mice, 4–6, 8, 35, 54, 61, 67, 68, 85–89, 165–177, 241, 242, 244, 245
Multi-functional, 94
Mungos mungo, 185
Mus domesticus, 4
Mustela eversmanni, 166
Mustela furo, 166, 170, 174
Mustela nivalis, 244
Mustela putorius, 166, 171
Mustela putorius furo, 245
Musteloidea, 94, 97, 167

N
Neighbours, 54, 94, 96, 100
Neisseria gonorrhoeae, 243
Nematode, 241, 244, 245
Neonatal, 118–120, 128, 130, 132, 133, 242
Neonate, 118, 120, 126
Neuroendocrine, 8, 241, 244
Neuron, 176, 221, 222
New Guinea, 72
Newt(s), 206–208, 212–215
Nile tilapia, 85, 244
Nipple, 118–123, 125, 126, 129–133
Nocturnal, 17, 39, 196
Nonanal, 31, 35, 62, 66, 67
Non-volatile, 28, 71, 78, 122, 167
Normosmic, 159
Nose-poking, 13, 20, 21
Note to self, 98
Nutrition, 153
Nutritional status, 81, 82, 85, 86, 110

O
Octadecadienoic acid, 34, 35, 223, 225, 233
Octanone, 30, 35
Odour, 3, 7, 21, 22, 27, 35, 67, 68, 71, 73–77, 81–89, 96, 108–113, 118, 120–122, 124–128, 130–133, 153–156, 158–160, 170–173, 175, 238–240, 242–246
Odour profile, 67, 88, 109–111, 241
Odour signature, 109, 112
Odour sources, 6, 76, 89, 122, 125, 126
Odour space, 107, 109–111
Oestradiol, 38, 39, 42, 44, 52, 53
Oestrogen, 42, 241
Oestrus, 6, 53, 68, 73, 97, 233, 242, 244
Oestrus cycle, 39
Olfaction, 60, 72, 86, 87, 90, 113, 120, 132, 133, 139, 140, 142, 143, 145–147, 165–167, 241, 244
Olfactometer, 208
Olfactory, 13–15, 22, 38, 39, 53, 60, 67, 71, 72, 75, 76, 89, 90, 95, 96, 100, 114, 115, 128, 132, 139, 140, 143, 144, 146, 153–155, 159, 165–168, 171, 173–177, 190, 194, 221, 222, 238–242, 244
Olfactory attractiveness, 81
Olfactory bulb, 14, 221, 222
Olfactory communication, 13, 14, 28, 39, 54, 94, 107, 108, 114, 115, 196, 197, 239, 246
Olfactory context, 114, 140
Olfactory cue, 54, 170, 171
Olfactory image, 139
Olfactory loss, 139
Olfactory receptor genes, 144, 222
Olfactory recognition, 239, 241
Olfactory sensitivity, 114
Olfactory sheath, 109

Olfactory signal, 96, 242
Olfactory stimulus/stimuli, 68, 85, 87, 121, 127, 128, 133, 142, 165–175, 177, 208, 216
Olfactory system, 222
Opioid, 228
Opposite-sex, 5, 6, 38, 53, 111
Oreochromis niloticus, 85, 244
OR genes, 144
Orientation, 94, 95, 98, 120–122, 125, 131
Otocyon megalotis, 185
Overmarking, 190, 194–197
Ovulation, 39, 73, 77, 174
Ownership, 95
Oxidative stress, 84
Oxytocin, 241

P
Panthera unica, 28, 182
Papilla mammae, 120
Paracynictis selousi, 185
Parasite control, 94
Parasites, 98, 99, 238–241, 245
Parasitic, 99, 238
Partner, 108, 112, 113, 140, 141, 144–147, 242, 244
Passive range exclusion, 100
Pathogen, 61, 118, 119, 140, 146, 198, 238, 240, 242–244
Payoff asymmetry hypothesis, 96
P-Cresol, 30, 33, 221, 223, 225, 231–233
Pentanol, 30, 35
Peptide, 175, 210
Perception, 81, 84, 89, 107–109, 114, 129, 142–147, 221
Perfume, 108, 111–113
Peromyscus bairoli, 244
Peromyscus manuculatus, 167
Personality, 108, 113, 147
Personality traits, 115
Pest control, 196, 197
Phenotype, 3, 4, 87, 124, 130
Phenylethylamine (PEA), 176
Pheromone(s), 28, 36, 61, 67, 68, 133, 141, 144, 167, 175, 190, 205–216, 221–223, 225–227, 231–234, 241, 246
Phodopus campbelli, 6
Phylogenetic tree, 224, 226, 227
Phylogeny, 206, 213, 224
Pig, 60–66, 68
Placement, 14, 23, 42, 76, 97, 99
Pleurodelinae, 207
Poison, 239
Polecat, 166
Polyplax serrate, 241
Population dynamics, 242
Predation, 99, 166, 196, 238, 239, 244, 245
Predator, 61, 95, 99, 165–168, 170–172, 174–177, 182, 183, 190, 194, 244, 245
Predator–prey interactions, 94
Predatory stimulus, 165, 166, 174
Preorbital gland, 4
Prey, 95, 99, 166, 168, 176, 182, 195, 198, 239, 244–246
Prey species, 99, 175, 242
Primates, 14, 94, 98, 119
Procyon lotor, 98
Pro-inflammatory, 240, 241
Propylthietane, 167
Protein(s), 4, 5, 54, 81, 85–87, 90, 167, 168, 175, 205, 207–210, 212, 215, 221–228, 230, 231, 233, 241, 242
Protozoan, 243
Provisioning, 40
Provitamin D, 82, 83
Pyrazine, 31

Q
Quality, 4, 7, 14, 15, 22, 54, 61, 81, 82, 84–86, 88–90, 108, 109, 112, 128, 139, 140, 145, 146, 153, 194, 197, 216, 223, 224, 231, 244

R
Raccoon, 98
Raccoon roundworm, 98
Random coils, 221, 228
Rat, 88, 175, 177
Rattus norvegicus, 4
Reactogenic, 128, 130, 131
Receiver, 3–5, 8, 9, 82, 97
Receptivity, 71, 205, 208, 212, 215, 244
Receptor, 144, 222, 225–227, 231–234
Receptor 1, 221, 223
Recognition, 6, 35, 81, 86–88, 166, 175, 192, 226, 234
Rectovaginal fistulae, 243
Red meat, 83
Reproduction, 97, 140, 141, 168, 205, 206, 208, 209, 216, 221, 223, 234, 238, 239, 242
Reproductive advertisement, 43, 95, 97
Reproductive behaviour, 14, 27
Reproductive output, 238, 242, 244
Reproductive state, 38, 39, 71
Reproductive success, 145, 242
Reptile, 165, 166, 177
Repugnant, 238

Resource, 14, 96–98, 195
Resource defence, 94–96, 100, 245
Resource use, 14, 15, 22
Responder, 38, 40, 41, 43, 44, 47, 48, 50–54
Response, 3, 5, 6, 8, 9, 14, 38, 42–44, 47, 48, 50, 52–54, 61, 67, 68, 73, 77, 82, 88, 120, 125, 127, 128, 132, 133, 144, 155, 176, 177, 192, 194, 197, 208, 223, 240, 241, 244
Rinechis scalaris, 168, 171, 175
Rival, 7
Rodent, 166, 168, 169, 176
Rodent predator allelochemicals, 166
Russia, 153, 154, 159, 160

S

Salamanders, 86, 176, 205–210, 213–215
Salamandridae, 205–207, 211
Salamandrinae, 207
Salamandrininae, 207
Saliva, 4, 6, 61, 67, 87, 175, 222, 223, 239
Same-sex, 5, 6, 8, 38, 53
Saturated fatty acid, 27, 34
Scanning, 13, 20, 75, 76
Scat, 17, 21, 22, 188, 189, 191
Scent, 3–9, 14, 15, 17, 20–23, 39–44, 53, 54, 71–75, 77, 87, 94–100, 121, 140, 142, 175, 182–185, 188–198, 212, 222, 239, 241, 242, 244, 245
Scent diffusion, 99
Scent dispenser, 183, 184, 194
Scent fence, 95, 96, 100
Scent gland, 122
Scent mark, 3, 7–9, 28, 71, 166, 182–186, 188–192, 194–197
Scent matching, 96
Scent playback, 41
Scent provisioning, 38
Scent war, 195
Schreckstoff, 244
Scrape, 17, 20, 28
Season, 6, 15, 16, 20–23, 27, 29, 32, 33, 35, 38, 41, 53, 71–77, 97, 206, 216, 244
Seasonal patterns, 17, 18
Sebaceous, 6, 120–122, 128, 142, 147
Sebum, 120–122, 126–128
Secondary metabolites, 35, 89
Secondary metabolites of plants, 27, 33
Secretion, 39, 54, 55, 83, 84, 97, 118, 122, 123, 126–128, 131, 132, 145, 167, 197, 210
Segment of odour space, 109
Selection pressure, 238, 242, 244
Selective advantage, 238
Self-advertisement, 100
Selous' mongoose, 185
Semiochemical, 122, 197, 222, 223, 225, 226, 233, 234, 245, 246
Sender, 3–6, 8, 9
Senescence, 39
Sequence analysis, 221, 224, 228, 229
Serpents, 167
Sett, 41
Sex pheromone, 205, 212, 233
Sex-specific, 53, 77, 97, 141, 234
Sex steroid, 38–40, 54
Sexual context, 3
Sexually transmittable infection, 243
Sexually Transmitted Diseases (STD), 49, 50, 242–244
Shark, 244
Short-beaked echidna, 13, 14
Short-term signal, 14
Signal, 3, 4, 14, 38, 54, 71, 81, 82, 96–99, 125, 140, 196, 210, 222, 241
Signal content, 94
Signal degradation, 97
Signal lifespan, 96
Signalling, 13, 22, 23, 54, 55, 97, 118, 120, 131, 209
Signal transduction, 221, 222
Signal validity, 97
Skin, 81–84, 108, 120–122, 126, 131, 134, 141, 167, 168, 170, 171, 175, 206, 213, 239–241, 243
Skin $_{p}$H, 241
Sleeping site, 96, 98
Slender mongoose, 185, 192, 196
Smell, 33, 36, 83, 84, 95, 99, 107, 109, 111, 112, 114, 139, 142, 144–146, 153, 154, 160, 238, 239, 241, 243, 244, 246
Smell of disease, 238, 239
Smell of injury, 238, 245
Snake, 165, 167, 168, 170, 171, 174–176
Snake shed, 171
Sniff duration, 38, 42–44, 47, 48, 50–53
Sniffing, 19, 40, 42, 45, 52, 71, 75, 76, 99, 167, 175, 184, 194
Snow leopard, 27–29, 33–36, 182, 189
Squamata, 166, 167
Social context, 3, 38, 81, 89
Social cues, 3, 4, 109, 110
Social group, 39–41, 54, 94
Social interaction, 146, 245
Socially relevant information, 107

Sodefrin, 205, 207, 210–212
Sodefrin Precursor-like Factor (SPF), 205–215
Soil, 13, 15–17, 41, 188
Solid-Phase Microextraction (SPME), 29, 30, 60, 62
Somesthesis, 120
Spatial, 3, 5, 8, 13, 14, 22, 54, 98, 100, 123, 131, 194
Spatial distribution, 13, 14, 94–100, 124
Species recognition, 81, 86
Spermatogenesis, 39
Spotted hyaena, 188
Status, 23, 27, 28, 36, 38–40, 60, 67, 68, 71–73, 97, 100, 107–109, 115, 129, 140, 142, 167, 242, 244, 245
Strangers, 54, 96
Stress, 97, 98, 131, 169, 171, 172, 174, 234, 238, 240, 242, 244, 245
Stress hormone, 242
Structural modeling, 221
Structure annotation, 224, 226
Subcaudal gland, 38, 39, 55, 97
Submandibular glands, 4
Suidae, 226, 227
Sulfanylalkanols, 241
Survey, 16, 18, 19, 77
Sweat, 6, 67, 84, 121, 122, 141, 145, 239, 243
Syphilis, 243
Syrian hamsters, 6

T
Tachyglossus aculeatus, 13, 18, 72
Tail-fan, 205
Tasmania, 14
Taste, 120, 153, 154
Temporal, 3, 5, 13, 15, 16, 22, 99, 100
Temporal usage, 94–99
Territory, 14, 22, 23, 27, 36, 54, 82, 86, 87, 95, 96, 98–100, 195, 197
Testosterone, 38, 39, 42–44, 53, 246
Thamnophis sirtalis, 78, 176, 244
Theory of proxemics, 139, 147
Three-Finger Protein (TFP), 209
Time-sensitive, 5
TM region, 228
TMT, 176
Tomcat thiol, 183, 184
Tongue flicking, 167
Toxoplasma gondii, 198, 246
Trade-offs, 3, 7
Trail, 97, 167, 171, 244
Trait, 4, 38, 40, 43, 53, 87, 113, 114, 130, 133, 134, 216
Transmembrane topology, 221
Treponema pallidum, 243
Trial, 42, 74, 240
Trichinella crassiceps, 242
Trichinella spiralis, 242
Trichomoniasis, 243
Trichostrongylus tenuis, 244
Trimesurus albolabris, 168, 171, 175
Type-1 vomeronasal receptors, 222

U
Unfamiliar, 8, 41, 45, 47, 48, 52, 54, 86, 87, 98, 121, 127, 128, 153, 158, 159
Ungulates, 14, 94, 168
UniGEOS scales, 144
Universal features, 139, 143
University of Pennsylvania Smell Identification Test (UPSIT), 153–156, 158–160
Urinary, 28, 33, 35, 36, 38, 42, 53–55, 60, 61, 67, 68, 88, 95, 175, 223, 241, 242
Urination, 27, 38–40, 94, 166, 184
Urine, 4–6, 13, 22, 23, 27–30, 32–36, 38–45, 47, 48, 50–54, 60–63, 66–68, 81, 82, 84–87, 89, 97, 168, 171, 172, 174–177, 182, 188, 191, 192, 194, 196–198, 222, 223, 239, 241–243, 246
Urodela, 205

V
Vegetation, 16, 18, 19, 22, 74, 183
Vertebrates, 78, 81, 82, 84, 86, 89, 107, 109, 111, 209, 239–241, 245
Video, 17, 20, 42, 60, 62, 76, 172, 182–185, 192, 194, 196
Video analysis, 172
Vigilance, 242
Vipera aspis, 168, 171, 175
Visit, 17, 20, 22, 54, 74, 75, 95–100, 188, 189
Vitamin D, 82, 83
Vitamin E, 82–84
Volatile, 28, 30, 32, 36, 53, 60, 61, 63, 66–69, 71, 78, 89, 97, 110, 122, 128, 131, 141, 167, 170, 172, 175, 239, 241
Volatile Organic Compounds (VOCs), 33, 36, 68, 108
Volatility, 14, 95, 96, 99
Volatilome, 61

Vole, 7, 245
Vomeronasal Organ (VNO), 53, 68, 97, 167, 175–177, 221, 222, 227, 242
Vomeronasal Sensory Neurons (VSNs), 222
Vomeronasal type-1 Receptor I (VN1R1), 221–223, 227
Vomit, 239
Vulpes chama, 185
Vulpes vulpes, 168, 171, 182, 245

W
Weasel, 166, 245

Y
Yellow mongoose, 185, 192
Y-maze, 212

The manufacturer's authorised representative in the EU is Springer Nature Customer Service Centre GmbH, Europaplatz 3, 69115 Heidelberg, Germany. If you have any concerns regarding our products, please contact ProductSafety@springernature.com

Printed and bound by CPI Group (UK) Ltd, Croydon, CR0 4YY

25/09/2026

02222869-0011